Sources and Studies in the History of Mathematics and Physical Sciences

Sources and Studies in the History of Mathematics and Physical Sciences was inaugurated as two series in 1975 with the publication in Studies of Otto Neugebauer's seminal three-volume History of Ancient Mathematical Astronomy, which remains the central history of the subject. This publication was followed the next year in Sources by Gerald Toomer's transcription, translation (from the Arabic), and commentary of Diocles on Burning Mirrors. The two series were eventually amalgamated under a single editorial board led originally by Martin Klein (d. 2009) and Gerald Toomer, respectively two of the foremost historians of modern and ancient physical science. The goal of the joint series, as of its two predecessors, is to publish probing histories and thorough editions of technical developments in mathematics and physics, broadly construed. Its scope covers all relevant work from pre-classical antiquity through the last century, ranging from Babylonian mathematics to the scientific correspondence of H. A. Lorentz. Books in this series will interest scholars in the history of mathematics and physics, mathematicians, physicists, engineers, and anyone who seeks to understand the historical underpinnings of the modern physical sciences.

More information about this series at http://www.springer.com/series/4142

Sally P. Ragep

Jaghmīnī's *Mulakhkhaṣ*

An Islamic Introduction to Ptolemaic Astronomy

Springer

Sally P. Ragep
Institute of Islamic Studies
McGill University
Montreal, QC, Canada

ISSN 2196-8810 ISSN 2196-8829 (electronic)
Sources and Studies in the History of Mathematics and Physical Sciences
ISBN 978-3-319-31992-6 ISBN 978-3-319-31993-3 (eBook)
DOI 10.1007/978-3-319-31993-3

Library of Congress Control Number: 2016962632

Printed on acid-free paper

This Springer imprint is published by Springer Nature
The registered company is Springer International Publishing AG
The registered company address is: Gewerbestrasse 11, 6330 Cham, Switzerland

to F. Jamil Ragep

A Prequel

Preface

Maḥmūd al-Jaghmīnī's thirteenth-century *al-Mulakhkhaṣ fī al-hayʾa al-basīṭa* provided an accessible introduction in the premodern period to Ptolemaic theoretical astronomy, for both specialists and the educated public throughout Islamic lands. It played a crucial role in the teaching, dissemination, and institutional instruction of Islamic astronomy; and the base Arabic text served as the starting point for at least sixty-one commentaries, supercommentaries, glosses, and translations (into Persian, Turkish, and Hebrew) that were composed and studied well into the nineteenth century and even beyond. The topics include basic astronomical definitions and concepts, parameters of the motions of the planets and the Earth's inhabited zone, and, above all, a structure or configuration (*hayʾa*) of the universe that offered a scientific account of God's creation.

The impact and longevity of the influence of the *Mulakhkhaṣ* are not in question, as evidenced by thousands of extant copies of the original and its various derivatives contained in repositories worldwide. However, the focus until now has been on the work itself, leaving unaddressed questions such as: why was the *Mulakhkhaṣ* commissioned; who was Jaghmīnī's target audience; and what kind of a society produced such a scholar? Moreover, ambiguity in the literature about the date for Jaghmīnī's *floruit* led to speculation that there were two Jaghmīnīs, a thirteenth-century scholar who composed the ubiquitous astronomical work *al-Mulakhkhaṣ*, and a fourteenth-century namesake who authored the equally popular medical treatise *al-Qānūnča*. Establishing that there was only one Jaghmīnī who composed a corpus of introductory scientific works during the late twelfth/early thirteenth centuries under the auspices of the Khwārizm Shāhs in Central Asia highlights that this period just before the Mongol invasions was not one of scientific stagnation, as is so often asserted. Rather, it indicates a continuity of scientific learning within Islamic lands and furthermore suggests a demand for works in the mathematical sciences and the desire of those societies to promote scientific education.

The fact that I refer to Jaghmīnī's *Mulakhkhaṣ* as an *Islamic* introduction to Ptolemaic astronomy, rather than simply an introduction to Ptolemaic astronomy, warrants some clarification. The commissioning of the *Mulakhkhaṣ* needs be situated within an Islamic context related to major and interconnected social, political, and religious transformations that were occurring during the late twelfth and early thirteenth centuries. Specifically, textual and conceptual transformations were altering the way the discipline of *hayʾa* (theoretical astronomy) was being taught, which were concurrent with institutional transformations that resulted in the codification and systematization of the teaching of both religious and non-religious subjects. In conjunction with this, the ʿulamāʾ were attempting to consolidate their position vis-à-vis the rulers and ruling elites, and one way of accomplishing this was to bring a substantial number of the public into contact with their understanding of Islam through teaching in the madrasas. It is my contention that the *Mulakhkhaṣ* fulfilled a growing demand for a simplified, user-friendly introductory textbook on theoretical astronomy; it was a work not just geared for a broad audience, but a treatise whose structure and content offered madrasa students a physical cosmography glorifying God's entire creation, both His celestial and sublunary realms.

Although my primary intention is to provide a critical edition and English translation of, and commentary on, this important and influential treatise for specialists in the field, anyone interested in learning the basics of Ptolemaic astronomy, and how it is presented to an Islamic audience, will benefit. Scholars engaged in the study of Islamic theoretical astronomy will be able to use the base text to trace textual and conceptual changes and developments that occurred over time and space through the ensuing commentaries and translations. In addition, the Arabic-English glossary of technical astronomical terminology enables those with a rudimentary knowledge of Arabic to read the edition and get a sense of Jaghmīnī's pedagogical style and erudition. I should add that I made great efforts to capture these features in the English translation so that a general reader could learn what constitutes an elementary introductory textbook on theoretical astronomy in Islamic lands; this will be useful for comparisons with other traditions, in particular that of the Latin West. An important point for comparative studies is that this "beginner" treatise is far from simple and requires at least some prior knowledge of astronomy and mathematics.

Jaghmīnī and his *Mulakhkhaṣ* play center stage in this book; however, it is my sincere hope that the issues raised, especially in the Introduction, will be useful for future research in a number of areas. It is noteworthy that after composing the *Qānūnča,* Jaghmīnī dedicated his *Mulakhkhaṣ* to a certain Badr al-Dīn al-Qalānisī, whose family hailed from Damascus and whose fame (as far as we know) was not in astronomy but in pharmacology. Among other things, this highlights ongoing scholarly pipelines throughout Islamic regions as well as the importance of avoiding the all too prevalent practice of examining scientific fields in isolation. It also serves to underscore the vibrant activities occurring in the various sciences during this understudied period.

I could not have written this book without the generosity and support of a great many people whom I am extremely grateful to acknowledge. With deep appreciation, I thank Len Berggren for his careful read, comments, and suggestions. I am indebted to Tzvi Langermann, Faith Wallis, and Robert Wisnovsky, for sharing their expertise and meticulous attention to detail; and to Rula Abisaab, who first introduced me to the Khwārizm Shāhs. I owe special thanks to Sajjad Nikfahm-Khubravan, Fateme Savadi, and Hasan Umut, all who assisted with translations, manuscript analyses, acquisitions, and proofreading. I am beholden to Sean Swanick and the late Stephen Millier (formerly of McGill's Islamic Studies Library), who tracked down crucial works for me; Robert Morrison, who brought the Hebrew translation of the *Mulakhkhaṣ* to my attention; Reza Pourjavady and Devin DeWeese, who provided me with various Persian sources; Taro Mimura, who, among other things, discovered a miniscule *Mulakhkhaṣ* text embedded within a codex margin; Issa Boulatta, who helped me uncover Jaghmīnī's poetical side; and the late Mercè Comes, Raine Daston, Adam Gacek, Judith Pfeiffer, Emilie Savage-Smith, Pouneh Shabani-Jadidi, all who supported me in innumerable ways.

This book would not have been possible without the assistance of many libraries that allowed me access to their collections: the Bašagić Collection of Islamic Manuscripts, University Library of Bratislava; the Bibliothèque nationale de France, Paris; the Süleymaniye Library and the Topkapı Sarayı Müzesi Library, Istanbul; the Staatsbibliothek, Berlin; the Forschungsbibliothek, Gotha; the Rare Book and Manuscript Library Collection, University of Pennsylvania; the Islamic Manuscripts Collection, Princeton University Library; Cambridge University Library; Oriental Collections, Leiden University; the Dār al-kutub, Cairo; and, various libraries throughout Iran, including those within Isfahan, Qum, and Tehran. I owe special thanks to the Süleymaniye Manuscripts Library and the Türkiye Yazma Eserler Kurumu Başkanlığı, Istanbul, and the Staatsbibliothek, Berlin, for permission to reproduce images from their collections. I am also indebted to the support of the Rational Sciences in Islam (RaSI) database research project housed at McGill University, and especially the Islamic Scientific Manuscript Initiative (ISMI) component.

Special heartfelt thanks go to İhsan Fazlıoğlu, who has generously shared his vast knowledge of the Islamic mathematical sciences over the years, and to Michael Powell and Angela Libby, both of whom graciously entered unwittingly into my Jaghmīnī world, and became invaluable complements to our family sphere. Finally, no words can convey what I owe to the late Mollie Palchik, for her unmitigated love and support that permeate still. And to Lina, Anwar, and Jamil—as boundless as I know the *Mulakhkhaṣ* tradition is, it will never be as extensive as my love for them.

Contents

Introduction

§ I.1 The Arabic Edition and English Translation of Jaghmīnī's *Mulakhkhaṣ*[1]

The *Mulakhkhaṣ fī al-hay'a al-basīṭa* was an extremely popular astronomical textbook that played a critical role in the teaching, dissemination, and institutional instruction of Islamic theoretical astronomy. It was composed by Maḥmūd ibn Muḥammad ibn ʿUmar al-Jaghmīnī in the early thirteenth century in the region of Khwārizm in Central Asia; and its study and use as a propaedeutic for more advanced teaching texts is evidenced by thousands of extant copies of the original and its numerous commentaries, supercommentaries, and glosses contained in research libraries and various other repositories throughout the world. I have identified fifty-seven treatises that were written to elucidate the *Mulakhkhaṣ*, and these span, conservatively speaking, at least seven centuries beyond Jaghmīnī's original composition date. Indeed, the *Mulakhkhaṣ* was also translated from its original Arabic into Persian, Turkish, and Hebrew, and continued to be taught in earnest well into the nineteenth century (and beyond).[2] The study of the *Mulakhkhaṣ* along with its commentaries was still relevant even after "European science" came on the scene, and it is noteworthy that concerted efforts were made to seek teaching approaches that could accommodate the older Islamic scientific traditions such as

[1] Throughout this book, all transcriptions of Arabic names follow the convention of the *Encyclopaedia of Islam*, 3rd ed. The first of two dates separated by a slash (/) is the hijra date; the second is the corresponding Common Era (CE) date. When only one date is given, it is the Common Era unless otherwise indicated. For purposes of alphabetizing in the bibliography, the Arabic article "al-" is ignored.

[2] See Appendix II for a list of 61 commentaries, supercommentaries, glosses, and translations on various aspects of the *Mulakhkhaṣ*. I should add that according to Zalkida Hadzibegovic, the *Mulakhkhaṣ* was still being taught in twentieth-century Bosnia (see "Compendium of the Science of Astronomy by al-Jaghmīnī Used in Bosnia for Teaching and Learning Planetary Motions," Oral Presentation at the GIREP-EPEC Conference 2007, Opatija, Croatia, 1–4).

S.P. Ragep, *Jaghmīnī's Mulakhkhaṣ*, Sources and Studies in the History of Mathematics and Physical Sciences, DOI 10.1007/978-3-319-31993-3_1

that of the *Mulakhkhaṣ* along with new (*jadīd*) scientific developments.[3] So Jaghmīnī's ubiquitous introductory textbook on theoretical astronomy provides us with a significant example with which to understand a vibrant, ongoing scientific educational tradition within Islam.[4] It is with this in mind that the first and foremost objective of this book is to present an Arabic edition along with an English translation of the *Mulakhkhaṣ*. Furthermore, establishing the base (or *matn*) text is a fundamental prerequisite for gaining better insights into the extensive, and rather daunting, commentary tradition that built upon the *Mulakhkhaṣ*.[5]

A high priority was to ensure that the Arabic edition was as close to Jaghmīnī's original version as possible. In other words, I was concerned that my edition not be contaminated with the interjections of later commentators and copyists.[6] For example, there was considerable tampering by later copyists and commentators with the

---

[3] For example, the Muslim Ottoman scholar al-Qūnawī (fl. 1857) attempted to reconcile the traditional and *jadīd* sciences by including an up-to-date version of the heliocentric system within the context of a traditional astronomical treatise for madrasa scholars; see Robert Morrison, "The Reception of Early Modern European Astronomy by Ottoman Religious Scholars," *Archivum Ottomanicum* 21 (2003): 187–95.

[4] Ḥasan al-Jabartī's (d. 1188/1774-75) circle of scholars provides an excellent example of the *Mulakhkhaṣ* still being studied in eighteenth-century Cairo, and at the Azhar. According to his famous son, the historian ʿAbd al-Raḥmān al-Jabartī (d. 1241/1825-26), his father Ḥasan was a member of the ʿulamāʾ and attracted students from all parts of the world, and his instruction included Jaghmīnī's *Mulakhkhaṣ* along with Qāḍīzāde's fifteenth-century commentary on it. See Ekmeleddin İhsanoğlu, "The Ottoman Scientific-Scholarly Literature," in *History of the Ottoman State, Society & Civilisation,* ed. Ekmeleddin İhsanoğlu, 2 vols. (Istanbul: IRCICA, 2002), 2:586–87; Jane H. Murphy, "Improving the Mind and Delighting the Spirit: Jabarti and the Sciences in Eighteenth-Century Ottoman Cairo," Ph.D. diss., Princeton University, 2006, 97–100; Boris A. Rosenfeld and Ekmeleddin İhsanoğlu, *Mathematicians, Astronomers and Other Scholars of Islamic Civilization and Their Works (7th–19th c.)* (Istanbul: IRCICA, 2003), 410 [hereafter cited as *MAMS*2]; *Osmanlı Astronomi Literatürü Tarihi* (History of Astronomy Literature during the Ottoman Period), ed. Ramazan Şeşen et al., 2 vols. (Istanbul: IRCICA, 1997), 2:479, no. 19 [hereafter cited as *OALT*]; and Cevat İzgi, *Osmanlı Medreselerinde İlim: Riyazî ilimler*, 2 vols. (Istanbul: İz, 1997), 1:386, ç6 [hereafter cited as *Riyazî ilimler*]. Furthermore, a century after Jabartī, a student of Muḥammad ʿAbduh (d. 1905) reported that Sayyid Jamāl al-Dīn al-Afghānī read the *Mulakhkhaṣ* in Cairo with his students (Thomas Hildebrandt, "Waren Ǧamāl ad-Dīn al-Afġānī und Muḥammad ʿAbduh Neo-Muʿtaziliten?" *Die Welt des Islams* 42, no. 2 [2002]: 215n22).

[5] Having the text of the *Mulakhkhaṣ* should also prove useful for other disciplines, since its influence extends well beyond the astronomical genre of *hayʾa* per se. For example, there is growing evidence to connect the content of the *Mulakhkhaṣ* with nautical cartography. Jaghmīnī is mentioned in several passages of the atlas of the Tunisian chart maker al-Sharafī (fl. 1551–79), who attributes his picture of the universe to the "cosmological scheme derived from Jaghmīnī's treatise on the fundaments of theoretical astronomy"; see Mónica Herrera-Casais, "The Nautical Atlases of ʿAlī al-Sharafī," *Suhayl* 8 (2008): 242, 242n49.

[6] This is a major problem of the German translation by G. Rudloff and Prof. Dr. Ad. Hochheim ("Die Astronomie des Maḥmûd ibn Muḥammed ibn 'Omar al-Ǵagmînî," *Zeitschrift der Deutschen Morgenländischen Gesellschaft* 47 [1893]: 213–75). Rudloff and

parameters for the climes. Thus given the enormous numbers of extant *Mulakhkhaṣ* witnesses, it certainly would be understandable if my goal of providing the "original" text would be met with skepticism. However, there are several reasons that I believe I have been able to reach a text very close to the author's original.

First of all, the advances in digital technology and information sharing meant that I had access to an enormous pool of extant *Mulakhkhaṣ* witnesses to review and analyze. I also have had, and continue to have, a strong network of support from colleagues worldwide who generously helped me obtain witness copies and shared valuable insights on topics related to my work. As a result of this access and networking, I was able to acquire a vast trove of manuscript witnesses, from which I identified three different versions of the preface of the *Mulakhkhaṣ*: one contained a dedication by Jaghmīnī to a certain Badr al-Dīn al-Qalānisī along with a dedicatory poem Jaghmīnī composed to him; a second version contained only the dedication (i.e., the poem was omitted); and a third version lacked both dedication and poem. As it turned out, it was this last stripped-down version that would become the most ubiquitous one; and in fact it is this preface that is contained in our earliest known *Mulakhkhaṣ* copy, dated 644/1246-47.[7] So apparently, within the relatively short period of forty years after the composition date of the *Mulakhkhaṣ* in 602-3/1205-6, the dedicatory material was removed from the preface.[8]

Although I knew that these earlier prefaces containing the dedication, with or without the poem, would prove quite significant for dating Jaghmīnī,[9] I also recognized that there was no guarantee that the contents of any given witness, whatever the preface, had not been changed given the tendency of certain copyists and commentators to modify parameters with "updated" ones. I was able to resolve this potentially serious problem based on the fact that certain parameters in the modified versions of the *Mulakhkhaṣ* came from Naṣīr al-Dīn al-Ṭūsī's *al-Tadhkira fī ʿilm al-hayʾa*, written over fifty years after the *Mulakhkhaṣ* in 659/1261.[10] Thus, I was

---

Hochheim unknowingly added numerous comments from al-Sayyid al-Sharīf al-Jurjānī's (d. 816/1413) commentary, one of the key witnesses they relied on for their translation.

[7] This was the discovery by Max Krause in 1936 of a copy of the *Mulakhkhaṣ* dated 644 hijra in the colophon of Istanbul, Süleymaniye Library, Laleli MS 2141, f. 81a; see "Stambuler Handschriften islamischer Mathematiker," *Quellen und Studien zur Geschichte der Mathematik, Astronomie und Physik,* Abteilung B, Studien 3 (1936), 509–10, no. 403.

[8] The three different preface versions and the dating of the *Mulakhkhaṣ* are discussed in more detail in the Commentary [Preface].

[9] In addition, these variant prefaces became a convenient tool for helping decide which witnesses to target and obtain for further examination, since repository catalogues often contain incipits within their witness descriptions.

[10] I discuss this in further detail in my commentary on the second clime (see II.1[4]). The argument for dating the later versions of the *Mulakhkhaṣ* rests on a scribal error that could only have come after the *Tadhkira* was written. F. Jamil Ragep points out that even as astute a commentator as al-Bīrjandī (d. 935/1528) was led astray by not realizing that in his dating of Jaghmīnī as post-Ṭūsī he was using values for the climes that had been altered; in Bīrjandī's defense he was writing some three centuries after the *Mulakhkhaṣ*'s composition ("On Dating Jaghmīnī and His *Mulakhkhaṣ*," in *Essays in Honour of Ekmeleddin İhsanoğlu*, ed. Mustafa Kaçar and Zeynep Durukal [Istanbul: IRCICA, 2006], 463).

able to ignore these witnesses for the edition. Given the relatively small numbers of remaining witnesses that had the unmodified parameters and the original preface (or, lacking this preface, was an early copy), it became a relatively straightforward task to establish the "original" version. Of course there will always be a few remaining ambiguous readings, and these are noted in the critical apparatus. A description of the manuscripts used for the edition and the editorial procedures employed can be found in *§ II.1: Editorial Procedures* and *§ II.2: Description of the Manuscripts.*

I should also mention that I made a concerted effort not to make Jaghmīnī appear more "erudite" than he actually was; in other words, my *modus operandi* was to provide the reader with the *Mulakhkhaṣ* as it is in the Arabic, which meant not "correcting" poorly composed sentences, inconsistent use of terminology, and so on. Indeed, these colloquial features bring out the orality of the text. However, I was confronted with yet another challenge: Jaghmīnī's plain or simple (*basīṭa*) introductory work was anything but simple-minded. And this was compounded by the fact that Jaghmīnī, as he put it, went "to great lengths to elucidate and illuminate the content"—some rather complex astronomical material—using "concise and succinct expressions."[11] This meant that I was charged with understanding and then explaining his often-pithy formulations that stand in marked contrast to the often overly elaborated discussions of other *hayʾa* writers. Moreover, Jaghmīnī did not limit his subject matter to straightforward basic definitions, rules, and parameters of the longitudinal motions of the planets and the Earth's inhabited zone: he also dealt with theories of the latitude of the planets, a subject known for its complexity,[12] and such difficult topics as the appearance of the sky in the arctic regions. Fortunately, I was greatly assisted by the following resources: (1) the *Mulakhkhaṣ* commentaries, whose authors often provided detailed explanations, along with clarifying examples, to shed light on Jaghmīnī's more obscure points or overly simplified statements. I relied on several, for each could provide a slightly different perspective on a given subject; my personal favorites were those of ʿAbd al-Wājid (d. 838/1435), Qāḍīzāde (d. after 844/1440), and Yūsuf ibn Mubārak al-Alānī (ca. 735/1334);[13] (2) the edition, translation, and study of the *Tadhkira,* Naṣīr al-Dīn al-Ṭūsī's major *hayʾa* work;[14] (3) al-Bīrūnī's *Tafhīm*, a reference of

[11] See *Mulakhkhaṣ*, Pref.[1] and II.3[11].

[12] Noel M. Swerdlow, "Ptolemy's Theories of the Latitude of the Planets in the *Almagest*, *Handy Tables*, and *Planetary Hypotheses*," in *Wrong for the Right Reasons*, ed. Jed Z. Buchwald and Allan Franklin, Archimedes 11: New Studies in the History and Philosophy of Science and Technology (Dordrecht; New York: Springer, 2005), 41–42.

[13] Jan Just Witkam has reflected on the importance of the "commentary culture" as it developed in an Islamic context and highlighted some reasons they were written ("Poverty or richness? Some ideas about the generation of Islamic texts revisited," 9–10; a paper presented at the Commentary Manuscripts [*al-Makhṭūṭāt al-Shāriḥa*] Conference, Bibliotheca Alexandrina, Alexandria, 7–9 March 2006; preprint [15 pp]: http://www.islamicmanuscripts.info/preprints).

[14] F. Jamil Ragep, *Naṣīr al-Dīn al-Ṭūsī's* Memoir on Astronomy (*al-Tadhkira fī ʿilm al-hayʾa*), 2 vols., Sources in the History of Mathematics and Physical Sciences 12 (New York: Springer-Verlag, 1993) [hereafter cited as *Tadhkira*]; a useful glossary of technical terms is included in 2:581–613. Ṭūsī and Jaghmīnī deal with much of the same astronomical subject

astronomical terms, concepts and explanations, even though it is ostensibly an "astrological" primer;[15] and (4) the availability of planetarium software which enabled me to see the movements of the constellations in the sky at various latitudes, and to determine the veracity of Jaghmīnī's statements.[16]

Jaghmīnī claimed that he was delighted in being entrusted with the lofty task of compiling an introductory book on the subject of *ʿilm al-hayʾa*.[17] His *Mulakhkhaṣ*, usually classified as an abridged (*mukhtaṣar*) beginner treatise on the subject of theoretical astronomy,[18] was composed for an early thirteenth-century audience, but it would continue to play a vital role in educating generations of students and individuals interested in learning about the structure of the universe. The Arabic edition and English translation of Jaghmīnī's treatise *al-Mulakhkhaṣ fī al-hayʾa al-basīṭa* presented in this book (eight centuries after the original composition) will, I hope, allow another group of readers to assess its significance.

§ I.2 The Dating of Jaghmīnī to the Late Twelfth/Early Thirteenth Century and Resolving the Question of Multiple Jaghmīnīs

Maḥmūd ibn Muḥammad ibn ʿUmar al-Jaghmīnī al-Khwārizmī wrote one of the most successful astronomical textbooks of all time. Given the extent of the influence of the *Mulakhkhaṣ* and its impact on the *hayʾa* tradition, it may seem surprising that there has been so little agreed-upon information about who Jaghmīnī was, the society that produced him, and the educational context in which his scientific textbooks were written. This has led to conflicting claims in numerous sources, some of which have placed Jaghmīnī in the early thirteenth century, others in the mid-fourteenth century. Though recently a number of historians have dated the *Mulakhkhaṣ* to the

---

matter, so their content overlap proved extremely helpful, especially since Ṭūsī provides far more elaborate explanations than Jaghmīnī.

[15] Abū Rayḥān Muḥammad ibn Aḥmad al-Bīrūnī's *Kitāb al-Tafhīm li-awāʾil ṣināʿat al-tanjīm* (=The Book of Instruction in the Elements of the Art of Astrology), trans. by R. Ramsay Wright (London: Luzac and Co., 1934) [hereafter cited as *Tafhīm*]. I discuss Bīrūnī's text as a reference in *I.3.3b: The Post-Moderns.*

[16] See *Mulakhkhaṣ*, II.2 on various locations having latitude. Frankly, I marveled at Jaghmīnī's accurate descriptions of what was occurring in the sky at these various latitudes. I used the open source Stellarium software: http://www.stellarium.org/, but Jaghmīnī obviously had to depend on other means; it would certainly be interesting to explore the tools scholars used to determine this information.

[17] *Mulakhkhaṣ*, Pref.[1].

[18] The sixteenth-century Ottoman encyclopedist Ṭāshkubrīzāde (1495–1561) listed Jaghmīnī's *Mulakhkhaṣ* along with four commentaries on it (by Faḍl Allāh al-ʿUbaydī, Kamāl al-Dīn al-Turkmānī, al-Sayyid al-Sharīf, and Qāḍīzāde al-Rūmī) under a separate category entitled "famous abridgments" (*Miftāḥ al-saʿāda wa-miṣbāḥ al-siyāda,* 3 vols. [Beirut: Dār al-kutub al-ʿilmiyya, 1985], 1:349).

early thirteenth century, ambiguity about him has continued, leading some to speculate that perhaps there were two Jaghmīnīs, one thirteenth-century Jaghmīnī whose work focused on astronomy, another who lived in the fourteenth century and wrote the popular medical treatise *al-Qānūnča*. Here we should emphasize what should be an obvious point: determining Jaghmīnī's dates, and whether the same person wrote the textbooks attributed to him, really does matter. If Jaghmīnī lived in the mid-fourteenth century, he would be coming after the Mongol invasions, the building of the Marāgha observatory, and the consolidation of the Islamic scientific, philosophical, and theological traditions in the late thirteenth/early fourteenth centuries. On the other hand, if he lived in the late twelfth/early thirteenth centuries, this would directly challenge the prevailing narrative that science declined in Iran and Central Asia immediately after Ghazālī (d. 1111),[19] and that there was a strong prejudice against teaching the exact sciences in religious institutions such as the madrasa.[20] Certainly, it would be noteworthy that during this alleged scientific Dark Age of the pre-Mongol period we have the example of at least one scholar composing in essence a corpus of elementary scientific textbooks. And this raises many questions, such as who was the target audience and where support might have come from within the context of this time and place. Therefore, establishing that there was one Jaghmīnī who flourished under the auspices of the Khwārizm Shāhs of Central

[19] Ghazālī has often been vilified as instigating scientific decline in Islam due to his fears that the teaching of science and especially philosophy in the madrasas could lead to heresy. Actually Ghazālī insisted on "not being overly overzealous in condemning all ancient science," especially its apodeictic parts such as the mathematical sciences, since this might lead to a mocking of Islam, especially by the young (F. Jamil Ragep, "Freeing Astronomy from Philosophy: An Aspect of Islamic Influence on Science," *Osiris* 16 [2001]: 54). Nevertheless, the view that "But for Al Ashʿarî and Al Ghazâlî the Arabs might have been a nation of Galileos, Keplers, and Newtons" continues to be perpetuated (see Aydın Sayılı [quoting E. Sachau], *The Observatory in Islam and its Place in the General History of the Observatory* [Ankara: Türk Tarih Kurumu Basımevi, 1960], 408). See Frank Griffel, "The Western Reception of al-Ghazālī's Cosmology from the Middle Ages to the 21st Century," *Dîvân* 16, issue 30 (2011/1): 33–62 (esp. Renan's views on Ghazālī); and F. J. Ragep, "When Did Islamic Science Die (and Who Cares)?" *Newsletter of the British Society for the History of Science* 85 (Feb. 2008): 1–3 (a rebuttal to the Nobel Laureate Steven Weinberg's claim that after Ghazālī "there was no more science worth mentioning in Islamic countries").

[20] Note that my focus here concerns the teaching of the mathematical sciences, especially theoretical astronomy. Historians of other disciplines, such as philosophy, have in recent years been more willing to accept the notion that they were allowed within the madrasa. As Sonja Brentjes points out: "Historical sources such as biographical dictionaries, study programs and historical chronicles leave no doubt that philosophical treatises by Ibn Sīnā (d. 428/1037), Faḫr ad-Dīn ar-Rāzī (d. 606/1209), Naṣīr ad-Dīn aṭ-Ṭūsī (d. 672/1274) or Ǧalāl ad-Dīn ad-Dawwānī (d. 907/1501) were studied at *madrasas* in Cairo, Damascus or even in cities of northern Africa" ("The Prison of Categories—'Decline' and its Company," in *Islamic Philosophy, Science, Culture, and Religion: Studies in Honor of Dimitri Gutas*, ed. Felicitas Opwis and David Reisman [Leiden: E. J. Brill, 2012], 131n2). Science is another matter. As noted by Ahmad S. Dallal: "Scholars of Islamic education mostly agree on the marginality of the sciences," and he goes on to remark that it is exceptional to find studies on the relationship between religious and scientific scholarship (*Islam, Science, and the Challenge of History* [New Haven: Yale University Press, 2010], 19, 184–85nn46–47).

Asia (470–628/1077–1231),[21] and who would play a major role in Islamic astronomy and medicine over the next seven centuries, would be a significant contribution to reaffirming the continuity of scientific traditions in Islam and the society's need to promote a scientific education.

I.2.1 A Man Who Should *Need No Introduction*

From the final *nisba* in his name, one can deduce that Jaghmīnī hailed from the region of Khwārizm; and indeed Qāḍīzāde al-Rūmī informs us "Jaghmīn is one of the villages in Khwārizm" in his *Sharḥ al-Mulakhkhaṣ*, the commentary he composed in 814/1412 and dedicated to Ulugh Beg in Samarqand. However, Khwārizm covers quite a bit of territory, so this is not very informative. Thus, it would seem that delving more deeply into the precise location of Jaghmīn was not of much concern to Qāḍīzāde, as well as the other commentators I checked, for they provide nothing more specific than this.[22] It is interesting that in the seventeenth century, the well-known Ottoman historian and bibliographer Kātib Çelebī (a.k.a. Ḥājjī Khalīfa) felt no compulsion to remove the cloud of obscurity surrounding Jaghmīnī's life and wrote in his *Kashf al-ẓunūn* that the *Mulakhkhaṣ* was "composed by the eminent [scholar] Maḥmūd ibn Muḥammad al-Jaghmīnī al-Khwārizmī, an author whose fame makes an identification unnecessary."[23] This sentiment is also attested by the fact that in the late thirteenth/early fourteenth century, the prominent scholar Quṭb al-Dīn al-Shīrāzī could mention, and even paraphrase, from the *Mulakhkhaṣ* without mentioning the name of the author or the source of the paraphrase in the introduction and explicit of his *Nihāyat al-idrāk fī dirāyat al-aflāk*, presumably with the expectation that the reader would recognize their provenance (assuming that Shīrāzī was not plagiarizing in his paraphrase).[24]

---

[21] For a nice family tree charting the reigns of the Khwārizm Shāhs, see Muḥammad ibn Aḥmad Nasawī, *Sīrat al-Sulṭān Jalāl al-Dīn Mankubirtī li-Muḥammad ibn Aḥmad al-Nasawī*, ed. Ḥāfiẓ Aḥmad Ḥamdī (Cairo: Dār al-fikr al-ʿarabī, 1953), intro., 2 [in Arabic].

[22] See Qāḍīzāde al-Rūmī, *Sharḥ al-Mulakhkhaṣ*, Istanbul, Süleymaniye Library, Ayasofya MS 2662, f. 2b: جغمين قرية من قرى خوارزم . A potential opportunity of pinpointing Jaghmīnī's location was lost in Jaghmīnī's discussion on the *qibla* bearing (see II.3[4]); here he compares the longitude and latitude of Mecca to "our locality" but unfortunately fails to be more specific (understandable since presumably his immediate audience knew where they were). On this point, Qāḍīzāde is once again content to reference Khwārizm (f. 62b); the other commentators I checked followed his example or omitted a location altogether. I have attempted to home in on Jaghmīn's location, but so far without much success. Given the lack of any record of a village, town, or region named Jaghmīn in the geographical sources, one speculative possibility is that Jaghmīnī's name designates a family and not a locale, or perhaps a Turkic tribal affiliation, e.g., he was one of the Jagh [Čagh]-mān.

[23] F. Jamil Ragep, "On Dating Jaghmīnī and His *Mulakhkhaṣ*," 464:

تاليف الفاضل محمود بن محمد الجغمينى الخوارزمى وهو مولف شهرته تغني عن تعريفه

[24] For Shīrāzī's direct reference to the *Mulakhkhaṣ* in his explicit, see F. Jamil Ragep, "Shīrāzī's *Nihāyat al-Idrāk*: Introduction and Conclusion," *Tarikh-e Elm* (Tehran, Iran) 11

It would certainly be understandable that many scholars such as Quṭb al-Dīn al-Shīrāzī (fourteenth century), Qāḍīzāde al-Rūmī (fifteenth century), and Kātib Çelebī (seventeenth century), as well as the literally dozens of commentators over the centuries, would have been more concerned with the *content* of the *Mulakhkhaṣ* than ascertaining the particulars of where and when Jaghmīnī lived. However, since definitively dating Jaghmīnī is vital to understanding the nature of his achievement, it is crucial to decipher how the conflicting information contained in the modern sources—which cite him alternatively as flourishing in the early thirteenth century and the mid-fourteenth century, or speculate that there are two Jaghmīnīs—first took hold and eventually became embedded in the literature. So what follows is an overview of the literature highlighting the main accounts of his life and work and the issues involved. I then provide evidence supporting my claim that there is one Jaghmīnī who wrote multiple works in the exact sciences and medicine, who flourished in the late twelfth/early thirteenth century in Khwārizm, most likely in the environs of Merv, and who was a witness to (and most likely victim of) the onslaught of the Mongol invasions into the region that put an end to the reign of the Khwārizm Shāhs. This evidence was collected from the *Mulakhkhaṣ* and some of Jaghmīnī's other works, and bolstered by primary and secondary sources.

I.2.2 Review of the Literature and the Tale of Two Jaghmīnīs

Textual analysis can be a rather complex and daunting endeavor in that it is often quite difficult to disentangle and decipher an original text from a contaminated one.[25] One may thus have sympathy for one's predecessors, but unfortunately errors, whether understandable or not, can become embedded in the literature and over the decades become increasingly difficult to eradicate. This has become the case with Jaghmīnī who has been confidently held to have flourished sometime in the fourteenth century—or even later.[26] Ironically, the basis for this confidence is the

---

(2013): 51 [Arabic], 55 [Eng. trans.]. Shīrāzī "borrows" the following from Jaghmīnī (50 [Arabic], 54 [Eng. trans.]) in his introduction:

ليكون اسمه دالّاً على معناه وظاهره مخبراً عن فحواه

"…so that its name will indicate its connotation and its literal sense will inform its signification" (*Mulakhkhaṣ*, Pref.[2]. See especially the preface variants for MSS B and L [*§ II.2: Description of Manuscripts*]).

[25] The fact is that any particular manuscript witness to a text might represent one of several versions; could be corrupted or an amalgam of different versions; and/or could intentionally have been modified over time (by the author himself or by others) due to updating. Indeed, it is not uncommon to find comments, emendations, and "corrections" added to texts. Furthermore, it was the habit of some "scholars and writers to leave blank spaces in their works for the later insertion, by themselves or others, of data which were not known to them at the time of writing" (Franz Rosenthal, "The Technique and Approach of Muslim Scholarship," *Analecta Orientalia* 24 [1947]: 30).

[26] Heinrich Suter informs us that The Cairo Khedieval Library catalogue contains statements that Jaghmīnī died in the ninth-century hijra, and composed the *Mulakhkhaṣ* in the year "808,

following statement, which appeared in the respected and widely used *Encyclopaedia of Islam, First Edition*, in 1913: "His date is not quite certain but it is very probable that he died in 745 (1344-45)."[27] Often repeated, this statement is still found today in numerous references, many quite reputable,[28] even though mounting evidence challenging this assertion had emerged as early as 1936, with the discovery by Max Krause of a copy of the *Mulakhkhaṣ* dated 644 H [1246-47 CE]).[29] In fact, two main Islamic reference resources currently list Jaghmīnī as flourishing in the fourteenth century, namely, the "al-Djaghmīnī" entry in *Encyclopaedia of Islam,* 2nd ed. (1965) (which simply repeats verbatim what appeared in the 1913 first edition);[30] and the *Encyclopædia Iranica* (2008) article entitled: "Jaġmini, Maḥmud b. Moḥammad b. ʿOmar (d. 1344), an astronomer from Jaġmin."[31]

1405/6" (see "Der V. Band des Katalogs der arabischen Bücher der viceköniglichen Bibliothek in Kairo," in *Historisch-literarische Abtheilung der Zeitschrift für Mathematik und Physik*, ed. Dr. O. Schlömilch and Dr. M. Cantor [Leipzig: Verlag von B. G. Teubner, 1893], vol. 38, no. 5, 162; and Heinrich Suter, "Zu Rudloff und Hochheim, Die Astronomie des Ǵagmînî," *Zeitschrift der Deutschen Morgenländischen Gesellschaft* 47 [1893]: 718). Carlo A. Nallino also points out that Cairo catalogues list the ninth-century hijra date for Jaghmīnī in several places ("Zu Ǵaġmînî's Astronomie," *Zeitschrift der Deutschen Morgenländischen Gesellschaft* 48 [1894]: 120). This could be the explanation for how 808 became the year currently listed for the completion of Jaghmīnī's *Mulakhkhaṣ* (without explanation) in Ismāʿīl Bāshā al-Baghdādī's *Hadiyyat al-ʿārifīn* (Istanbul, 1955), vol. 2, col. 410 [in Arabic].

27 Henrich Suter, "al-Djaghmīnī," in *Encyclopaedia of Islam, First Edition* (Leiden: E. J. Brill, 1913), 1:1038.

28 Let me point out that this error of referencing an eighth-/fourteenth-century Jaghmīnī occurs in non-Western sources as well as Western ones, such as: The Majlis Library Catalogue (*Fihrist-i Kitābkhānah-yi Majlis-i Shūrā-yi Millī, kutub-i khattī*), ed. ʿAbd al-Ḥusayn Ḥāʾirī (Tehran, 1347 H. Sh./1968-69), vol. 10, part 1, 512; Abū al-Qāsim al-Qurbānī's *Zindagī-nāmah-yi rīyāḍī-dānān dawrah-yi Islāmī* (Tehran, 1365/1986), 219–20 (no. 69); and Halil Inalcık's *The Ottoman Empire: The Classical Age 1300–1600*, trans. Norman Itzkowitz and Colin Imber (London: Weidenfeld and Nicolson, 1973), 176n*.

29 Krause, "Stambuler Handschriften islamischer Mathematiker," 509–10. The 644 hijra copy date that Krause mentions, still the oldest one to date, is found in Laleli MS 2141, f. 81a; for the entire colophon, see *§ II.2: Description of the Manuscripts*, no. 4. See also Plate 6.

30 In his defense, Heinrich Suter (d. 1922) had died long before Krause's 1936 discovery, and well before the printing of the entry in *Encyclopaedia of Islam,* 2nd ed. [hereafter cited as *EI*2] (Leiden: E. J. Brill, 1965), 2:378; it was incumbent upon Juan Vernet, who is listed as co-author, to revise the date. This responsibility also applies to others, such as Fuat Sezgin, who lists "Maḥmūd b. Muḥammad b. ʿUmar AL-ǦAĠMĪNĪ" as "probably 745/1345" in the important bio-bibliographical resource *Geschichte des arabischen Schrifttums*, vol. 5: *Mathematik* (Leiden: E. J. Brill, 1974), 115 (no. 56) [hereafter cited as *GAS,* 5].

31 Lutz Richter-Bernburg suggests here that since "Nothing specific is known about his life, but it would seem plausible (but no more) to speculate that the author of *al-Qānunja* was a linear descendent of his earlier namesake…" ("Jaġmini, Maḥmud," *Encyclopædia Iranica*, vol. 14, fasc. 4, 373; originally published Dec. 15, 2008; online version last updated April 10, 2012: http://www.iranicaonline.org/articles/jagmini-mahmud).

The fact that many sources cited Jaghmīnī as flourishing circa 618/1221[32] did not sway those committed to a fourteenth-century Jaghmīnī to reevaluate their position. In those cases, the 618 H date was either ignored altogether or mitigated by the suggestion of the possibility of there being *two* Jaghmīnīs: an early thirteenth-century astronomer/mathematician (fl. ca. 618), the one who authored the *Mulakhkhaṣ*; and, a fourteenth-century physician who wrote the *Qānūnča* (the "little Qānūn"),[33] an abridged treatise of Ibn Sīnā's medical text *al-Qānūn fī al-ṭibb*. Almost without exception, references to the Jaghmīnī who authored the *Qānūnča* state that he flourished in the eighth/fourteenth century.[34]

How did the two-Jaghmīnī narrative take hold? The short answer is most references that have cited a 618/1221-22 date[35] for Jaghmīnī provided no information as to its source, and this ambiguity led some to question the trustworthiness of the date, but not enough to rule it out completely even after the fourteenth-century option emerged as a contender. Two different dates, hence two Jaghmīnīs, seemed a logical compromise to many; however, this was not the conclusion of Henrich Suter (d. 1922) who, as we shall see, insisted on one fourteenth-century Jaghmīnī who authored both the *Mulakhkhaṣ* and the *Qānūnča*. The longer answer follows.

Rudloff and Hochheim, in the introduction to their 1893 German translation of Jaghmīnī's *Mulakhkhaṣ*, bemoan the fact that "one searches in vain for any notes from which conclusions can be drawn concerning the date of birth, place of residence, and life circumstances of the author of the following treatise."[36] Now it is

---

[32] For a list of some of the more prominent references that cite the 618/1221 date, see fn. 55.

[33] Jaghmīnī's choice of *Qānūnča* for the title is interesting, since the medical treatise is written in Arabic, but the diminutive suffix "che" is found in Persian. As far as I know, Jaghmīnī never wrote scientific texts in Persian, though he does reference the two holidays of Nayrūz and Mihrjān in the *Mulakhkhaṣ* (see II.2[2]). So this title may be an indication of Jaghmīnī's background or perhaps some playful tribute by him acknowledging the wealth of medical literature written in Persian during the twelfth century.

[34] A prominent example is the date "d. 1344/745H" currently listed for Jaghmīnī online at the bio-bibliographies on the *Islamic Medical Manuscripts at the National Library of Medicine* website (with text written by Emilie Savage-Smith, The Oriental Institute, Oxford University): http://www.nlm.nih.gov/hmd/arabic/bioJ.html). Whereas Savage-Smith points out that there is conflicting evidence about when Jaghmīnī lived and raises the possibility of two Jaghmīnīs in her description, many other sources do not: A. Z. Iskandar simply lists "d. A.H. 745/A.D. 1344" in his *A Catalogue of Arabic Manuscripts on Medicine and Science in the Wellcome Historical Medical Library* (London: The Wellcome Historical Medical Library, 1967), 56. And a recent edition of Jaghmīnī's *Qānūnča fī al-ṭibb* currently bears both the date 751 H [!] on the book cover *and* 745 H on the inside title page (ed. and Persian trans. by Ismāʿīl Nāẓim [Tehran: Tehran University of Medical Sciences, 2012]). I found one exception in which Jaghmīnī (author of the *Qānūnča*) is listed as flourishing circa 618 H in *Fihris al-makhṭūṭāt al-muṣawwara,* ed. Ibrāhīm Shabbūḥ (Cairo, 1959), vol. 3, pt. 2, 145 (no. 186).

[35] Exactly what this date refers to in the literature is ambiguous. At times it is a death date, at others a date for the composition of the *Mulakhkhaṣ*.

[36] G. Rudloff and Prof. Dr. Ad. Hochheim, "Die Astronomie," 213. Note all English translations of the German here are mine.

evident that Rudloff and Hochheim are unfamiliar with Islamic history; for example, they openly admit that they are unaware of the identity of al-Shāfiʿī (d. 204/820) and Abū Ḥanīfa (d. 150/767), the famous founders of two Sunnī legal schools. Both are cited in the *Mulakhkhaṣ* (II.3[2]) in the context of Jaghmīnī distinguishing between their opinions regarding the determination of prayer times using shadow lengths. In addition, Rudloff and Hochheim were unacquainted with the renowned Islamic scholar al-Sayyid al-Sharīf al-Jurjānī (d. 816/1413), except for the fact that he was the author of the *Mulakhkhaṣ* commentary they used for their German translation. But despite these shortcomings, Rudloff and Hochheim discovered an important piece of information in Prof. Josephus Gottwaldt's 1855 Library Catalogue of Kazan, which simply states that Jaghmīnī died in 618 H, and they present this date in their introduction.[37] (Keep in mind that their spade work occurred some five years before the initial publication of Carl Brockelmann's seminal *Geschichte der arabischen Litteratur,* vol. 1 in 1898.[38]) But since Prof. Gottwaldt provided no indication as to how he came by this date, it opened the door for speculation that Jaghmīnī may or may not have lived then. Nonetheless, Rudloff and Hochheim upheld this date based on their translation of the *Mulakhkhaṣ*, concluding that Jaghmīnī was a scholar who "delivers through his presentation, a luminous picture of the ideas of those Arabs of the thirteenth century, who dedicated to astronomy a purely scientific interest."[39] With qualifications, their assessment has merit. Nevertheless, one must say that their German translation is not always reliable, since it is based on an amalgam of manuscripts, all obtained from the Gotha Library,[40] that were often interlaced with commentary notes. Of the four main witnesses that Rudloff and Hochheim used for their final translation: one had a late copy date of 1137 H; two others were defective; and the last one was a copy of Jurjānī's fifteenth-century commentary.[41] Although it is highly questionable

[37] Rudloff and Hochheim, "Die Astronomie," 213. I was able to check the Library Catalogue of Kazan and verify that indeed it states "618 (1221)" without qualification. See Josephus M. E. Gottwaldt, *Opisanie arabskich rukopisej prinadležavšich bibliotekě Imperatorskago kazangskago universiteta* (Kazan, 1855), 245 (entry for the *Mulakhkhaṣ* [no. 169] under the category of mathematics) [in Russian].

[38] Carl Brockelmann, *Geschichte der arabischen Litteratur*, 2 vols. plus 3 supplements (Weimar: Verlag von Emil Felber, [vol. 1] 1898; Berlin: Verlag von Emil Felber, [vol. 2] 1902; Leiden: E. J. Brill, [suppl. 1] 1937, [suppl. 2] 1938, [suppl. 3] 1942) [hereafter cited as *GAL*].

[39] Rudloff and Hochheim, "Die Astronomie," 215.

[40] Although Rudloff and Hochheim restricted their translation to witnesses from the Gotha Library Oriental collection, it evidently housed "the largest collection held at German libraries during the nineteenth century." It began with an expedition sent to the Middle East in 1802, specifically charged with acquiring Oriental books and manuscripts; and this apparently "created a need for specialists who were able to read, evaluate, and catalogue the collection"; see Ursula Woköck, *German Orientalism. The Study of the Middle East and Islam from 1800 to 1945* (London/New York: Routledge, 2009), 92 and 130. Perhaps this may help contextualize why Suter trusted their catalogue information [see below].

[41] For Rudloff and Hochheim's description of these four manuscripts, see "Die Astronomie," 216–18. For the catalogue descriptions, see Wilhelm Pertsch, *Die orientalischen*

whether Rudloff and Hochheim could distinguish between the words of Jaghmīnī and those of Jurjānī or other commentators and interpolators, or understand the subtleties of the text they were dealing with, they did, as noted, correctly place Jaghmīnī in the seventh/thirteenth century. This fact did not go unchallenged by Suter, whose review of their work appeared alongside their translation within the same journal issue in 1893. He contended that their date was "a little too early, although I can find no compelling evidence for a later lifetime."[42]

Suter's opinion was formulated on information gathered from the mathematical and astronomical parts of the published catalogues of the Cairo Khedieval Library.[43] Unlike Rudloff and Hochheim, Suter *was* familiar with Jurjānī and the composition date of his *Mulakhkhaṣ* commentary (813/1410-11); and he was also aware of other *Mulakhkhaṣ* commentators such as Qāḍīzāde al-Rūmī and Kamāl al-Dīn al-Turkmānī. He believed that Turkmānī's commentary was especially significant for dating Jaghmīnī since it was written in 755 H. Armed with this date, Suter confidently concluded that any claim that Jaghmīnī was "a scholar of the 9th century H" is highly improbable;[44] but he also states that Jaghmīnī "with near certainty…flourished in the first half of the 8th century H."[45] This was an assertion that Suter would tenaciously champion in publications throughout his career.[46]

The crucial piece of "new evidence" that Suter relied on to support his claim that "with near certainty" Jaghmīnī flourished in the "first half of the eighth century H" was a reference he found in the 1881 Gotha Library catalogue; Wilhelm Pertsch informs us here that codex Gotha 1930, folio 1b has a marginal note that states that Maḥmūd ibn ʿUmar al-Jaghmīnī, the author of *al-Qānūnča fī al-ṭibb*, an abridgement of Ibn Sīnā's work, died in the year 745 H.[47] Actually, Pertsch's comment is not found in his description of Gotha 1930, but rather in his summary of codex

Handschriften der Herzoglichen Bibliothek zu Gotha (Gotha: Friedrich Andreas Perthes, 1881), vol. 3, part 3, 46–48 (nos. 1385, 1386, 1387, 1388).

[42] Suter, "Zu Rudloff und Hochheim, Die Astronomie des Ǵagmînî," 718.

[43] Suter, "Zu Rudloff und Hochheim," 718; and Suter, "Der V. Band des Katalogs der arabischen Bücher der viceköniglichen Bibliothek in Kairo," vol. 38, no. 5, 161 and 162.

[44] Suter, "Zu Rudloff und Hochheim," 718–19. C. A. Nallino is also familiar with the ninth-century dating of Jaghmīnī (also based on Cairo catalogues), but is less willing than Suter to reject it; however, Nallino is unaware of Turkmānī's commentary ("Zu Ǵaġmînî's Astronomie," 120).

[45] Henrich Suter, "Zur Frage über die Lebenszeit des Verfassers des Mulaḫḫaṣ fî'l-hei'a, Maḥmûd b. Muḥ. b. 'Omar al-Ǵaġmînî," *Zeitschrift der Deutschen Morgenländischen Gesellschaft* 53 (1899): 540.

[46] Suter: "Der V. Band des Katalogs der arabischen Bücher der viceköniglichen Bibliothek in Kairo," 161 and 162; "Zur Frage über die Lebenszeit des Verfassers des Mulaḫḫaṣ fî'l-hei'a," 539–40; "Die Mathematiker und Astronomen der Araber und ihre Werke," *Abhandlungen zur Geschichte der mathematischen Wissenschaften mit Einschluss ihrer Anwendungen* 10 (1900): 164 (no. 403); and "al-Djaghmīnī," in *Encyclopaedia of Islam, First Edition*, 1:1038.

[47] Suter, "Zur Frage über die Lebenszeit des Verfassers des Mulaḫḫaṣ fî'l-hei'a," 539–40 (no. 2).

Gotha 1928 (Section XIX. Medicin, 468).[48] Nevertheless, the marginal note is actually contained in Gotha 1930 (which I checked).[49] It is not at all clear from Suter's article whether he actually examined Gotha 1930 or was relying entirely on Pertsch's catalogue for his information. This is unfortunate for had he checked the codex he would have been alerted to the fact that there were several errors in marginal notes pertaining to this particular witness making their reliability suspect.[50]

According to Suter, the emergence of *Mulakhkhaṣ* commentaries around the fourteenth century strongly supported the 745 H [1344-45 CE] catalogue date, and thus strengthened his dating claim. Suter's heavy reliance on the dates of *Mulakhkhaṣ* commentaries (especially that of Kamāl al-Dīn al-Turkmānī) to bolster his argument may explain why he never considered the two-Jaghmīnī option. Indeed, Suter has just *one* entry for Jaghmīnī, the scholar who authored both the *Mulakhkhaṣ* and the *Qānūnča*, in his seminal work listing 600 Islamic astronomers and mathematicians and their works.[51] Suter never mentions the 618 H date in this entry; and he does not suggest it as an option in his supplement to this work.[52] Furthermore, he fails to cite other references that list Jaghmīnī twice due to the contrary information, such as we find in Brockelmann, who lists Jaghmīnī under the

---

[48] Wilhelm Pertsch, *Die orientalischen Handschriften*, vol. 3, part 3, 468–69 (no. 1928) and 469–71 (no. 1930).

[49] I examined the witness *Sharḥ Qānūnča,* Gotha Ms. orient. A 1930 (which bears a copy date of 949 H/1542 CE [f. 144b]), and the marginal note on f. 1b reads as follows:

اختصره من القانون الكبير لابن سينا توفي محمود المذكور سنة ٧٤٥

"He abridged it from the great *Qānūn* [Canon] of Ibn Sīnā, the aforementioned Maḥmūd died in 745" [the aforementioned Maḥmūd being Maḥmūd ibn ʿUmar al-Jaghmīnī].

So indeed, according to this statement, Jaghmīnī died in 745.

[50] A. Z. Iskandar, who made a careful examination of Gotha MS 1930, pointed to several unreliable marginal dates and notes in the witness, to wit: the commentary is actually by ʿAlī b. Kamāl al-Dīn Maḥmūd al-Astarābādī al-Makkī (as noted on f. 144b), but is misattributed by the annotator (on f. 1a) to Muḥammad b. Muḥammad al-ṭabīb al-Miṣrī, whose name does not appear anywhere in the text (unfortunately, this misattribution is then given by Pertsch [*Die orientalischen Handschriften*, vol. 3, part 3, 469]); and the date of al-Miṣrī's death is given by the annotator as the year 801 (f. 1a), which is impossible since the work is dedicated to Sultan Bāyazīd II (f. 2a) who reigned 886–918/1481–1512. However, this error is not reported by Pertsch (ibid.). See Iskandar, "Commentaries on K. Qānūnča," in *A Catalogue of Arabic Manuscripts on Medicine and Science*, 58–59.

[51] Suter, "Die Mathematiker und Astronomen," 164–65, no. 403. This reference work (written in 1900 on Islamic authors and their works on the exact sciences) was foundational for many subsequent resources: M. Krause used Suter's author numbers for his 1936 "Stambuler Handschriften islamischer Mathematiker" as did G. P. Matvievskaya and B. A. Rosenfeld, *Matematiki i astronomi musulmanskogo srednevekovya i ikh trudi (VIII–XVII vv.)* [Mathematicians and Astronomers of the Muslim Middle Ages and Their Works (VIII–XVII centuries)], 3 vols. (Moscow: Nauka, 1983).

[52] Suter, "Nachträge und Berichtigungen," *Abhandlungen zur Geschichte der mathematischen Wissenschaften mit Einschluss ihrer Anwendungen* 14 (1902): 177.

categories of both astronomy and medicine but with different dates.[53] Ironically, Pertsch is among those who listed Jaghmīnī twice; so in effect, Suter based his claim on dating Jaghmīnī circa 745 H from information obtained from the Gotha catalogue but ignored the information from the same catalogue, which also dated him as flourishing circa 618 H.[54]

Suter aside (and the sources that followed him perpetuating the 745 date), the vast majority of references to Jaghmīnī the astronomer cite the date 618 H [1221-22 CE] as either the year of Jaghmīnī's death or the date of the composition of the *Mulakhkhaṣ*. In either case, one typically finds that this date is stated without qualification, except it is not uncommon that sources simply reference other sources that provide no evidence for the date.[55]

One of the most frequently cited sources for Jaghmīnī is Ḥājjī Khalīfa's *Kashf al-ẓunūn*; and this is particularly noteworthy because the printed editions of his work (at least the two that I have been able to check) omit Jaghmīnī's dates twice, i.e., in the listings for the *Mulakhkhaṣ* and the *Qānūnča*.[56] Nevertheless, Ḥājjī Khalīfa may have been the original source of the 618 H date. A viable explanation for this is that the date was contained in one of the several manuscript versions of the *Kashf al-ẓunūn*.[57] In support of this view, the title page of a manuscript copy of

---

[53] See Brockelmann, *GAL*1:473 (no. 5) for Jaghmīnī on astronomy (d. after 618/1221); and *GAL*1:457 for Jaghmīnī on medicine (745/1344). Brockelmann repeats this bifurcation in suppl. 1:826, 865. After Brockelmann we find other prominent sources following suit; see, for example, Charles A. Storey, *Persian Literature: A Bio-Bibliographical Survey,* 2 vols. (London: Luzac and Co., 1927–72), vol. 2, pt. 1 (A. Mathematics. B. Weights and Measures. C. Astronomy and Astrology. D. Geography), 50 (no. 88) [for astronomy]; and Storey, vol. 2 (E. Medicine), pt. 2, 219 (no. 377) [for medicine]. As the basis for dating Jaghmīnī 745/1344-45, Storey refers to the marginal note on folio 1b in Gotha 1930 listed in Pertsch's *Die orientalischen Handschriften.*

[54] Cf. Pertsch, *Die orientalischen Handschriften,* vol. 3, pt. 3, 46, no. 1385 (Section XIV: Astronomie und Astrologie) and 468–69, nos. 1928–1930 (Section XIX: Medicin). Suter was obviously aware of the two separate listings for Jaghmīnī since he includes codices listed for Jaghmīnī on astronomy (bearing the 618 H date) in his own list (but without comment); one example is Gotha, no. 1385 (see Suter, "Die Mathematiker und Astronomen," 164).

[55] Some of the more prominent references citing the 618 H date are: Brockelmann, *GAL*1:473 (no. 5), *GAL* suppl. 1:865; İzgi, *Riyazî ilimler*, 1:370, esp. n1010; David A. King, *A Survey of the Scientific Manuscripts in the Egyptian National Library* (Winona Lake, Indiana: Eisenbrauns, 1986), 150 (G17; 1.2.7) [hereafter cited as *Survey*]; Krause, "Stambuler Handschriften islamischer Mathematiker," 509 (no. 403) [Krause does not state a year but refers to Brockelmann, *GAL*1:473]; Rudloff and Hochheim, "Die Astronomie," 213; Matvievskaya and Rosenfeld, *Matematiki i astronomi*, 2:368; Rosenfeld and İhsanoğlu, *MAMS*2, 198 (no. 547); and Khayr al-Dīn al-Ziriklī, *Kitāb al-Aʿlām,* 8 vols. (Beirut: Dār al-ʿilm, 1980), 7:181.

[56] See Ḥājjī Khalīfa, *Kashf al-ẓunūn ʿan asāmī al-kutub wa-ʾl-funūn*, 2 vols. (Istanbul, 1941, 1943), vol. 2, cols. 1819–1820 [for astronomy] and vol. 2, col. 1311 [for medicine]; and also Gustavus Flügel, *Lexicon Bibliographicum et Encyclopædicum a Mustafa ben Abdallah. Katib Jelebi dicto et nomine Haji Khalfa celebrato compositum*, 7 vols. (Leipzig and London, 1835–58), 6:113–14 (no. 12886) [for astronomy], 4:495–96 (no. 9347) [for medicine].

[57] The fact that witness copies vary for a title (each witness being unique and thus potentially containing valuable information) highlights the important and complex issue of establishing

Qāḍīzāde's commentary on the *Mulakhkhaṣ* (Cairo, Dār al-kutub, Taymūr Riyāḍa 338, f. 1b) contains a note stating that Jaghmīnī "completed it in the year 618," and that this information was obtained from the *Kashf al-ẓunūn.*[58] This then could have been the basis for Gottwaldt's entry in his 1885 catalogue as well as for other sources that subsequently repeated the date.

I.2.3 Evidence Shedding New Light

The origins of the date 618 H [1221-22 CE] for Jaghmīnī's date of death would be interesting to resolve;[59] and an historiographical analysis of the literature regarding Jaghmīnī's dates is undeniably important for many reasons, among which are the insights one gets from tracing the nachleben of faulty assumptions. Nevertheless, my primary concern here is to remove some of the obscurity surrounding Jaghmīnī's life and works. So the remaining parts of this section provide conclusive evidence to support the contention that there was only one Jaghmīnī who wrote multiple scientific works (and in particular the *Mulakhkhaṣ* in 602-3 H [1205-6 CE]), and who flourished in the late twelfth/early thirteenth centuries during the extremely tumultuous period that witnessed the end of the Khwārizm Shāhs.[60] As mentioned earlier, my assertions are based on evidence gleaned directly from within the *Mulakhkhaṣ* itself, several of Jaghmīnī's other scientific treatises, and primary and secondary sources that provide valuable supplementary information for buttressing my claims. Needless to say, there are hundreds (if not thousands) of still-to-be-read manuscripts that need to be examined for future research;[61] and undoubtedly these contain information that will broaden, alter, and enrich our spectrum of knowledge.

---

the veracity of information, especially if there are conflicting claims. Suter seemingly believed that multiple versions of a primary source such as *Kashf al-ẓunūn* made using it suspect (see "Zu Rudloff und Hochheim," 719; and Suter, "Zur Frage über die Lebenszeit des Verfassers des Mulaḫḫaṣ fī'l-hei'a," 539); and he had few qualms about relying on a single secondary source, namely, the Gotha catalogue.

[58] F. J. Ragep provides the Arabic text of this passage along with an English translation in "On Dating Jaghmīnī and His *Mulakhkhaṣ*," 464–65.

[59] This year may simply have surfaced based on the assumption that Jaghmīnī died with the Mongol invasions.

[60] For historical overviews that provide insights into the complex alliances that were being formed among the peoples of this region during this period, see C. Edmund Bosworth, "The Political and Dynastic History of the Iranian World (A.D. 1000–1217)," in *The Cambridge History of Iran*, vol. 5, *The Saljuq and Mongol Periods*, ed. J. A. Boyle (Cambridge: Cambridge University Press, 1968), 185–95 (Section XIII: K͟hurāsān in the Second Half of the $6^{th}/12^{th}$ Century, and the Expansion of the K͟hwārazm-S͟hāhs); and W. Barthold, *Turkestan Down to the Mongol Invasion*, 2nd ed. (London: Luzac and Co., 1958), 323–80 (Ch. III: The Qarā-Khiṭāys and the Khwārazm-Shāhs).

[61] My colleague Sajjad Nikfahm-Khubravan has made a preliminary listing of over 1,000 extant witnesses of the text of the *Mulakhkhaṣ* and its commentaries; just to provide perspective, there are about 300 extant copies of Qāḍīzāde's commentary in Istanbul libraries alone that are not included in his list (*OALT*, 1:8–20).

To begin with, the original version of Jaghmīnī's *Mulakhkhaṣ* contains a dedication and a poem that Jaghmīnī composed for the Imām Badr al-Dīn Muḥammad ibn Bahrām al-Qalānisī (Pref.[1]),[62] who Jaghmīnī informs us proposed that he compile a work on the subject of *ʿilm al-hayʾa* (i.e., an epitome of theoretical astronomy that provides a configuration [or *hayʾa*] of the Universe). Jaghmīnī also dedicated a short treatise on planetary sizes and distances to Badr al-Dīn al-Qalānisī, a subject he did not include within the *Mulakhkhaṣ*.[63] Since *hayʾa* works often devote a section to this topic, perhaps Jaghmīnī recognized the omission and tried to rectify it by composing this brief astronomical work as a kind of appendix. In any event, Jaghmīnī's presumed oversight is our gain since this work on sizes and distances provides important confirmation of the dedicatee's name, which is stated in the explicit to this work (copied as is from the Cairo witness):

تمت الرسالة التى افادها الامام الجغمينى الخوارزمى حين فرغ من تاليف الملخص فى الهيئة واهداها
الى الامام بدر الدىن القلانسى والله اعلم

"The treatise is completed, which the Imām al-Jaghmīnī al-Khwārizmī put forth at the time he completed the work *al-Mulakhkhaṣ fī al-hayʾa*, and he dedicated it to the Imām Badr al-Dīn al-Qalānisī, and God is all-knowing." (Cairo, Dār al-kutub, ṬJ 429, f. 4b)[64]

So who was Badr al-Dīn al-Qalānisī? Although his life is not well known,[65] there is a substantial number of sources that specifically reference Badr al-Dīn

[62] The *Mulakhkhaṣ* has three different preface versions; for more details on this, see Commentary [Preface] and *§ II.2: Description of Manuscripts.*

[63] Jaghmīnī's *Mulakhkhaṣ* is not the only elementary *hayʾa* text that lacks a discussion of planetary sizes and distances; *al-Tabṣira fī ʿilm al-hayʾa* of Kharaqī (fl. mid-twelfth century, Merv) is another example.

[64] I know of two extant copies of this treatise, both of which I have consulted. The first is listed in David A. King's *Survey,* 150 (G17, 1.2.7): Cairo, Dār al-kutub, ṬJ [طلعت مجاميع] 429, 2, ff. 4a–4b. King also provides the colophon in his *A Catalogue of the Scientific Manuscripts in the Egyptian National Library* (Cairo: General Egyptian Book Organization, 1986), vol. 2, 21 (2) [in Arabic]; however, King misread al-Qalānisī as "al-Falāsitī (?)". The witness is described by King as unique, but I was able to identify another witness of it from an online image contained in the Bratislava collection, whose catalogue description stated "no title" for a Jaghmīnī text (TG 15; Ordinal Number 291: http://retrobib.ulib.sk/Basagic/EN/291.htm). I am grateful to Mr. Sajjad Nikfahm-Khubravan for bringing this image to my attention. This witness is missing a folio, but fortunately the extant folio (f. 33a) contains the dedication to Badr al-Dīn, since it is written at the beginning of the text:

من فواىد الامام الچغمينى الخوارزمىّ حين فرغ من تأليف الملخص واهداه الي الامام بدر الدين القلانسىّ

[65] This paucity of information on the life of "Mohammed ben Bahram ben Mohammed Bedr eddin el Calanisy Essamarcandy" was expressed by Lucien Leclerc in his *Histoire de la médecine arabe: exposé complet des traductions du grec; Les sciences en Orient, leur transmission à l'Occident par les traductions latines,* 2 vols. (Paris: E. Leroux, 1876), 2:128. The sentiment was echoed over a century later by Irene Fellmann, who translated into German Qalānisī's pharmaceutical work, and also reviewed the author and his work in her

Muḥammad ibn Bahrām ibn Muḥammad al-Qalānisī as the author of a pharmaceutical treatise (in 49 chapters) entitled *Aqrābādhīn al-Qalānisī* (composed ca. 590/1194),[66] and who flourished in the late sixth/twelfth to early seventh/thirteenth centuries.[67] There are also references to Qalānisī found in other early thirteenth-century medical sources, such as the pharmacological treatise by Najīb al-Samarqandī, who is reported to have died in the city of Herat in the wake of the Mongol invasion of 619/1222,[68] and al-Suwaydī (600–90/1204–92), who hailed from Damascus and was a contemporary of Ibn Abī Uṣaybiʿa.[69] Some references

---

introduction to *Das Aqrābāḏīn al-Qalānisī: Quellenkritische und begriffsanalytische Untersuchungen zur arabisch-pharmazeutischen Literatur* (Beirut, 1986), 1.

[66] In addition to Leclerc and Fellmann, see Ibn Abī Uṣaybiʿa, *ʿUyūn al-anbāʾ fī ṭabaqāt al-aṭibbāʾ* (2 editions): ed. A. Müller, 2 vols. plus corrections (Cairo: al-Maṭbaʿa al-Wahabiyya, 1299/1882, Königsberg, 1884), 2:31; and ed. Nizār Riḍā (Beirut: Dār maktabat al-ḥayāh, 1965), 472. Also see: Brockelman, *GAL*1:489 (no. 23); suppl. 1:893 (no. 23); A. Z. Iskandar, *A Catalogue of Arabic Manuscripts on Medicine and Science,* 79–80; Iskandar, "A Study of Al-Samarqandī's Medical Writings," in *Le Muséon Revue d'Études Orientales* 85 (Louvain, 1972), 452 (esp. n7); ʿUmar Riḍā Kaḥḥāla, *Muʿjam al-muʾallifīn: tarājim muṣannifī al-kutub al-ʿArabiyya,* 15 vols. (Beirut: Dār iḥyāʾal-turāth al-ʿarabī, 1980), 9:122; Manfred Ullmann, *Die Medizin im Islam.* Handbuch der Orientalistik (Leiden: E. J. Brill, 1970), 307–8; Lutz Richter-Bernburg, "Medical and Veterinary Sciences, Pt. One: Medicine, Pharmacology and Veterinary Science in Islamic Eastern Iran and Central Asia," in *History of Civilizations of Central Asia,* vol. 4, *The Age of Achievement: A.D. 750 to the End of the Fifteenth Century. Part Two: The Achievements*, ed. C. Edmund Bosworth and M. S. Asimov (Paris: UNESCO Publ., 2000), 310; and *Fihris al-makhṭūṭāt al-muṣawwara,* vol. 3, pt. 2, 24 (no. 25).

[67] Only Richter-Bernburg questions this as a "dubious date"; however, he provides neither reason nor alterative ("Medical and Veterinary Sciences," 310n42).

[68] A. Z. Iskandar pointed out that the marginal notes to two medical works by Najīb al-Samarqandī contain quotes attributed to Qalānisī; see codex Coll. 1062, MS. Ar. 73 [=UCLA Ar. 73] ("A Study of Al-Samarqandī's Medical Writings," 452, esp. n7). I was able to check this (here I am indebted to E. Savage-Smith for graciously allowing me to consult her copy); however, whether Samarqandī is actually quoting Qalānisī within his treatises, as well as the dating of these marginal notes, needs further careful examination. For more on Najīb al-Samarqandī, see Tarabein Chérif, "Contribution à l'histoire de la pharmacie arabe. Étude particulière du manuscrit intitulé: Al-Nadjibiate Al-Samarkandiate" (Ph.D. diss., Strasbourg University, 1952), intro., 6; Storey, *Persian Literature*, vol. 2, pt. 2, 215 (no. 368); Ibn Abī Uṣaybiʿa, *ʿUyūn al-anbāʾ*, Beirut ed., 472, Cairo-Königsberg ed., 2:31; and Ullmann, *Die Medizin im Islam,* 170, 278, 294, 308, 339.

[69] Leclerc states that Badr al-Dīn al-Qalānisī was among the numerous sources cited by Abū Isḥāq Ibrāhīm b. Muḥammad ʿIzz al-Dīn b. Ṭarkhān al-Suwaydī in his medical treatise on remedies entitled *al-Tadhkira al-hādiya* (*Histoire de la médecine arabe,* 2:128, 199–202 [on "Soueidy"]). For more on Suwaydī, see Brockelmann [listed as "ʿIzzaddīn a. Isḥāq Ibr. b. M. b. Tarḫān b. *as-Suwaidī* al-Anṣārī"], *GAL*1:493 (no. 38), *GAL* suppl. 1:900 (no. 38); Albert Dietrich, "al-Suwaydī," in *EI2* (1997), 9:909–10; Ibn Abī Uṣaybiʿa, *ʿUyūn al-anbāʾ*, Beirut ed., 759–61, Cairo-Königsberg ed., 266–67; Ullmann, *Die Medizin im Islam*, 284–85, 291; and *Islamic Medical Bio-Bibliographies at the National Library of Medicine*: http://www.nlm.nih.gov/hmd/arabic/bioS.html.

add the *nisba* al-Samarqandī to Badr al-Dīn al-Qalānisī's name,[70] which is noteworthy since the Banū Qalānisī hailed from a prominent Damascene family. So Badr al-Dīn would seem to have been an émigré to Central Asia from Damascus, which highlights connections between the two regions during this time.[71] Moreover, during this period the Qalānisī family was known to have "gradually evolved into a family of Shāfiʿī scholars and qāḍīs" from a family of government bureaucrats during a period that witnessed attempts to "professionalize" the ʿulamāʾ and codify law.[72] This professionalization of the ʿulamāʾ extended well beyond regulating salaries, and also included attempts to standardize their training and practice. Consequently, there was an upsurge in the number of teaching institutions that were constructed, and with them a proliferation of positions, accompanied by a growing demand for standardized textbooks.[73] Evidence indicates that this demand was not just restricted to the subject of religious law, and also that this phenomenon was not confined to Damascus alone.

So it should not surprise us that Jaghmīnī dedicated works to a scholar-Imām. On the other hand, it may seem somewhat odd that Badr al-Dīn al-Qalānisī, whose scholarly pursuits seem to focus on medicine, singled out Jaghmīnī to compose a work on astronomy. A possible explanation, admittedly somewhat speculative, is that Badr al-Dīn may have been a teacher or mentor of a younger Jaghmīnī. Here, we should keep in mind that twelfth-century Central Asia was a hub of scholarly activity (a further indication that this period was not one of scientific stagnation), "remarkable for the development of a vernacular medical and scientific literature,"[74] and would have been a locus for those seeking a proficient scientific education. In support of this, we have growing indications that Masʿūd ibn Maḥmūd al-Shīrāzī

---

[70] Brockelmann originally listed him as Badraddīn M. b. Bahrām al-Qalānisī (*GAL*1:489), and then later modified it by changing the name to Badr ad-Dīn M. b. Bahrām al-Qalānisī as-Samarqandī (*GAL* suppl. 1:893).

[71] Furthermore, the sources on al-Suwaydī report that he traveled between Damascus and Egypt; so presumably Badr al-Dīn's work disseminated both westward and eastward from Damascus.

[72] Joan E. Gilbert provides valuable information about the Banū Qalānisī residing in Damascus between 468/1076 and 736/1335 in "The Ulama of Medieval Damascus and the International World of Islamic Scholarship" (Ph.D. diss., University of California, Berkeley, 1977), ProQuest (7812573). According to Gilbert, the family was emblematic of the major political and social changes that were occurring in Damascus during the twelfth and thirteenth centuries (206–8, 222–25). It would seem that the Qalānisī family wore many "hats," figuratively and literally, since the family name *qalānisī* is a *nisba* for small cap/hat makers. Also, see Joan E. Gilbert, "Institutionalization of Muslim Scholarship and Professionalization of the ʿUlamāʾ in Medieval Damascus," *Studia Islamica* 52 (1980): 113–26.

[73] Gilbert informs us that: "By degrees specialized buildings replaced common teaching sites such as mosques, private homes, shops, libraries and gardens, and served not only as places of instruction and devotion but also as residents for professors and students" ("The Ulama of Medieval Damascus," 59).

[74] Edward G. Browne, *Arabian Medicine, Being the Fitzpatrick Lectures Delivered at the College of Physicians in November 1919 and November 1920* (Cambridge: Cambridge University, 1921), Lecture IV, 98.

(who was Quṭb al-Dīn al-Shīrāzī's father) had pursued studies in Khurāsān during this period, and not just in medicine (for which he is most famous).[75] Badr al-Dīn's *Aqrābādhīn al-Qalānisī* is a work full of quotations that "attest to his wide reading in the field; besides Ibn Sīnā, a whole range of authors, of whom al-Bīrūnī is the latest datable one, is represented."[76] So it is not inconceivable that Badr al-Dīn's medical knowledge had an influence on Jaghmīnī, directly or indirectly. In any event, Jaghmīnī did compose the *Qānūnča* (which lacks a dedication), and this concise elementary textbook on medicine became extremely popular (comparable to the *Mulakhkhaṣ*).[77] Perhaps in recognition of Jaghmīnī's success in adeptly writing a medical textbook, Badr al-Dīn was hopeful that he could write another primer, this one on the subject of theoretical astronomy.

This scenario is based on the following assumptions: that there is only one Jaghmīnī (who authored both the *Qānūnča* and the *Mulakhkhaṣ*); that he flourished in the late twelfth/early thirteenth centuries; and that Jaghmīnī composed the *Qānūnča* not in the fourteenth century (as Suter et al. have claimed), but prior to his composing the *Mulakhkhaṣ*. What follows is evidence to support all of these assumptions.

I.2.3a Dating the *Qānūnča*

A *Qānūnča* manuscript, recently discovered, states that it was copied on 12 Ramaḍān 601 H [=3 May 1205 CE], in the city of Konya (lit., ق [=Qūniya قونيه]).[78]

[75] See Junayd ibn Maḥmūd Junayd Shīrāzī, *Taẕkirah-yi Hazār Mazār: Tarjumah-yi Shadd al-Izār: Mazārāt-i Shīrāz* (Shiraz, 1364 H. Sh./1985-86), 109–11; and al-Jaghmīnī, *Talkhīṣ kitāb Ūqlīdis*, ed. Ḥusaynī al-Ishkavarī (Qum, 2006), 246 (where there is a marginal note in this mathematical treatise indicating that Jaghmīnī may have been a teacher of Masʿūd ibn Maḥmūd al-Shīrāzī). In his autobiography, Quṭb al-Dīn al-Shīrāzī (d. 710/1311) informs us that his father was considered to be the "Hippocrates of his age and the Galen of his day" (62; 83). Quṭb al-Dīn also tells us that he traveled to Khurāsān to engage with scientists and benefit from their knowledge; perhaps he was following in his father's footsteps (see Kaveh Farzad Niazi, "A Comparative Study of Quṭb al-Dīn Shīrāzī's Texts and Models on the Configuration of the Heavens" [Ph.D. diss., Columbia University, 2011], ProQuest [3479090], 82–85; and Niazi, *Quṭb al-Dīn Shīrāzī and the Configuration of the Heavens: A Comparison of Texts and Models,* Archimedes 35: New Studies in the History and Philosophy of Science and Technology [Dordrecht: Springer, 2014], 62–64, 67–68).

[76] See Richter-Bernburg, "Medical and Veterinary Sciences," 310.

[77] A. Z. Iskandar reports that two witnesses of Jaghmīnī's *Qānūnča* specifically state that the textbook was used in schools of medicine "…in all countries, and indeed, students were as familiar with it as with the midday sun …" (*A Catalogue of Arabic Manuscripts on Medicine and Science*, 56–57).

[78] See Istanbul, Süleymaniye Library, Ayasofya MS 3735, f. 25a. I have İhsan Fazlıoğlu to thank for his assistance in deciphering the ambiguous < ق > in the colophon. This can be understood as the city of Konya since another colophon in the same codex (and in the same hand) states explicitly that the work was copied in city of Qūniya (بمدينة قونيه) and completed

That this treatise was copied in Anatolia in the early seventh/thirteenth century highlights the point that scientific texts were disseminating westward to lands that would later become part of the Ottoman Empire; and it also indicates that this specific treatise was in circulation by 601/1205. As far as I know, this *Qānūnča* witness is the oldest one to date, evidently copied during Jaghmīnī's lifetime. This should effectively put to rest the purported fourteenth-century date for the *Qānūnča* as well as the two-Jaghmīnī hypothesis unless one wishes to maintain that there were two Maḥmūd ibn Muḥammad ibn ʿUmar al-Jaghmīnī al-Khwārizmī's living in Khwārizm at the same time and that both were writing scientific textbooks.[79]

I.2.3b Dating the *Mulakhkhaṣ*

There is strong evidence that the *Mulakhkhaṣ* was composed in 602-3 H (=1205-6 CE); this places its composition as being after the *Qānūnča,* based on the above extant witness dated 601 H. The 603 H date is provided by Jaghmīnī himself in the *Mulakhkhaṣ* in his chapter on planetary motions, and specifically within the discussion of the parameters for the apogee and nodes (I.5[26]). Here Jaghmīnī states: "As for the position of the apogees, they are for the beginning of the year 1517 of Dhū al-Qarnayn [the two-horned, i.e., the era of Alexander the Great]: …" Jaghmīnī did not select this date arbitrarily; rather it was chosen because 1517 was *his* current year and he was providing the students with updated positions. (More speculative is that Jaghmīnī was also using this as an exercise in calendar conversion.) In any event, in support of my update claim, I was able to calculate, using the positions provided by Jaghmīnī and Battānī for their apogees and the Alexandrian years between them (1517 minus 1191=326), a constant value for the motion of the apogees, namely, 1 degree per 66 years, which is exactly the value given by Jaghmīnī for the precessional motion of the stars and apogees (see Commentary, I.5[26]). It would have been very odd indeed for Jaghmīnī to use a date that was not his own, given that he uses it to report the position of the planetary apogees. We can thus conclude that 1517 of the Alexander era, which converts to the year 1205-6 CE [or 602-3 H], is the date of composition of the *Mulakhkhaṣ*.[80]

---

on 20 Ramaḍān (i.e., eight days later) in the previously-mentioned year (سنه المذكور), i.e., 601 (f. 40a).

[79] Furthermore, Ayasofya MS 3735 is not the only *Qānūnča* bearing a thirteenth-century copy date; for another example, see Princeton, Garrett 3559Y, which bears a date of the middle of Ṣafar 680/1281 in the colophon (f. 57a).

[80] An even more precise calculation of this date is 1517 years from Monday, 1 October –312, or in other words 1 October 1205 [=16 Ṣafar 602 H]. See Commentary, I.5[26] for more information on this calendar conversion, the term Dhū al-Qarnayn, and variant readings of the year 1517. I am not alone in asserting that Jaghmīnī's use of the date 1517 Dhū al-Qarnayn indicates when he lived; see Hanif Ghalandari, "Chaghmīnī," *The Great Islamic Encyclopedia* (Tehran, 1390 H. Sh./2012), 19:356–57 [in Persian]; and Farīd Qāsimlū,

I.2.3c Further Evidence for Dating Jaghmīnī

Additional support that Jaghmīnī flourished during this time comes from another of his compositions, this one being a short astrological treatise entitled *Fī quwā al-kawākib wa-ḍaʿfihā* (The Strengths and Weaknesses of the Planets). This work contains a discussion similar to the one Jaghmīnī put forth in the *Mulakhkhaṣ* on the positions of the apogees for each of the planets; however, his listing of parameters here are based on the planetary positions for the beginning of the year 1516 of Dhū al-Qarnayn.[81] So presumably this work was composed one year earlier than the *Mulakhkhaṣ* in 1204-5 CE [or 601-2 H].[82] This work is also very important for dating Jaghmīnī because one of the two extant witnesses (Paris, BnF, Ms ar. 2589, f. 174b [=f. 27b]) states that the work is dedicated to "our teacher Shihāb al-Dīn, may God prolong his life."[83]

Identifying this Shihāb al-Dīn with nothing but the abbreviated form of the name is not an easy task; there were several Shihāb al-Dīns who lived in this region during this period.[84] But fortunately a fuller version of his name—Shihāb al-Dīn Abū Saʿd ibn ʿImrān al-Khwārizmī al-Khīwaqī—is provided by Jaghmīnī himself in another work that he dedicated to him, namely, a mathematical treatise entitled *Talkhīṣ kitāb Ūqlīdis* (Epitome of Euclid's book [i.e., the *Elements*]),[85] in which we learn from the colophon that it was completed on Sunday, 22 Ṣafar 615 H (=Saturday-Sunday, 19-20 May 1218 CE).[86]

"Chaghmīnī," *Encyclopaedia of the World of Islam* (Tehran, 1387 H. Sh./2008-9), 12:61 [in Persian]. See also Arash Abutorabi Hamedani (ed.) in the introduction to the printed Persian commentary by Muḥammad ibn ʿUmar al-Andiqānī of Jaghmīnī's *al-Mulakhkhaṣ* (*Tarjumah-yi al-Mulakhkhaṣ fī al-Hayʾa,* in *Nāmah-yi Maʿānī, Yādnāmah-yi Ustād Aḥmad Gulčīn Maʿānī (Memoirs of Master Aḥmad Gulčīn Maʿānī)* [Tehran, 1383 H. Sh./2004-5], 866 [in Persian]).

81 There are two extant copies of this work: Paris, BnF, MS ar. 2589, ff. 174b–176b [=Arabic-script numbering: ff. 27b–29b]; and a witness that has been published with the *Talkhīṣ kitāb Ūqlīdis*, 249–53. Jaghmīnī specifically mentions the year 1516 of Dhū al-Qarnayn on 250 of the facsimile, and on f. 174b [=f. 27b] of BnF, MS ar. 2589; and in both witnesses the numbers for the year are not alphanumerical but are clearly written out in words.

82 Jaghmīnī's use of these two successive dates was also noted by Ghalandari, "Chaghmīnī," 19:356–57; and Qāsimlū, "Chaghmīnī," 12:61.

83 Obviously being able to connect a specific date 601-2/1204-5 with the dedicatee Shihāb al-Dīn is extremely valuable information; it is also significant that the statement informs us that Shihāb al-Dīn is still alive. This information, however, is only contained in the Paris manuscript; the Qum facsimile substitutes the word *fulān* [meaning "unspecified person"] in its place (see *Talkhīṣ kitāb Ūqlīdis*, 249).

84 See Barthold, who lists four Shihāb al-Dīn's, all flourishing in the late twelfth/early thirteenth centuries in this region (*Turkestan*, General Index). I cannot resist pointing out that Suter should have been aware of Jaghmīnī's dedicatee since he lists Paris MS 2589 as a witness for this work in "Die Mathematiker und Astronomen," 164–65.

85 See *Talkhīṣ kitāb Ūqlīdis*, 16.

86 See *Talkhīṣ kitāb Ūqlīdis*, 246.

In addition to Shihāb al-Dīn al-Khīwaqī's *nisba,* which indicates that his family hailed from Khīwa in the heart of Khwārizm, it turns out that there is much information available about him, from both primary and secondary sources, especially in comparison with the information we have regarding Badr al-Dīn al-Qalānisī. This is due to Shihāb al-Dīn's eminence as a scholar as well as his important role as advisor to the Khwārizm Shāh ʿAlāʾ al-Dīn Muḥammad (r. 596–617/1200–20). In fact, Muḥammad ibn Aḥmad al-Nasawī (fl. 639/1241), in his *Sīrat al-Sulṭān Jalāl al-Dīn Mankubirtī* (a biography of the Khwārizm Shāh who reigned 617–28/1220–31) devotes an entire chapter to Shihāb al-Dīn in which he describes his departure from Khwārizm to Nasā during the crumbling of the Khwārizm dynasty just prior to the arrival of the Mongols circa 618/1221.[87] Furthermore, there are also other sources that specifically mention Shihāb al-Dīn al-Khīwaqī and that provide insightful information about the period; among these, several were written by contemporary historians.[88]

With specific regard to Jaghmīnī's patrons, we have a situation parallel to that of Badr al-Dīn in that Shihāb al-Dīn is another example of a dedicatee who is recognized as a highly esteemed scholar/Shāfiʿī Imām.[89] In addition, both Badr al-Dīn

[87] See Nasawī, *Sīrat al-Sulṭān Jalāl al-Dīn,* 109–15, Ch. 23: On the Arrival of Shihāb al-Dīn al-Khīwaqī to Nasā from Khwārizm (=*Histoire du sultan Djelal ed-Din Mankobirti, prince du Kharezm par Mohammed en-Nesawi,* French trans. Octave Houdas [Paris: Leroux, 1895], 82–89, Ch. 22 [=Ch. 23 in Arabic]). Nasawī entered the service of Jalāl al-Dīn in 1223. The valuable detailed information he provides presumably is due to having the "home court advantage" of writing from the perspective of a native Khurāsānian and living there during this period (see Barthold, *Turkestan,* 38–39).

[88] See ʿAlāʾ al-Dīn al-Juwaynī (d. 681/1283), *Tārīkh-i Jahān-gushā* [in Persian] (=*The History of the World-Conqueror by ʿAla-ad-Din ʿAta-Malik Juvaini. Translated from the text of Mirza Muhammad Qazvini,* trans. John A. Boyle, 2 vols. [Manchester: Manchester University Press, 1958], 1:322–23); Ibn al-Athīr, *al-Kāmil fī ʾl-Taʾrīkh* [in Arabic] (Beirut: Dār ṣādir, 1966), 12:362–63 (=*The Chronicle of Ibn al-Athīr for the Crusading Period from* al-Kāmil fīʾl-Taʾrīkh. Part 3, *The Years 589–629/1193–1231: The Ayyūbids after Saladin and the Mongol Menace,* trans. D. S. Richards [Aldershot, Hants, England; Burlington, Vt.: Ashgate, 2008], 206 (within the "Account of the Tatars' irruption into Turkestan and Transoxiana and what they did," 204–10); Minhāj Sirāj Jūzjānī (born 589/1193) wrote his *Ṭabaqāt-i Nāṣirī* [in Persian] in 658/1260 (=*A General History of the Muhammadan Dynasties of Asia, including Hindūstān, from A.H. 194 (810 A.D.) to A.H. 658 (1260 A.D.) and the Irruption of the Infidel Mughals into Islām,* 2 vols., trans. Major H. G. Raverty (London, 1881), esp. 252–78; and Yāqūt ibn ʿAbd Allāh al-Ḥamawī (d. 626/1229), who was in Merv just prior to its destruction (616/1220), and reports on the extensive endowed libraries and collections of the city (*Muʿjam al-buldān li-l-Shaykh al-imām Shihāb al-Dīn Abī ʿAbd Allāh Yāqūt ibn ʿAbd Allāh al-Ḥamawī al-Rūmī al-Baghdādī,* 5 vols. (Beirut: Dār ṣādir, 1957), 5:114. See also J. A. Boyle, "Dynastic and Political History of the Īl-Khāns," in *The Cambridge History of Iran,* vol. 5, *The Saljuq and Mongol Period,* ed. J. A. Boyle (Cambridge: Cambridge University Press, 1968), 306; and S. M. Stern, "Petitions from the Ayyūbid Period," *Bulletin of the School of Oriental and African Studies, University of London* 27, no. 1 (1964): 15–16. Barthold provides a nice overview of many of these contemporary historians (*Turkestan,* 31–37; on Shihāb al-Dīn, see 376, 404–5, 429).

[89] Nasawī informs us that "Regarding the science of law, [Shihāb al-Dīn] combined knowledge of lexicography, medicine, and dialectic, and other sciences. Eloquent and versed

al-Qalānisī and Shihāb al-Dīn had government affiliations. In the case of Shihāb al-Dīn, these are more pronounced; it is reported that he held the status of a trusted advisor (*wakīl*) to the Khwārizm Shāh ʿAlāʾ al-Dīn Muḥammad himself, who "consulted him in all serious circumstances and yielded to his decision in important matters."[90] These scholars seem to have used their positions with governmental connections to promote scholarly activities (especially the teaching of the sciences).[91] It is worth noting that the Khwārizm Shāh ʿAlāʾ al-Dīn Muḥammad also had a close relationship with Fakhr al-Dīn al-Rāzī (d. 606/1210),[92] so presumably he was highly receptive to supporting scholarly endeavors. Indeed, it is stated that Shihāb al-Dīn was directly responsible for establishing numerous Islamic institutions throughout the region and filling their libraries with extensive collections. It is certainly conceivable that he used his position to promote the teaching of the sciences. Nasawī informs us that Shihāb al-Dīn was charged with teaching in five *madrasa*s and had built a library in a Shāfiʿī mosque in Khwārizm that had no equal "either before or since."[93] If this were true then it would surpass Yāqūt al-Ḥamawī's citing of 12,000

---

in various languages, he was also a man of good counsel. Mars had bought happiness from him, Mercury had benefited from his lessons, the finest man was the slave of his wisdom and the greatest thinker was the servant of his ideas" (*Sīrat al-Sulṭān Jalāl al-Dīn*, 109 [=Houdas, *Histoire du Sultan,* 82]). Cf. Ibn Athīr, *al-Kāmil fī 'l-taʾrīkh,* 12:362–63 (=Richards, *The Chronicle of Ibn al-Athīr,* 206).

90 Nasawī, *Sīrat al-Sulṭān Jalāl al-Dīn,* 109 (=Houdas, *Histoire du Sultan*, 82). The position of *wakīl* meant he "was by no means a subordinate official whose function was literally to carry the decision of the sultan to the chancery…it is obvious that it was an honorary duty attributed to high-ranking courtiers" (S. M. Stern, "Petitions from the Ayyūbid Period," 16). Barthold also mentions Shihāb al-Dīn's position of *wakīl* at the Khwārizmian court, and points out that in the twelfth century, Khwārizm Shāh ʿAlāʾ al-Dīn Muḥammad's "bold reform" (379) transferred power from the Imperial wazīr alone to a mandatory unanimous decision by six *wakīl*s (*Turkestan*, 376–80).

91 Carla L. Klausner stresses that a major innovation of the Seljuks was their attempt to establish close links between the central government and the ʿulamāʾ through state support for the madrasa system of education (*The Seljuk Vezirate: A Study of Civil Administration 1055–1194* [Cambridge, MA: Harvard University Press, 1973], 22). This echos Aydın Sayılı's assertion that the Seljuk system of civil administration looked to the Shāfiʿī madrasas as training grounds for "judges, lawyers, and administrators, secretaries, ministers, ambassadors, political advisers, in short, the personnel for all public and private functions" ("The Institutions of Science and Learning in the Moslem World," Ph.D. diss., Harvard University, 1941, 23).

92 ʿAlāʾ al-Dīn Muḥammad was a patron of Fakhr al-Dīn, and also entrusted him with tutoring his children (see See Frank Griffel, "On Fakhr al-Dīn al-Rāzī's Life and the Patronage He Received," *Journal of Islamic Studies* 13, no. 3 [2007]: 316–17, 331–34). Cf. Fathalla Kholeif, who suggests that this relationship may have started earlier with Fakhr al-Dīn being a tutor to a young Muḥammad during the reign of his father ʿAlāʾ al-Dīn Tekish (567–96/1172–1200) (*A Study on Fakhr al-Dīn al-Rāzī and His Controversies in Transoxiana* [Beirut: Dar El-Machreq Editeurs, 1984], 19).

93 Nasawī, *Sīrat al-Sulṭān Jalāl al-Dīn,* 109–10 (=Houdas, *Histoire du Sultan*, 83, 84). See also Barthold, *Turkestan*, 429.

volumes (in one library alone) among the multitude of scholarly books he scouted that were located throughout Merv.[94]

The dating for Jaghmīnī's two dedications to Shihāb al-Dīn (601-2/1204-5 and 615/1218) span some thirteen years; these dates not only indicate a rather long-standing relationship between the two, but also fall within the long reign of the Khwārizm Shāh ʿAlāʾ al-Dīn Muḥammad (596–617/1200–20). So the composition dates of several of Jaghmīnī's treatises in conjunction with the *floruits* of his two dedicatees all support the contention that Jaghmīnī flourished during his reign and that of the Khwārizm Shāh ʿAlāʾ al-Dīn Tekish (r. 567–96/1172–1200). Where Jaghmīnī lived throughout this period, though, is not at all clear. The last composition date we have for him is 615/1218, which as mentioned earlier is given for his *Talkhīṣ kitāb Ūqlīdis*. If he managed to evade the ensuing massacres that occurred in the cities of Bukhara (616–17/1219–21) and Samarqand (617/1220-21),[95] he would have witnessed the ushering in of the reign of Jalāl al-Dīn, which occurred in 617. However, it is equally possible that ultimately he became a victim of one of these many raging battles that ravaged the regions of Khurāsān and Khwārizm and destroyed major centers of learning, such as the cities of Merv and Gurgānj (in 617–18/1220–21), where most likely Jaghmīnī was residing; hence we have a viable explanation for the 618 H death date for him that surfaced (unreferenced) in the Islamic reference sources.[96]

We know more specifically about the fate of Shihāb al-Dīn; it is reported that his ill-fated advice to the Khwārizm Shāh ʿAlāʾ al-Dīn Muḥammad eventually led to his fleeing to the city of Nasā,[97] along with his son Tāj al-Dīn, where Nasawī writes that they both perished circa 1220. Nasawī also informs us of Shihāb al-Dīn's

---

[94] See Yāqūt, *Muʿjam al-buldān,* 5:114; Svat Soucek provides an English translation of these relevant parts in *A History of Inner Asia* (Cambridge: Cambridge University Press, 2000), 114–15 ("The conquering Mongols").

[95] Accounts vary whether it was 616 or 617 H [1220] for the capture of Bukhara. However, all the sources agree on the ensuing devastation; Juwaynī describes how Chingiz Khān "martyred the whole of the inhabitants, put the ʿUlamāʾ to the sword, and gave the libraries of books to the flames." He then marched towards Samarqand and captured it on 617/1220 (*Ṭabaqāt-i Nāṣirī*, Raverty [trans.], 274–75); cf. Ibn al-Athīr, *al-Kāmil fī ʾl-taʾrīkh,* 12:361–68 (=Richards, *The Chronicle of Ibn al-Athīr*, 204–10); and Juwaynī, 1:75–84 (=Boyle, *History of the World*, 1:97–109 [XVI: Of the Capture of Bukhara]).

[96] Ibn al-Athīr, *al-Kāmil fī ʾl-taʾrīkh,* 12:389–95, esp. 394–95 [On the Destruction of Khwārazm] (=Richards, *The Chronicle of Ibn al-Athīr*, 224–28, esp. 227–28); Juwaynī, 1:97–101 and 119–32 (=Boyle, *History of the World*, 1:123–28 [Of the Fate of Khorazm] and 1:153–68 [XXVII: Of Merv and the Fate Thereof]); and Barthold, *Turkestan*, 436–37.

[97] Nasā, also the hometown of al-Nasawī and where Shihāb al-Dīn is buried, is situated in Khurāsān [near modern-day Ashgabat, Turkmenistan] and was considered a five-day journey westward from Merv, two days from Sarakhs, one day from Abīvard, and six or seven days from Nīsābūr. See Yāqūt, *Muʿjam al-buldān,* 5:282; and "Nésâ and Nisa" in Houdas, *Histoire du Sultan*, 458. See also V. Minorsky [C. E. Bosworth], "Nasā, Nisā" in *EI2* (1993), 7:966–67.

valiant attempts to preserve what he considered the most valuable books, but concludes that ultimately they were lost.[98]

Destroyed perhaps, but their contents were not all completely lost; this in light of the extant scientific works that date from this tumultuous period composed by Jaghmīnī among other writers. Conceivably many of these works were able to circulate to safer lands due to having been copied (possibly multiple times) either before the eye of the storm actually hit the region or between waves of attacks. In any event, some twenty-five years after the devastation, specifically in 644 H [1246-47 CE], we find a copy of the *Mulakhkhaṣ* surfacing (=MS L); and shortly thereafter, we find two extant copies of a treatise that Jaghmīnī composed on arithmetic, both bearing colophon dates from the seventh/thirteenth century, and one explicitly stating that it was completed in the Ṣadriyya madrasa in Khwārizm in 661/1263.[99] One is reminded of Mark Twain's 1897 retort upon reading of his demise: "...the report of my death was an exaggeration."; so too are the false proclamations concerning the demise of Islamic science during this period.

I.2.4 So What's in a Date?

Pinpointing that Jaghmīnī flourished in the late twelfth/early thirteenth centuries, and resolving once and for all that there was only *one* of him—one scholar who authored both the *Mulakhkhaṣ* and the *Qānūnča*—is not insignificant. More is at stake than just finally putting to rest repeated errors contained within the secondary sources; determining that Jaghmīnī flourished during the period of the reign of the Khwārizm Shāhs in the region of Central Asia *prior* to the Mongol invasions has a major impact on how we view the development of scientific inquiry within Islamic society during the premodern period. That he was writing elementary textbooks on a variety of scientific topics such as astronomy, medicine, and mathematics also raises other questions, such as the makeup of his target audience. This demand for scientific textbooks within Islamic lands is clearly a strong indication that science had not dwindled to a handful of individuals, nor was dependent on these few to keep the scientific torch burning.[100]

This also means that we should consider that the massive scientific efforts occurring in thirteenth-century Marāgha under the directorship of Naṣīr al-Dīn al-Ṭūsī (597–672/1201–74) was a remarkable resuscitation by Ṭūsī (and others) of a

---

[98] Nasawī, *Sīrat al-Sulṭān Jalāl al-Dīn,* 110–11, 115 (=Houdas, *Histoire du Sultan*, 84, 88–89); Barthold, *Turkestan*, 424, 429–30. One can certainly sympathize with Nasawī's anguish regarding the loss of massive numbers of scholarly works; clearly, there was no suitcase large enough to contain them all.

[99] Specifically, Tehran, University of Tehran, Central Library and Documentation Center, MS 6911, p. 12, states it was completed Monday, at noon, the beginning of Rabīʿ II 660 [=probably, 27 February 1262]; and Princeton, Princeton University, Garrett MS 502H, f. 51a states it was completed in Khwārizm, at the Ṣadriyya madrasa at the end of Shawwāl 661 [early Sept. 1263].

[100] See F. J. Ragep, "When Did Islamic Science Die (and Who Cares)?" 1–3.

well-established mathematical tradition within the fabric of Islamic society that had been interrupted, but not curtailed or terminated, due to the disruptions resulting from the Mongol invasions and political events of the late twelfth/early thirteenth century. Quṭb al-Dīn al-Shīrāzī provides a historical summing up of the *hay'a* literature up to his time in his major astronomical work *Nihāyat al-idrāk fī dirāyat al-aflāk* (the first version completed in 680/1281); it should not go unnoticed that included in his list are a number of pre-Mongol treatises, Jaghmīnī's *al-Mulakhkhaṣ* included.[101]

The *hay'a* literature, this rich corpus of works on theoretical astronomy that Jaghmīnī inherited and built upon (and was ultimately disseminated through generations), is the focus of the next section.

§ I.3 An Overview of Summary Accounts of Astronomy Before the *Mulakhkhaṣ*

Jaghmīnī's elementary astronomical work *al-Mulakhkhaṣ fī al-hay'a al-basīṭa* came on the scene in the early thirteenth century; it would become one of the most popular textbooks on theoretical astronomy ever written in Islamic lands and would play a critical role in its development. As with other *hay'a* texts, Jaghmīnī's stated aim was to introduce the reader to the entirety of the cosmos, which included both the celestial and sublunar realms (see [Preface] and Introduction). He makes it clear that the cosmos, or "World" (*al-ʿālam*), is composed of bodies, these bodies being the subject of his treatise. However, since this does not exactly correspond to "astronomy," either in the modern sense or even in the sense in which the term *astronomia* was used in Hellenistic Greece, this section explores how *ʿilm al-hay'a* (the science of *hay'a*) came into being in an Islamic context, and how Jaghmīnī's text fits into this genre, both in content and historically.[102]

---

[101] See F. J. Ragep, "Shīrāzī's *Nihāyat al-Idrāk*: Introduction and Conclusion," 51 [Arabic], 55 [Eng. trans.]. Shīrāzī specifically cites *al-Mulakhkhaṣ* as one of "the books…set forth and composed in this discipline." This is yet another indication that Jaghmīnī flourished prior to the *Nihāya*'s composition date (i.e., 680/1281).

[102] My goal is not to provide a general survey of *hay'a* literature, but rather to highlight *hay'a* works prior to Jaghmīnī's *Mulakhkhaṣ* that were mainly used as introductory texts for teaching purposes. For overviews of *hay'a*, see F. Jamil Ragep: "Astronomy," *Encyclopaedia of Islam, THREE* [hereafter cited as *EI*3], ed. Gudrun Krämer et al., Brill Online, 2014. Reference. McGill University. 03 March 2014 http://referenceworks.brillonline.com/entries/encyclopaedia-of-islam-3/astronomy-COM_22652 (esp. part 1, "Theoretical astronomy and cosmology"); "*Hay'a*," in *Encyclopaedia of the History of Science, Technology, and Medicine in Non-Western Cultures*, ed. Helaine Selin (Dordrecht: Kluwer Academic Publishers, 1997), 395–97; and *Tadhkira*, 1:33–41 ("The *Tadhkira* as Genre"). See also Y. Tzvi Langermann, *Ibn al-Haytham's "On the Configuration of the World"* (New York: Garland [Harvard Dissertations in the History of

I.3.1 The Meaning of ʿIlm al-hayʾa

Jaghmīnī's *al-Mulakhkhaṣ* is part of a genre of astronomical literature termed *ʿilm al-hayʾa,* which developed at least as early as the eleventh century in eastern Islam and replaced *ʿilm al-nujūm* (the science of the stars), or sometimes "astronomia" as transliterated from the Greek, as the general term for the discipline of astronomy.[103] Naṣīr al-Dīn al-Ṭūsī (d. 672/1274) provides us with what would become the classical definition of the discipline: "The subject of astronomy is the simple bodies, both superior and inferior, with respect to their quantities, qualities, positions, and intrinsic motions."[104] This new delineation of astronomy focused on topics related to the configuration (*hayʾa*) or structure of the universe as a coherent whole, in other words its subject matter dealt with both the upper bodies of the celestial region ("cosmo-graphy") and the lower bodies of the terrestrial realm ("geo-graphy"). According to Qāḍīzāde, a fifteenth-century commentator on the *Mulakhkhaṣ*, this definition was a way "modern" Islamic astronomers (in which he includes Jaghmīnī) differentiated their science from that of the ancient Greeks in that it brought together the unchanging realm of the celestial aether and the ever-changing realm of the four elements, the world of generation and corruption, into a single discipline.[105] And although one finds topics dealing with the inhabited world included within Greek astronomical works, indeed a prominent example being

---

Science], 1990), 25–34 ("Predecessors and the *hayʾah* tradition"); and David Pingree, "*ʿIlm al-hayʾa,*" in *EI2* (1971), 3:1135–38.

[103] E.g., in the tenth century *ʿilm al-nujūm* is still being used in Islamic reference books by al-Fārābī in his *Enumeration of the Sciences*, Abū ʿAbd Allāh al-Khwārizmī in his *Mafātīḥ al-ʿulūm*, and the Ikhwān al-Ṣafāʾ in Epistle 3 of their *Rasāʾil* as the general term for astronomy (with the latter two designating *ʿilm al-hayʾa* as a branch). However, *ʿilm al-hayʾa* becomes the general term in Ibn Sīnā's (d. 1037) *Aqsām al-ʿulūm al-ʿaqliyya* (Classification of the Rational Sciences), and it becomes synonymous with astronomy in most accounts of the discipline after this time (see Ragep, *Tadhkira*, 1:34–37).

[104] See Ragep, *Tadhkira*, I.Intr.[2] (1:90–91), and 1:38 (on "All simple bodies as the subject matter of astronomy"). Some seventy-five years later, one finds a similar definition of the discipline by the Egyptian encyclopaedist Ibn al-Akfānī (d. 749/1348): "the science from which one learns the situations of the lower and upper simple bodies, their forms, their positions, their magnitudes, the distances between them, the motions of the orbs and the planets and their amounts. Its subject is the aforementioned bodies from the point of view of their quantities, positions, and inherent motions" (Jan J. Witkam, *De Egyptische Arts Ibn al-Akfānī* [Leiden: Ter Lugt Pers, 1989], 408).

[105] Qāḍīzāde (*Sharḥ al-Mulakhkhaṣ*, Istanbul, Ayasofya MS 2662, f. 2b) informs us:

ويمكن أنْ يكون المراد بهيئة العالم علم الهيئة الذي يبحث فيه عن أحوال **الأجرام البسيطة العلوية والسفلية من حيث الكمية والكيفية والوضع والحركة اللازمة لها وما يلزم** منها. وإنّما أطلقنا القول في البسائط السفلية لانّ المتأخّرين ومنهم المصنّف تعرّضوا لها مطلقاً وإنْ لم يتعرّض صاحب المجسطي منها إلاّ لكرة الأرض والماء معاً.

"…[Also] it is possible that [Jaghmīnī's] intention by *hayʾa* of the world is the science of *hayʾa* in which one studies the states of the **superior and inferior simple bodies, with respect to their quantity, quality, position, and intrinsic motion** and what pertains to them. Moreover, we have used the term "the lower simples" without qualification because the

Ptolemy's *Almagest*, Book II,[106] it is significant that Islamic astronomers saw themselves as doing something new and considerably expanded.[107]

One major consequence of the recategorization of the discipline from *ʿilm al-nujūm* to *ʿilm al-hayʾa* is that *hayʾa* is no longer a subdivision of astronomy but becomes the term for the field in general. *Hayʾa basīṭa* (plain *hayʾa*) then becomes one branch of the discipline, which provides a general overview of cosmography but devoid of geometrical proofs and complex mathematical derivations;[108] Jaghmīnī's *al-Mulakhkhaṣ fī al-hayʾa al-basīṭa* falls into this category. Thus, the genre of *hayʾa* literature is recognized as a strictly mathematical discipline with an emphasis on transforming mathematical models of celestial motion into physical bodies as a way of providing a picture of the universe as a whole; and its focus addresses the external aspects of cosmology, in other words issues related to "how" the celestial and terrestrial realms operate the way they do, and not with dealing with questions of "why." The fact that *hayʾa* works do not discuss subjects related to the "causes" of natural phenomena and matters of Aristotelian metaphysics is quite significant; however, it should be duly noted that this is not because these issues are unimportant, but rather because the internal aspects of cosmology, or matters related to natural philosophy, were dealt with elsewhere.[109]

Some modern studies of the discipline of *ʿilm al-hayʾa* maintain that Islamic astronomers regarded the universe "purely as a mathematical construct having no necessary physical counterpart," at least until Ibn al-Haytham came on the scene in the eleventh century.[110] This interpretation categorizes *ʿilm al-hayʾa,* at least before

---

Moderns, among whom is the author [i.e., Jaghmīnī], deal with it without restriction even though the author of the *Almagest* only presents the sphere of the Earth and water together [i.e., without the other lower simple elements]."

Note that Qāḍīzāde's definition for the subject matter of the discipline of astronomy [in bold] is strikingly similar to that of Ṭūsī's (*Tadhkira*, I.Intr.[2]) quoted above. Qāḍīzāde suggests this definition could also apply to Jaghmīnī, since Jaghmīnī does not provide one in the *Mulakhkhaṣ* within his explanation of *hayʾa* and discussion of the simple bodies (see *Mulakhkhaṣ*, Intr.[1]).

[106] Moreover, Ptolemy is the authority Jaghmīnī relies on for matters pertaining to the terrestrial region (*Mulakhkhaṣ*, II.1[2]).

[107] This was pointed out by F. J. Ragep, *Tadhkira*, 1:38; and Ragep, "Astronomy in the Fanārī-Circle: The Critical Background for Qāḍīzāde al-Rūmī and the Samarqand School," in *Uluslararası Molla Fenârî Sempozyumu (4–6 Aralık 2009 Bursa)* (International symposium on Molla Fanārī, 4–6 December 2009 Bursa), ed. Tevfik Yücedoğru et al. (Bursa, 2010), 165–76.

[108] Ragep, *Tadhkira*, 1:35, 36–37.

[109] Jaghmīnī informs us that *why* simple bodies are spherical when left unimpeded and in their natural state is "shown in another science" (*Mulakhkhaṣ*, Intr.[1]). Likewise, Ṭūsī explicitly states that there is a demarcation of how to deal with the same subject matter between disciplines and that the science of *hayʾa* relies on principles "proved in another science and are taken for granted in this science" (see *Tadhkira,* Intr.[1], 1:90–91).

[110] See Pingree, "*ʿIlm al-hayʾa*," 3:1135–36. Carlo Nallino takes a similar position: "…Like Ptolemy, the most ancient Arabic astronomers neglect to define the idea of the celestial spheres and limit themselves to considering them in the mathematical aspect of ideal circles

Ibn al-Haytham, as dealing with the universe as a nonrealistic geometric structure, one endorsed by Ptolemy himself, in which the models contained in the *Almagest* were mathematical devices or fictions designed to account accurately for observations (i.e., "to save the phenomena") and for their predictive ability.[111] This view of *ʿilm al-hayʾa*, which reduces the debate to an either/or situation (i.e., geometrical constructs versus physical realities), significantly ignores that most Islamic astronomers believed that the mathematical models needed to be compatible with a physical representation of the universe (as Ptolemy himself had shown in his *Planetary Hypotheses*).[112] It also assumes that physical bodies were only first introduced with Ibn al-Haytham's *On the Configuration of the World*, certainly not a clear-cut conclusion.[113]

Unlike works falling under the rubric of *ʿilm al-nujūm*, another significant feature of *hayʾa* works was the exclusion of topics on astrology, especially those espousing predictive capabilities related to future events; and this dissociation of

---

representing the movements of the heavenly bodies. The Aristotelian conception of solid spheres was introduced for the first time into a purely astronomical treatise by Ibn al-Haitham" ("Sun, Moon, and Stars (Muhammadan)," in *Encyclopædia of Religion and Ethics*, ed. James Hastings [Edinburgh: T. & T. Clark; New York: Charles Scribner's Sons, 1921], 12:99). In defense of these earlier scholars, a few Islamic astronomers did focus on imaginary circles rather than solid spheres, a point made by al-Kharaqī (d. 553/1158) who claims that Ibn al-Haytham was one of the first to emphasize real spheres; see F. J. Ragep, *Tadhkira*, 1:33 and 2:454.

[111] An assertion that Ptolemy's geometrical models were only mathematical fictions with no basis in reality ignores or downplays his great cosmological work *The Planetary Hypotheses*; this is discussed below in *I.3.2: Ancient Forebears* (see Proclus). G. E. R. Lloyd's seminal article, "Saving the Appearances," provides an adept analysis of the "instrumentalist" and "realist" debate and its repercussions on the interpretation of ancient Greek science. Lloyd includes a discussion of Pierre Duhem (d. 1916), the foremost proponent of the instrumentalist view, whose insistence that Ptolemy was an instrumentalist (despite evidence to the contrary) was intertwined with upholding a methodological approach for the development of the history and philosophy of science, one I might add not favorable to Arabs (*Methods and Problems in Greek Science: [Selected Papers]* [Cambridge: Cambridge University Press, 1991], 248–50). For more on Duhem and his ramifications, see F. J. Ragep, "Duhem, the Arabs, and the History of Cosmology," *Synthese* 83 (1990): 201–14; and Ragep, "Freeing Astronomy," 51–52, esp. n9. And for evidence that this debate still continues, see Peter Barker, who argues that Peurbach's introduction of Ptolemaic geometrical models as physically real corporeal orbs, rather than mathematical fictions, was innovative and a new departure rather than a culmination of the old *theorica* tradition ("The Reality of Peurbach's Orbs: Cosmological Continuity in Fifteenth and Sixteenth Century Astronomy," in *Change and Continuity in Early Modern Cosmology*, ed. Patrick J. Boner, Archimedes 27: New Studies in the History and Philosophy of Science and Technology [Dordrecht; New York: Springer, 2011], 7–32).

[112] For G. E. R. Lloyd's stance on the *compatibility* of the two options, see "Saving the Appearances," 250. See also, Ragep, "Duhem, the Arabs, and the History of Cosmology," 210.

[113] Langermann argues that this is a misconception and that "it is quite clear…that Ibn al-Haytham does not regard himself to be the first person to address the problem of the physical description of the heavens" (*Ibn al-Haytham's "On the Configuration of the World,"* 25).

ʿilm al-hayʾa from astrology had important ramifications, a prominent one being that it helped to secure *hayʾa* a niche within Islamic religious circles. It should not be surprising that a strictly scientific discipline based on mathematics and observations would be far less objectionable to certain Islamic theologians (mutakallims) than one that seemingly limited God's omnipotence, with claims of a parallel ability to make judgments by tapping into the powers of the stars. George Saliba has maintained that the need to demarcate astronomy from astrology gave birth to the genre of *ʿilm al-hayʾa*, this being a corpus of literature within a strictly Islamic context distinct from, and free of the stigma attached to, the Greek astronomical tradition that had been appropriated into Islamic society with the ninth-century translation movement.[114]

However, one should not conclude that any "Islamic" corpus of scientific and philosophical works totally eliminated Greek or any other "foreign" elements;[115] and it would also be misguided to assume that there was a strict demarcation with no overlap in subject matter between these (or other) disciplines.[116] Undoubtedly, the role of the astrologer was multifaceted within medieval Islamic society, and the practice of astrology was widespread and quite popular in some circles. However, any discipline, perhaps especially a scientific one, that incorporated tenets of Aristotelian natural philosophy and/or relied on Greek, Indian, and Persian sources attracted its share of critics as well as adamant supporters.[117] Few could deny the allure of a discipline that dangles "the promise of predictive power over a full scale of phenomena ranging from cosmic events to the outcome of a battle or the length

---

[114] See George Saliba: "Islamic Astronomy in Context: Attacks on Astrology and the Rise of the *Hayʾa* Tradition," *Bulletin of the Royal Institute for Inter-Faith Studies* 2, no. 1 (Spring/Summer 2002): 25–27, 42; "The Development of Astronomy in Medieval Islamic Society," in *A History of Arabic Astronomy: Planetary Theories during the Golden Age of Islam* (New York: New York University Press, 1994), 53–61, 65; and "Arabic versus Greek Astronomy: A Debate over the Foundations of Science," *Perspectives on Science* 8, no. 4 (2000): 328–29, 330.

[115] Saliba specifically states, "the attack on astrology did not entail a rejection of the foreign sciences altogether" ("The Development of Astronomy in Medieval Islamic Society," 56; and "Astrology/Astronomy, Islamic," in *A History of Arabic Astronomy*, 66–81).

[116] Many Islamic scholars straddled multiple subjects, and there could be considerable overlap in disciplines. So, for example, we find that in his fifteenth-century commentary to Ṭūsī's "theological" work, the *Tajrīd al-ʿAqāʾid*, ʿAlī Qushjī discusses astronomy and puts forth the "case that astronomy should dispense with its dependence upon Aristotelian physics" and that the "the Earth's rotation is a possibility" (see F. J. Ragep, "ʿAlī Qushjī and Regiomontanus: Eccentric Transformations and Copernican Revolutions," *Journal for the History of Astronomy* 36, no. 4 [2005]: 360–61 and Ragep, "Freeing Astronomy," 61–63).

[117] Saliba provides an overview of the social status of the astrologer between the ninth and eighteenth centuries that includes a detailed examination of the pros and cons of astrology ("The Role of the Astrologer in Medieval Islamic Society," *Bulletin d'Études Orientales* 44 [1992]: 45–67). For a discussion of astrology as a scientific discipline and some accepted methods of argumentation, see Charles Burnett, "The Certitude of Astrology: The Scientific Methodology of al-Qabīṣī and Abū Maʿshar," *Early Science and Medicine* 7, no. 3 (2002): 198–213.

of an individual's life";[118] nevertheless, opponents of the practice of astrology and alchemy—and they ranged from Hellenized philosophers to religious adherents—found much fault among the practice itself and its practitioners, not the least of these being claims of special abilities for interpreting God's divine will.[119] Therefore, astronomical treatises that lacked the taint of astrology, such as *hayʾa* works, were presumably far less objectionable for inclusion within religious institutions.[120] On the other hand, content contained in *hayʾa* works was highly indebted to the scientific works of the Greeks and other "foreign" sources, with the result that making them more suitable for a broad Islamic audience presented a challenge. So how did Islamic scholars adapt or reformulate subject matter into an astronomy that was "distinctly Islamic"?[121]

When Badr al-Dīn al-Qalānisī requested that Jaghmīnī compose an elementary *hayʾa basīṭa* textbook sometime in the late twelfth/early thirteenth centuries, he clearly felt that there was a pressing need for an abridged version on the subject matter of *ʿilm al-hayʾa*; however, it was a genre that had been several centuries in the making and so by then had become an established discipline. Jaghmīnī then was confronted with a rather daunting task (*Mulakhkhaṣ,* II.3[11]) in that by that period he had inherited quite an extensive corpus of sources as well as pedagogical styles to choose from.[122] So what follows is an overview of some formative sources from late antiquity up until the late twelfth century that were inherited by Muslim scholars

---

118 See A. I. Sabra, "Configuring the Universe. Aporetic, Problem Solving, and Kinematic Modeling as Themes of Arabic Astronomy," *Perspectives on Science* 6, no. 3 (1998): 289.

119 For a scathing critique against both astrology and alchemy, see Ibn Khaldūn, *The Muqaddimah: An Introduction to History*, trans. Franz Rosenthal, 3 vols. (Princeton, NJ: Princeton University Press, 1967), 3:258–67 and 3:267–81 (respectively).

120 That a *hayʾa* work such as the *Mulakhkhaṣ* was viewed as one "dedicated purely to interests of science" was duly noted by Rudloff and Hochheim. Although they were writing from a nineteenth-century perspective, both assumed that Jaghmīnī's exclusion of astrological discussions indicated that "he must have looked down on it with contempt" and that "Jaghmīnī's abstinence is all the more to be admired since astrological ambition of the time held a lot of attraction for the easily aroused imagination of the Oriental, and furthermore under favorable circumstances brought a lot of profit" ("Die Astronomie des Maḥmûd ibn Muḥammed ibn ʿOmar al-Ǵagmînî," 215–16). Their views probably indicate more about the attitudes of nineteenth-century German scholars than late twelfth-century Islamic ones. Also recall that Jaghmīnī composed a separate work on astrology (see *I.2.3c: Further Evidence for Dating Jaghmīnī* and Appendix I, no. 4).

121 Saliba: "The Development of Astronomy in Medieval Islamic Society," 65.

122 Liba Taub points out that it is often neglected that authors writing on scientific, mathematical, and medical subjects had numerous options available to them to convey their ideas and information. Her focus is scientific texts, and she explores how mathematical ones display a "variety of forms, or genres, including, but not limited to, poetry, dialogue, lecture, question-and-answer text, letter, biography, recipe, epitome, encyclopedia and commentary" in "On the Variety of 'Genres' of Greek Mathematical Writing: Thinking about Mathematical Texts and Modes of Mathematical Discourse," in *Writing Science: Medical and Mathematical Authorship in Ancient Greece*, ed. Markus Asper (Berlin/Boston: Walter de Gruyter, 2013), 333–34.

or written during the early Islamic period that could arguably have been at Jaghmīnī's disposal, either directly or indirectly, and found suitable to use and modify for his brand of theoretical astronomy so as to comply with Badr al-Dīn's lofty proposal that he compose an elementary introduction to *ʿilm al-hayʾa*.

I.3.2 Ancient Forebears

I.3.2a Ptolemy's Predecessors

According to Otto Neugebauer, the eminence of the scientific works of the Alexandrian Claudius Ptolemy in the second century CE would cause "an almost total obliteration of the prehistory of the Ptolemaic astronomy."[123] Indeed, once Ptolemy came on the scene, he undoubtedly had a profound impact on theoretical astronomy, including the development of the *hayʾa* tradition that became dependent on his works. Nevertheless, before moving on to this, it is worth highlighting the emergence of "a particular literary *topos*, the introduction," beginning in the first century BC as a way for writers to present views of celestial science (*astrologia* in Latin; ἀστρολογία in Greek) to their readers.[124] According to James Evans and J. Lennart Berggren, the demand for popular surveys in the Hellenistic period led to the production of "comprehensive astronomy textbooks written at an elementary level"; and both view these works as forming a Greek genre of elementary textbooks on astronomy, even though they duly point out that the works "differ markedly in tone" as well as content and period composed.[125] Some examples included in their "corpus" are Geminus's *Introduction to the Phenomena* (also referred to as the *Isagoge*) (first century BCE), Theon of Smyrna's *Mathematical Knowledge Useful for Reading Plato* (second century CE), and Cleomedes's *Meteora* (third to fourth centuries CE). Although the titles alone indicate their diversity, they all provide general descriptive overviews of astronomy (some more than others) and cover a variety of topics and basic concepts on celestial science.

However, it is unclear whether these works were actually intended as elementary astronomical textbooks. Indeed, the huge discrepencies between these works—with respect to content, structure, and literary style—raise serious questions about what it means to lump together general works dealing with astronomical topics (admittedly with some overlap) and refer to them as a genre of "elementary textbooks" or "introductions," especially when they clearly contain significant differences and

[123] Otto Neugebauer, *A History of Ancient Mathematical Astronomy*, 3 parts (Berlin; New York: Springer-Verlag, 1975), 1:5 [hereafter cited as *HAMA*].

[124] Alan C. Bowen, "Three Introductions to Celestial Science in the First Century BC," in *Writing Science*, 299–300, 319, 326–27.

[125] James Evans and J. Lennart Berggren, *Geminos's Introduction to the Phenomena: A Translation and Study of a Hellenistic Survey of Astronomy* (Princeton, NJ: Princeton University Press, 2006), 8, 10.

may have been written centuries apart.[126] Thus Alan Bowen understandably underscores the need for "a more carefully thought out notion of what an introduction is," especially without real evidence to support the claim that these works were intended to form part of a curriculum of study.[127] Bowen points out that the lack of standardization between these textbooks makes it unclear what exactly was being taught, who the targeted audience was, and what a more advanced study of these topics would have entailed.[128]

On the other hand, the discrepancies between these works provide insights into the range of topics that were of interest, the level of proficiency, the influences at work, sources used, and so on during this time.[129] For example, it is interesting that within Geminus's hodge-podge of topics that purportedly cover "all important branches of Greek astronomy," he omits planetary theory, but includes discussions on the limitations of weather prognostication and the astrological doctrine of the "aspects" according to which Babylonian astrologers calculated the zodiacal signs' influence on human affairs—indications that the role and veracity of astrological theory and practice were major concerns during this period.[130] Also, several of these works include causal explanations to account for various aspects of celestial science,[131] which distinguishes them from *hayʾa* treatises that sought to weed out philosophical issues in order to confine the subject matter to dealing with only the external aspects of the celestial bodies. Finally, we can note that Babylonian astronomy/astrology made inroads into Greek celestial science during this period, although the depth of its penetration seems to be a matter of disagreement.[132] Nevertheless, their employment of the Babylonian sexagesimal system (dating back to

---

[126] Evans and Berggren, *Geminos's Introduction to the Phenomena*, 8. Evans and Berggren seemingly find it unproblematic to refer to Geminus's work as an introduction even though they acknowledge, "we cannot be sure that *Introduction to the Phenomena* is the title that Geminos himself gave it," 3.

[127] See Bowen, "Three Introductions to Celestial Science in the First Century BC," 303n11, 319–20. Cf. Evans and Berggren, *Geminos's Introduction to the Phenomena*, 8.

[128] Bowen, "Three Introductions to Celestial Science in the First Century BC," 318–19.

[129] For example, unlike Geminus (and others), Jaghmīnī does not employ literary references in the *Mulakhkhaṣ* (unless one counts his dedicatory verse to Badr al-Dīn [see Preface]), and we know that he composed poetry from another of his works (see *Talkhīṣ kitāb Ūqlīdis*, 247–49).

[130] See Evans and Berggren, *Geminos's Introduction to the Phenomena,* 2, 8–9, 105–6; 220–22 (Ch. XVII [15–23]: The Stars Indicate But Do Not Cause The Weather); and 125–36 (Ch. II [Aspects of the Zodiacal Signs]); Neugebauer, *HAMA*, 2:581–88; and D. R. Dicks, "Geminus," in *Dictionary of Scientific Biography* [hereafter cited as *DSB*] (New York: Charles Scribner's Sons, 1972), 5:345–46.

[131] Bowen discusses Geminus's "emphasis" on causation in "Three Introductions to Celestial Science in the First Century BC," 320–26. See also Ragep, *Tadhkira*, 1:39–40; and Evans and Berggren, *Geminos's Introduction to the Phenomena,* 49–51.

[132] Neugebauer is impressed with how far Babylonian astronomy had penetrated into early Greek mathematical astronomy" (*HAMA,* 2:579), but Evans and Berggren conclude (without explanation) that "...this material was still being absorbed and adapted by the Greeks" (*Geminos's Introduction to the Phenomena,* 22; cf. Geminus's references to the Babylonians,

Eratosthenes circa 250 BCE) would be the hallmark of a sound astronomical textbook in that it indicates a concern for precision. Indeed, it became *the* notation for Ptolemaic parameters and subsequent *hayʾa* works (including the *Mulakhkhaṣ*), and was a system so widely used by Islamic astronomers that it became known as "the astronomers' arithmetic."[133]

I.3.2b Ptolemy[134]

In the second century BCE, the Alexandrian Claudius Ptolemy proposed a coherent picture of the universe consisting of contiguous or nested planetary spheres around an immobile, spherical Earth; each sphere contained embedded within it additional non-concentric eccentric spheres and epicycles whose various combinations of motions accounted for perceived observations. This so-called "Ptolemaic system" transformed ancient mathematical astronomy; the nesting principle for the orbs would become the cornerstone of *hayʾa*,[135] and thus was crucial for its development and consequently its textbook tradition.

Though certainly not a "user-friendly" textbook, Ptolemy's great comprehensive compilation of Greek mathematical astronomy, *Mathematike Syntaxis*, also commonly known as *The Almagest* (*al-Majisṭī*), supplanted most of the work of his scientific predecessors; and it became the standard textbook on astronomy for more "advanced" students in Alexandria (and presumably Athens and Antioch). Ptolemy assumes that the student is familiar with elementary geometry as well as some basic terminology and concepts, at least schooled enough to have "already made some progress in the field."[136] The work, in thirteen books, provides geometric models,

---

13–15, 125, 192n4, 228–29). See also, Bowen, "Three Introductions to Celestial Science in the First Century BC," 306–8, 316–18, 322–26.

[133] See J. Lennart Berggren, *Episodes in the Mathematics of Medieval Islam* (New York: Springer-Verlag, 2003), 41. See also Neugebauer, *HAMA,* 2:590–93; and G. J. Toomer, *Ptolemy's Almagest*, trans. and annot. G. J. Toomer (Princeton, NJ: Princeton University Press, 1998), 6–7.

[134] For information on Ptolemy, see: Bernard R. Goldstein, "The Arabic Version of Ptolemy's Planetary Hypotheses," *Transactions of the American Philosophical Society*, n.s., 57, no. 4 (1967): 3–55 at 3–4; Langermann's *Ibn al-Haytham's "On the Configuration of the World,"* 15–25; Neugebauer, *HAMA*, 2:834–38 ("Biographical and Bibliographical Data" and "The Almagest"), 2:900–26 ("Planetary Hypotheses" and "Canobic Inscription"), and 2:926–41 ("Additional Writings of Ptolemy"); Alexander Jones, ed., *Ptolemy in Perspective: Use and Criticism of His Work from Antiquity to the Nineteenth Century,* Archimedes 23: New Studies in the History and Philosophy of Science and Technology (Dordrecht; New York: Springer, 2010), xi–xv (Introduction), 217–29 (Bibliography); Liba Taub, *Ptolemy's Universe: The Natural Philosophical and Ethical Foundations of Ptolemy's Astronomy* (Chicago: Open Court, 1993); G. J. Toomer, "Ptolemy," in *DSB* (1975), 11:186–206 (an excellent overview).

[135] Ragep, *Tadhkira*, 2:517.

[136] See *Almagest*, Book I [Preface]; and Toomer, *Ptolemy's Almagest*, 1–2, 6, 37, 37n13. Ptolemy's assumption is that the student has already studied the works of Euclid, and, likely,

along with quantitative parameters, to account for the celestial motion of each of the heavenly bodies (the Sun and Moon, each of the upper and lower planets, and the Fixed Stars), each contained within its own sphere. Ptolemy also provides tables to calculate positions of the heavenly bodies and other phenomena, and values demarcating the climata.[137]

Ptolemy continued to develop and modify his astronomy throughout his career. For example, in the *Almagest* Ptolemy is still uncertain about the order of the spheres (especially for Venus and Mercury) and their distances (*Almagest*, IX.1), and he provides absolute distances only for the Moon (through parallax) and the Sun (through eclipses) based on Earth radii.[138] However, Ptolemy revisits and rectifies these concerns in his later two-part work, the *Planetary Hypotheses* (*Kitāb al-iqtiṣāṣ* or *Kitāb al-manshūrāt*); there he provides absolute distances of the celestial bodies from the Earth (in Earth radii and stades, based on the assumption that the Earth's circumference is 180,000 stades) and sizes so that "these bodies may be fitted together to form a coherent, unified structure, or *hay'a*."[139] Ptolemy's

---

the so-called "Middle Books," a corpus of pre-Ptolemaic Greek texts that would later be recommended for study before tackling the *Almagest*. They were sometimes grouped together under the title of the "little or small astronomy collection." Translated into Arabic in the ninth century, they included: "the *Data*, the *Optics*, the *Catoptrica* and the *Phenomena* of Euclid [fl. ca. 300 BCE]; the *Spherics*, *On Habitations* and *On Days and Nights* of Theodosius [d. ca. 90 BCE]; *On the Moving Sphere* and *On Risings and Settings* by Autolycus [d. ca. 290 BCE]; *On the Sizes and Distances of the Sun and Moon* by Aristarchus of Samos [d. ca. 230 BCE]; *On the Ascensions of Stars* of Hypsicles [d. ca. 120 BCE]; and the *Spherica* by Menelaus" (Régis Morelon, "General Survey of Arabic Astronomy," and "Eastern Arabic Astronomy between the Eighth and the Eleventh centuries," in *Encyclopedia of the History of Arabic Science*, vol. 1, *Astronomy–Theoretical and Applied*, ed. Roshdi Rashed [London: Routledge, 1996], 7, 18–19, 21, 55n6). See also Toomer, "Ptolemy," 187–88.

137 See Jones, *Ptolemy in Perspective*, xi–xii. Ptolemy would also later compile his astronomical computations into a separate work entitled the *Handy Tables*. The Ptolemaic parameters for planetary motions (from his works and tables) greatly influenced the *zīj* literature, which Jaghmīnī refers to in the *Mulakhkhaṣ* (I.2[10]; II.3[7]). For example, the *zīj* of al-Battānī (whom Jaghmīnī mentions (II.3[9])) shows strong Ptolemaic influence (see E. S. Kennedy, "A Survey of Islamic Astronomical Tables," *Transactions of the American Philosophical Society*, n.s., 46, pt. 2 [1956]: 132–33). For the *Handy Tables*, see Anne Tihon and Raymond Mercier, *Ptolemaiou Procheiroi Kanones: Les Tables Faciles de Ptolémée. Volume 1a: Tables A1–A2* (Louvain-la-Neuve: Université catholique de Louvain, Institut Orientaliste, 2011–).

138 *Almagest*, V.13–16, 19. See Neugebauer, *HAMA*, 2:917–22; and Toomer, "Ptolemy," 191–94.

139 Ragep, *Tadhkira*, 2:500. In the *Planetary Hypotheses*, Ptolemy assumes the following order of the celestial bodies: Moon, Mercury, Venus, Sun, Mars, Jupiter, Saturn, Fixed Stars, with the Earth at the center; each planet, as well as the Fixed Stars, is contained within a physical, geocentric sphere, and all these spheres are contiguously fitted exactly together without a void. See Willy Hartner's seminal article, "Mediaeval Views on Cosmic Dimensions and Ptolemy's *Kitāb al-Manshūrāt*," in *Mélanges Alexandre Koyré*, 2 vols. (Paris: Hermann, 1964), 1:254–82; Hartner's work was superseded by Goldstein's "The Arabic Version of Ptolemy's Planetary Hypotheses," 3–55 (which includes an English translation and commentary of Book I, second part, as well as an Arabic facsimile of the entire work (British

Almagest and *Planetary Hypotheses* together provide both the geometrical modeling and the physical structure for a unified celestial and sublunar cosmography, fundamental for any *hayʾa* work.

Although Ptolemy himself states in the introduction to the *Planetary Hypotheses* that he modified (simplified as well as improved) some of the parameters with respect to the *Almagest*, not all *hayʾa* works include the new, improved parameters. Jaghmīnī, for example, opts in the *Mulakhkhaṣ* (for reasons unknown) to use the Ptolemaic values of the *Almagest* rather than those of the *Planetary Hypotheses.*[140] Jaghmīnī also omits any discussion of sizes and distances in the *Mulakhkhaṣ*, which is typically contained in a *hayʾa* work. However, it would be unwarranted to assume that this is due to his being unaware of the *Planetary Hypotheses*. More likely, this is a case in which Jaghmīnī considered that pedagogically the subject of sizes and distances was inappropriate for an elementary textbook. In fact, this is Qāḍīzāde's assessment, namely, that the topic was omitted due to its difficulty.[141]

On the other hand, Jaghmīnī does present "updated" information from Ptolemy's *Geography*, which he wrote after the *Almagest.*[142] In the *Geography,* which was translated into Arabic in the ninth century,[143] Ptolemy showcases topics of the terrestrial realm, in contrast to the relatively minor role they played in the *Almagest*;

Museum MS. arab. 426 [Add. 7473], ff. 81b–102b)). See also Neugebauer, *HAMA*, 2:919–22 (who includes tables comparing the parameters found in the *Almagest* and the *Planetary Hypotheses* with Proclus's *Hypotyposis* and his Commentary to the *Timaeus* along with the values of Thābit b. Qurra); Olaf Pedersen, *A Survey of the Almagest* (Odense: Odense University Press, 1974), 393–97; and Toomer, "Ptolemy," 197.

[140] A specific example of this is Jaghmīnī's parameters for the maximum inclination of the inclined orbs from the zodiacal orb (see *Mulakhkhaṣ*, I.5[16]); Jaghmīnī gives for Mars 1;0 and for Mercury 0;45 [*Almagest*], not 1;50 and 0;10 respectively [*Planetary Hypotheses*]. See also Neugebauer, *HAMA*, 2:907–9; Pedersen, *A Survey of the Almagest,* 392–93; and Swerdlow, "Ptolemy's Theories of the Latitude of the Planets in the *Almagest*, *Handy Tables*, and *Planetary Hypotheses*," 68 (Swerdlow provides here a convenient table of the inclinations for the three Ptolemaic works and the modern values); and Neugebauer, *HAMA*, 2:907–9.

[141] See Qāḍīzāde, *Sharḥ al-Mulakhkhaṣ* (Istanbul, Ayasofya MS 2662, f. 4a) where he states that the difficulty *(ṣuʿūba)* of the subject is the reason for its omission. See also Ragep, *Tadhkira*, 2:500n1. Here one should also keep in mind that Jaghmīnī wrote a separate short treatise on sizes and distances. It is interesting, though, that Jaghmīnī omits all discussion of sizes and distances in the *Mulakhkhaṣ*, but includes in this "elementary" textbook the parameters for planetary latitudes, a subject known for its complexity (according to Swerdlow, "Ptolemy's Theories of the Latitude of the Planets in the *Almagest*, *Handy Tables*, and *Planetary Hypotheses*," 41–42).

[142] Actually, Ptolemy mentions that its publication will be forthcoming in Book II.13 of the *Almagest*.

[143] Ibn al-Nadīm reports that Ptolemy's "Geography of the Inhabited Lands and a Description of the Earth" was a book in eight sections, which was translated several times in the ninth century. He further comments that: "Al-*Kindī* made a bad translation of it and then *Thābit* [ibn Qurrah] made an excellent Arabic translation. It is also extant in Syriac" (see Ibn al-Nadīm, *Kitāb al-Fihrist*, ed. Riḍā Tajaddud [Beirut: Dār al-Masīra, 1988], trans. as *The Fihrist of al-Nadīm: A Tenth-Century Survey of Muslim Culture* by Bayard Dodge, 2 vols.

and he provides "for the first time a mathematically clear theory of geographical mapping" along with a grid of coordinates, reckoned in degrees.[144] Included in this is Ptolemy's latest information on the borders of the *oikoumenê*;[145] and since this is a subject dealt with extensively in *hay'a* textbooks, i.e., directly related to matters of *hay'at al-arḍ*, any Ptolemaic modifications made here would have been of great concern. This is reflected in the *Mulakhkhaṣ*; though Jaghmīnī does not cite the *Geography* specifically, he states: "Ptolemy, after writing the *Almagest,* claimed that he found habitation below the equator to a distance of 16;25 [degrees]" (see II.1[2] and Commentary).[146]

Some have suggested that Ptolemy's *Almagest* and *Planetary Hypotheses* be "linked" with his great astrological work the *Tetrabiblos* as together providing a better understanding of his cosmology;[147] however, there is no evidence that the *Tetrabiblos* played a part in the *Mulakhkhaṣ*. On the other hand, there could have been some valuable borrowing of information between the disciplines of astronomy and astrology due to the overlap of topics, basic terminology and concepts, and parameters.[148]

[New York: Columbia University Press, 1970], 2:640; cf. the more recent critical edition by Ayman Fu'ād Sayyid, 2 vols. in 4 [London: Al-Furqān Islamic Heritage Foundation, 2009], 2/1:216). [Hereafter cited as *Fihrist*. Citations below will refer to the Dodge trans. and the Sayyid Arabic ed.] See also J. Lennart Berggren and Alexander Jones, *Ptolemy's Geography: An Annotated Translation of the Theoretical Chapters* (Princeton, NJ: Princeton University Press, 2000), 50–52 (on early readers and translators). The line of textual transmission of this treatise has not been a straightforward one. Florian Mittenhuber points out that of the fifty-three preserved Greek manuscripts, none were written before the late thirteenth century ("The Tradition of Texts and Maps in Ptolemy's *Geography*," in *Ptolemy in Perspective,* 95).

144 Neugebauer, *HAMA*, 2:934.

145 Neugebauer points out that Ptolemy relied on Hipparchus for many of his basic assumptions in the *Almagest*, and so the various geographical data found in his *Geography* indicate different stages of his development (*HAMA*, 2:939–40). See also Berggren and Jones, *Ptolemy's Geography*, 21–22, 64–77.

146 Cf. Berggren and Jones, *Ptolemy's Geography,* 110 (Book 7); cf. Toomer, *Ptolemy's Almagest*, 82–83 (II.6[1]).

147 Both Toomer and Taub seem to be advocating this position: Toomer points out that Ptolemy regards "the *Tetrabiblios* as the natural complement to the *Almagest*: as the latter enables one to predict the positions of the heavenly bodies, so the former expounds the theory of their influences on terrestrial things" ("Ptolemy," 198); and Taub has asserted that Ptolemy's "detailed demonstration of the planetary order in the *Planetary Hypotheses* served to fortify the foundation of the physical claims in the *Tetrabiblos*" (*Ptolemy's Universe*, 132–33). However, Neugebauer concluded that, "On the whole, the *Tetrabiblios* stands alone" among Ptolemy's works (*HAMA*, 2:897).

148 Jaghmīnī was certainly familiar with the subject of astrology, and composed a short treatise entitled *Fī quwā al-kawākib wa-ḍa'fihā* (The Strengths and Weaknesses of the Planets) (see *I.2.3c: Further Evidence for Dating Jaghmīnī* and Appendix I ("Jaghmīnī's Works," no. 4).

I.3.2c The Ptolemaic Aftermath: Theoretical Astronomy with—and without—Him

In the fifth century, the neo-Platonist Proclus wrote the *Hypotyposis*, a textbook on Ptolemaic theoretical astronomy that has been described as "the first and last summary of the contents of the Almagest from antiquity."[149] Proclus, a director of the "Academy" in Athens, demonstrates a remarkable knowledge of astronomy as well as pedagogical acumen. However, Proclus's *Hypotyposis* was written within a philosophical milieu, and there are differences when compared with some *hayʾa* compositions.[150] Far more than an overview of the *Almagest*, this work provides a detailed examination of the celestial realm as well as instructions on the use and construction of astronomical instruments.[151] Proclus presents ten problems by which he criticizes various attempts by astronomers to account for the irregular movements of the heavenly bodies;[152] and he specifically problematizes the status of epicycles and eccentrics as either geometrical fictions or physical realities, raising objections to *both* options but without choosing sides.[153] Ultimately, Proclus (the Platonist) and Ptolemy (the Mathematician) both believe in the regularity of celestial motion, and seek simple solutions to account for the problem of perceived irregularites.[154] Given his neo-Platonic bent, Proclus's ability to distinguish himself

---

149 Neugebauer, *HAMA*, 3:1036.

150 For example, Proclus's *Hypotyposis* lacks any discussion of the terrestrial realm, which is included in Ptolemy's *Almagest* and a prominent feature of most *hayʾa* works. On the other hand, Ibn al-Haytham, like Proclus, omitted this topic altogether in his *On the Configuration of the World*; and furthermore, whereas Ibn al-Haytham does not discuss the sizes and distances of the planets, Proclus does. For comparisons of Ptolemy and Proclus, see Hartner, "Mediaeval Views on Cosmic Dimensions and Ptolemy's *Kitāb al-Manshūrāt*," 323–40; and Neugebauer (*HAMA*, 2:920–91), who includes tables of comparative values for Ptolemy, Proclus, and Thābit ibn Qurra (2:920–22).

151 Proclus discusses the use and construction of Ptolemy's instrument for determining the obliquity of the ecliptic (*Almagest*, I.12 [Toomer, 61–63]) and also Ptolemy's "ringed" or spherical astrolabe [armillary sphere] (*Almagest*, V.1 [Toomer, 217–19]). See *Procli Diadochi Hypotyposis Astronomicarum Positionum*, ed. Carolus Manitius (Stuttgart: B. G. Teubner, 1974 [original Leipzig: B. G. Teubner, 1909]), 41–55 and 199–213, respectively. Cf. Neugebauer, *HAMA*, 2:1036.

152 Proclus clearly does not believe the *Almagest* is definitive. For example, in point nine we see that he disagrees with Ptolemy regarding the movement of the Fixed Stars; Proclus believes there is no movement, whereas Ptolemy states they move one degree per 100 years (*Almagest,* VII.2 [328]) (see *Procli Diadochi Hypotyposis*, 235).

153 Lloyd's overview of Proclus's *Hypotyposis* highlights key points regarding Proclus's position on astronomy, including how he has been misrepresented as being an instrumentalist. Lloyd asserts that Proclus attempted to reconcile the complex movements of the heavenly bodies with his desire to uphold Plato's "authority," but charged astronomers with not making clear enough "those things that it is possible to grasp" (263) ("Saving the Appearances," 256–64).

154 Ptolemy explicitly states his position on the meaning of "simplicity" in *Almagest*, XIII.2: "Let no one, considering the complicated nature of our devices, judge such hypotheses to be

from Ptolemy and ultimately accept an attitude of agreeing to disagree with him, by raising objections to difficulties contained in the *Almagest* without rejecting Ptolemy outright, strikes me as remarkable and similar to later medieval Islamic astronomers working within the *hayʾa* tradition of "reforming" Ptolemaic astronomy rather than "overthrowing" it.[155]

In addition to Proclus's *Hypotyposis,* mention should be made of several other works sometimes listed as "introductions" to the *Almagest* but which are actually commentaries. One is by Pappus (fl. 320); another is by Theon of Alexandria (fl. Alexandria, second half of the fourth century), who tells us in the preface that he composed the work for his students;[156] and a third is an anonymous work attributed to Eutocius, who also authored a commentary on Apollonius's *Conics* and is considered to be the head of the Alexandrian school between Ammonius and

overelaborated. For it is not appropriate to compare human [constructions] with divine, nor to form one's beliefs about such great things on the basis of very dissimilar analogies. For what [could one compare] more dissimilar than the eternal and unchanging with the ever changing, or that which can be hindered by anything with that which cannot be hindered even by itself? Rather, one should try, as far as possible, to fit the simpler hypotheses to the heavenly motions, but if this does not succeed, [one should apply hypotheses] which do fit" (Toomer, *Ptolemy's Almagest*, 600).

[155] By "overthrowing," I am thinking here of the radical reaction against Ptolemaic astronomy found in twelfth-century Andalusia, whereby Islamic scholars rejected it in search of a purer version of Aristotelian cosmology, one free of eccentrics and epicycles, in which planetary motions of spherical bodies with embedded planets occur in uniform, circular motion within homocentric nested spheres about a stationary Earth. Proclus's approach seems more aligned with those Islamic astronomers who attempted to reconcile inconsistences, the form of argument found in the so-called *Shukūk* [Doubts] literature, in which difficulties or objections were raised against ancient authorities. (The term *shukūk* [doubts]) is here meant in the sense of the Greek *aporia*, i.e., not simply for an error to be deleted or corrected, but a difficulty, puzzle or problem to be defined before requiring a particular solution.) See A. I. Sabra: "The Andalusian Revolt Against Ptolemaic Astronomy: Averroes and al-Biṭrūjī," in *Transformation and Tradition in the Sciences: Essays in Honor of I. Bernard Cohen,* ed. Everett Mendelsohn (Cambridge: Cambridge University Press, 1984), 133–35; "Configuring the Universe," 290–1, 297–300; and "Reply to Saliba," *Perspectives on Science* 8, no. 4 (2000): 343.

[156] Theon's commentary on the *Almagest*, a work characterized by Gerald Toomer as "never critical, merely exegetic," suggests a redaction of his Alexandrian lectures; of the original thirteen books, Book XI is lost and only a fragment of Book V survives, but these parts are probably extant in other works ("Theon of Alexandria," in *DSB* [1976], 13:321–22). The work is listed in Ibn al-Nadīm as an "Introduction to 'Almagest' [*Introductio in Almagestum*] with an ancient translation" (*Fihrist*, Dodge 2:641; Sayyid 2/1:217), and also as an introduction in Sezgin, *Geschichte des arabischen Schrifttums*, vol. 6: *Astronomie bis ca. 430 H.* (Leiden: E. J. Brill, 1978), 102 [hereafter cited as *GAS*, 6]. The content of Pappus's commentary, in which only books 5 and 6 are extant, indicates that Theon built on his work. Pappus's work is listed by Ibn al-Nadīm as: "A commentary on Ptolemy's book about finding the plane," and it was translated into Arabic by Thābit ibn Qurra (642). See also Neugebauer, *HAMA*, 2:965–69.

Olympiodorus.[157] All three of these works contain some overlap, and deal with content concerned with explanations of mathematical computations.[158] In other words, their orientation is quite technical, and so one would presume that the target audience would have been rather limited. However, apparently this topic found resonance with some Islamic scholars, such as al-Kindī, who was well known for his attempts to make difficult Greek subject matter more comprehensible for a broader audience.[159] According to Franz Rosenthal, Theon's commentary on the *Almagest* was a major source for al-Kindī's *Kitāb fī al-ṣināʿa al-ʿuẓmā,* a work dealing with the first eight chapters of Book I of the *Almagest,* and from which "Ptolemy's original ideas are often given precedence, but on the whole, Theon's text is followed faithfully."[160]

None of these early Greek works are comparable to Proclus's *Hypotyposis* in providing the reader, in both scope and explanatory detail, a general background of the Ptolemaic system. However, Proclus's work is not listed in the literature as having been translated into Arabic[161] or into Latin (*in toto* or in parts). We do know,

---

[157] Joseph Mogenet considers Eutocius as being the anonymous author of a work he entitles, "*L'introduction à l'Almageste*" (Mémoires de la classe des lettres, Collection in-8°, 2e Série, vol. 51, fasc. 2 [Bruxelles: Palais des Académies, 1956]); however, Wilbur Knorr skeptically views Eutocius's authorship as only a "possibility" (156), in *Textual Studies in Ancient and Medieval Geometry* (Boston: Birkhäuser, 1989), 155–211 (Ch. 7: On Eutocius: A Thesis of J. Mogenet).

[158] See Neugebauer, *HAMA*, 2:1042–43. Regarding these three works, G. J. Toomer concluded that "there is no doubt that they are derived from the same work"; he based this in part on the fact that all three works contain content from the mathematician Zenodorus concerning isoperimetric problems (see "The Mathematician Zenodorus," *Greek, Roman and Byzantine Studies* 13 [1972]: 177, and n1; and Toomer, "Theon of Alexandria," 321). Although Zenodorus's name is not mentioned by Ptolemy in the *Almagest*, extensive excerpts of his proofs of propositions are used by Theon in discussing Ptolemy's section on the sphericity of the heavens (*Almagest*, I.3; see Toomer, *Ptolemy's Almagest*, 39–40, 40n25).

[159] See Franz Rosenthal, "Al-Kindī and Ptolemy," *Studi Orientalistici in Onore di Giorgio Levi Della Vida,* Pubblicazioni dell'Istituto per l'Oriente, 52 (Rome, 1956), 2:440, 444. I return to al-Kindī's *Almagest* commentary later as well as one by Abū Jaʿfar al-Khāzin, whose *Tafsīr al-Majisṭī* also dealt with isoperimetric problems.

[160] Rosenthal is rather explicit in asserting that the "context leaves no doubt that Theon's *Commentary* is al-Kindī's source" and that in parts of the work "al-Kindī follows Theon almost literally but expands the discussion in some places" ("Al-Kindī and Ptolemy," 2:446, 449n2, 450).

[161] Ibn al-Nadīm does include "Diadochus *Proclus*, the Platonist"; however, the *Hypotyposis* is not listed among his works translated into Arabic (*Fihrist*, Dodge 2:607–8; Sayyid 2/1:173). The work was printed in Basel in the sixteenth century: *Procli Diadochi Hypotyposis astronomicarum positionum*, ed. Simon Grynaeus (Baseleae, 1540). However, Franz Rosenthal includes Proclus among "the proud list of names of writers part of whose work has been preserved only in Arabic"; and he points out that "Often, the original text of eminent authors proved hard to understand, and paraphrases and elaborations were easier to master. This happened to the famous Neo-Platonists, Plotinus and Proclus" (*The Classical Heritage in Islam* [Berkeley: University of California Press, 1975], 11–12).

however, that Proclus had students and successors;[162] and so one would presume his astronomical knowledge (with a Ptolemaic bent) would have influenced future generations of scholars.[163]

In ninth-century Western Europe there is "no knowledge of works by Hipparchus, Ptolemy, or Theon of Alexandria";[164] the textbooks used for teaching astronomy were basically Roman, which included Latin translations and commentaries of a few Greek works. The overwhelming consensus by modern historians is that without the knowledge of the principal astronomical works of Greek antiquity, especially *sans* Ptolemy, the teaching of theoretical astronomy and planetary theory was a challenging endeavor; indeed, many portray this period as one of scientific stagnation.[165] Originally, Roman astronomy relied on "odds and ends" of ancient Greek astronomy for pedagogical purposes; but this virtually made

---

[162] It is known that Proclus's student Ammonius had students who included Philoponus, Asclepius, Olympiodorus, Damascius, and Simplicius; and that Olympiodorus's pupil Stephanus of Alexandria left Athens/Alexandria a century later for Constantinople. For a brief survey of some of the key scholars of the Academy in Athens, and the school of Alexandria, see Neugebauer, *HAMA*, 2:1031–54; and also David Pingree, "The Greek Influence on Early Islamic Mathematical Astronomy," *Journal of the American Oriental Society* 93, no. 1 (Jan.-Mar., 1973): 32–34.

[163] The library of Cardinal Bessarion is said to have housed the largest collection of Greek manuscripts in fifteenth-century Italy; it included several Greek *Almagest*s, Proclus's *Hypotyposis*, Theon of Alexandria's commentary on the *Almagest*, Theon of Smyrna's *Mathematical Knowledge Useful for Reading Plato*, and other hard-to-find Greek works. See Michael H. Shank. "The Classical Scientific Tradition in Fifteenth-Century Vienna," in *Tradition, Transmission, Transformation*, ed. F. Jamil Ragep and Sally P. Ragep (Leiden: E. J. Brill, 1996), 128–29.; and Shank, "Regiomontanus and Astronomical Controversy in the Background of Copernicus," in *Before Copernicus: The Cultures and Contexts of Scientific Learning in the Fifteenth Century*, ed. Rivka Feldhay and F. Jamil Ragep (McGill-Queens University Press, in press).

[164] See Bruce Eastwood, *Ordering the Heavens: Roman Astronomy and Cosmology in the Carolingian Renaissance* (Leiden: E. J. Brill, 2007), 10. Stephen C. McCluskey states that in the Latin West, Ptolemy's name "remained little more than a name, often confused with the Ptolemaic rulers of Egypt" (*Astronomies and Cultures in Early Medieval Europe* [Cambridge: Cambridge University Press, 1998], 20).

[165] In his inimitable way, Neugebauer proclaimed, "Ptolemy had no successor," and also deemed the extent of extant Greek scientific works at the time of the Roman period as "rather sad" (*HAMA*, 1:5). But even more graphic demonstrations of this overall sentiment of stagnation were employed by Henry Smith Williams, who intentionally left blank pages for his entire chapter entitled "Astronomy in the Medieval Period ("The Christian World—Twelve Centuries of Progress [325–1543, A.D.]" to indicate "astronomical progress" (*The Great Astronomers* [New York: Newton Publishing Co., 1932], 99–102); and also by Carl Sagan, whose timeline of the development of Western civilization after the Greeks left a millennium gap (ca. 500–1500) in the middle, describing the period as a "poignant lost opportunity for the human species" (*Cosmos* [New York: Random House, 1980], 335).

learning astronomy "along Greek lines impossible."[166] In sum, there was no single astronomical textbook for teaching astronomy; rather, what emerged was a corpus of Roman works, compiled during the first to the fifth centuries, that had piecemealed bits of Greek astronomy, and whose astronomical topics as well as focus varied greatly. It is recognized that foremost among these works were:[167] Pliny the Elder's (first century CE) detailed encyclopedic *Historia naturalis*, specifically Book II (on celestial phenomena) and Book VI (on terrestrial matters);[168] Macrobius's *Commentarii in somnium Scipionis* (fifth century), a broad cosmological overview connecting the celestial and terrestrial realms;[169] and Martianus Capella's *De nuptiis Philologiae et Mercurii* (fifth century), Book VIII, which provided some elementary astronomical concepts and data.[170]

---

166 See Olaf Pedersen, "The *Corpus Astronomicum* and the Traditions of Medieval Latin Astronomy: A Tentative Interpretation," in *Colloquia Copernicana III*, ed. Owen Gingerich and Jerzy Dobrzycki (Wrocław: Ossolineum, 1975), 62.

167 See Eastwood, *Ordering the Heavens* (on Pliny the Elder, 95–178 [Ch. 3]; on Macrobius, 31–94 [Ch. 2]; and on Martianus Capella, 179–311 [Ch. 4]). Cf. Neugebauer, *HAMA*, 2:1028–30; McCluskey, *Astronomies and Cultures*, 16–17; and, Pedersen, "The *Corpus Astronomicum*," 60–62.

168 According to Eastwood, Pliny the Elder's ambitious 37-volume *Natural History*, which claims to cover all human knowledge, is a "gold-mine of information" (*Ordering the Heavens*, 178). Book II alone contains 109 chapters, with topics ranging from eclipses to why the sea is salty. Bernard R. Goldstein and Alan C. Bowen highlight Pliny's passage on eclipses (II.12) to assert that he has been an underappreciated source for our knowledge of pre-Ptolemaic Greco-Latin astronomy. Their intent was to counter claims such as Kepler's, who stated that Pliny led "both himself and the reader astray by the obscurity of his words" ("Pliny and Hipparchus's 600-year Cycle," *Journal for the History of Astronomy* 26, no.2 [1995]: 155–58). Actually, I share Kepler's view; the scope and magnitude of Pliny's work certainly make it difficult for any reader to distinguish the gold from fool's gold (but of course, one could consult XXXIII.43: "Touchstones for Testing Gold").

169 Whereas Pliny presents a wide-range of astronomical topics, Macrobius's commentary on Cicero's dream (written some four centuries later), provides a broad picture of a Platonic cosmos of mathematically-harmonious ordered spheres (with Venus and Mercury above the Sun). Excerpts from both works were used in schools, though apparently not heavily glossed. Striking is Macrobius's theme of relating order of the cosmos with order in the soul; discussions include corresponding zones of the heavens and the Earth, a human soul that migrates between the two realms (one that both ascends and descends) in pursuit of eternal rewards, and what the stars indicate, but do not cause (Eastwood, *Ordering the Heavens,* 19, 27, 59–60, 66–67). See also McCluskey, *Astronomies and Cultures*, 117–19.

170 The title of this 9-volume work is an allegory for the marriage of elegance and wisdom, uniting to combine respectively the *trivium* (grammar, rhetoric, and logic) and the *quadrivium* (arithmetic, geometry, music, and astronomy); Book VIII is devoted to "Lady" astronomy, and in the ninth century, ten astronomical diagrams were appended to it. In addition to providing elementary terminology, Capella seems to be grappling with explaining planetary irregularities, such as the varying lengths of daylight throughout the year and the different lengths of the four seasons. Unlike both Pliny and Macrobius, he also asserts (without reference) that the paths of Venus and Mercury are around the Sun, not the Earth, whereas the Sun, Moon and three other planets circle around the Earth. See Eastwood, *Ordering the*

Theoretical astronomy was a topic dealt with only peripherally in these Roman sources.[171] Consequently, one finds a range of competing and often contradictory astronomical theories; so, for example, we find different scenarios for the sequencing of the planets.[172] Accompanying this is a general lack of technical accuracy and mathematical explanations,[173] i.e., epicycles and eccentrics make only cameo appearances in Roman sources. It seems that far more important for a Roman audience was presenting a general cosmological description, often enhanced with literary references,[174] of a relatively miniscule Earth encompassed by homocentric spheres—in short, a geocentric universe that was ordered and regular. But as Ptolemy wisely forewarned: "It is possible for many people to possess some of the moral virtues even without being taught, whereas it is impossible to achieve theoretical understanding of the universe without instruction."[175]

Roman astronomy then, in roughly the ninth century, provides us with an alternative account of an astronomical education, one that developed in the main without the benefit of ancient Greek sources for theoretical guidance.[176] This would have

---

Heavens, 12–14, 20–21, 244–59, 303; Gerd Graßhoff, "Natural Law and Celestial Regularities from Copernicus to Kepler," in *Natural Law and Laws of Nature in Early Modern Europe: Jurisprudence, Theology, Moral and Natural Philosophy*, ed. Lorraine Daston and Michael Stolleis (Farnham: Ashgate, 2008), 144–46; and McCluskey, *Astronomies and Cultures*, 120–22.

171 Note that since theoretical astronomy is my primary focus, I do not deal with the teaching of computus. See Eastwood, who seemingly concurs with my assessment in stating that the "separate concerns of astronomy and computus are far more numerous than the overlaps" (*Ordering the Heavens,* 10–12).

172 A striking example of this is that the three most popular Roman sources for teaching astronomy each presented a different order for the sequence of the planets: Pliny the Elder held that Venus and Mercury circled the Earth below the Sun; Macrobius maintained that these two planets circled the Earth above the Sun; and, Martianus Capella asserted that both did not enclose the Earth, but had circumsolar motions.

173 Neugebauer provides us with what he refers to as some of the more "absurd parameters" regarding sizes and distances found in Roman sources (*HAMA*, 2:723–24, 1029–30); and he singles out their oft-repeated postulate that all seven planets move with equal speed in their respective orbits. See also Pedersen, who bemoans the "non-mathematical character" of astronomical works of popularization ("The *Corpus Astronomicum*," 61–62, 65).

174 McCluskey concludes that "literary presentation was more important than rigorous demonstration, philosophical significance more important than mathematical precision" (*Astronomies and Cultures*, 117). Cf. both Eastwood and Pedersen, who intentionally omit those literary sources referencing astronomy and cosmology in their surveys of popular pedagogical astronomical texts (Pedersen, "The *Corpus Astronomicum*," 60; and Eastwood, *Ordering the Heavens*, 13).

175 *Almagest*, I.1 [Preface] (Toomer, *Ptolemy's Almagest*, 35).

176 Pedersen attributes "the disappearance of the Greek tongue," and thus the inability to comprehend Greek sources, as "the decisive factor" in stunting the development of early medieval astronomy in the West; according to him, "the West was left with a small number of works written by Latin authors of minor scientific importance and inferior quality compared with Ptolemy or his Greek commentators" ("The *Corpus Astronomicum*," 59–60).

been in sharp contrast to Islamic scholars who had amassed a huge corpus of ancient Greek philosophical and scientific texts by this same period. As Franz Rosenthal has aptly stated, it is indisputable that "Islamic civilization as we know it would simply not have existed without the Greek heritage."[177] Thus, it is highly unlikely that Islamic astronomers would have relied on Roman sources for astronomical knowledge.[178]

I.3.3 Islamic Forebears

Many Islamic scholars writing on theoretical astronomy supported Ptolemy's view, as stated in the *Almagest* preface, that two of Aristotle's three divisions of theoretical philosophy (theology and physics) should "be called guesswork," and that only the third division of "mathematics can provide sure and unshakeable knowledge to its devotees, provided one approaches it rigorously."[179] By the ninth century, Ptolemy's *Almagest* had been translated no less than five times,[180] along with the translation of his *Planetary Hypotheses*, other ancient Greek scientific works, and those of other cultures. However, the translation of the *Planetary Hypotheses* into Arabic deserves special mention given its significant role in putting forth the physical component for the picture of the universe, i.e., the so-called "Ptolemaic system" of nested orbs along with absolute distances and sizes of the planets. The *Planetary Hypotheses* complemented the mathematical models of the *Almagest* and handed Islamic astronomers the roadmap to a complete cosmographical system.[181]

---

[177] Rosenthal, *The Classical Heritage in Islam*, 14.

[178] Neither Ibn al-Nadīm's *Fihrist* nor Sezgin's *GAS*, 6 include Roman authors in their listings of works translated into Arabic.

[179] See *Almagest* I.1: Relation of astronomy to philosophy (*Ptolemy's Almagest*, 36).

[180] To summarize, these *Almagest* translations include a lost Syriac version translated from the Greek as well as translations from Greek into Arabic, including two different versions for the Caliph Maʾmūn (one by al-Ḥasan ibn Quraysh and another by al-Ḥajjāj ibn Maṭar) and a later translation by Isḥāq ibn Ḥunayn for Abū al-Ṣaqar ibn Bulbul, which was revised by Thābit ibn Qurra. Ibn al-Nadīm reports attempts at translating the *Almagest* into Arabic even earlier in the eighth century due to the interest of Yaḥyā ibn Khālid, the Barmakid vizier to the Caliph Hārūn al-Rashīd. Ibn al-Nadīm also adds al-Nayrīzī to the list of translators, stating this version was corrected by Thābit (*Fihrist*, Dodge 2:639–40; Sayyid 2/1:215). See Sezgin, *GAS*, 6:83–96; Toomer, *Ptolemy's Almagest*, 2–3; and Morelon, "Eastern Arabic Astronomy," 21–23. For comparison, Latin versions of the *Almagest* only became available around the twelfth century, with scholars such as Gerard of Cremona (d. 1187) translating from the Arabic; only later were there translations from the original Greek.

[181] The entire two books of the *Planetary Hypotheses* are extant in Arabic translation (by an unknown translator with corrections by Thābit ibn Qurra); only the first part of Book I is extant in Greek; and there is a fourteenth-century Hebrew translation based on the Arabic version. The work was plagued by a series of mishaps, which eventually led to an English translation, and commentary on just the supposedly "lost" Book I, part 2 by B. Goldstein ("The Arabic Version of Ptolemy's Planetary Hypotheses," 3–4). (The entire text has yet to

Islamic scholars took Ptolemy's notion of the advancement or progress of astronomy through inquiry as a mandate. This great ancient "authority" had made it quite explicit that he was not the final word on the subject, but merely had recounted "everything useful for the theory of the heavens" up until his time in the second century. In other words, Ptolemy had done his part by updating the three elapsed centuries since Hipparchus's observations, and it was incumbent on Islamic astronomers to continue the struggle for astronomical advancement some seven centuries later, by correcting results and striving for greater accuracy, the advantage that long intervals of time provide to test and improve upon past observations.[182] Needless to say, elementary Islamic astronomical textbooks on *hayʾa* were the beneficiaries of this directive.

I.3.3a The Moderns[183]

It is not uncommon to find references in *hayʾa* works referring to the opinions of the "Moderns" (see, for example, *Mulakhkhaṣ,* I.2[6]), which originally meant those Islamic scholars who flourished in the ninth century (or later) and provided "updated" information on ancient authorities; in the case of *hayʾa,* this usually meant Ptolemy. This new information was the fruit of concerted efforts by many individual scholars, but it was also due to sponsored scientific endeavors by various patrons that included ʿAbbāsid caliphs like al-Maʾmūn (r. 813–33).[184] So some three

---

be critically edited, though there are partial translations into German and French.) Also see Neugebauer, *HAMA*, 2:900–1, 918–19; Ragep, *Tadhkira*, 1:27n7; and Sezgin, *GAS*, 6:94–95.

[182] See *Almagest*, I.1 [Preface]; VII.1 and 3; XIII.11 [Epilogue] (Toomer, *Ptolemy's Almagest*, 37, 37n11, 321, 329, 647 respectively). A. I. Sabra reiterates the point that, "Islamic astronomers must have derived much hope and encouragement from the fact that their observational activities were taking place at a time sufficiently remote from Ptolemy's to allow for obtaining significant results, the intervening period being significantly longer than the one that had separated Ptolemy's own observations from, say, those of Hipparchus" ("Configuring the Universe," 289). The mandate for scientific advancement so dearly upheld by Islamic astronomers seems to contrast sharply with what was occurring in the Latin Middle Ages; perhaps that is why McCluskey stresses that "our question should not be what contributed to progress in astronomy, for episodes of progress were few. Instead, we will ask what forestalled the decline of astronomy and shaped the continuation and renewal of astronomical practice and knowledge from the fourth to the thirteenth centuries" (*Astronomies and Cultures,* ix).

[183] The contrast between "ancients" and "moderns" is commonplace among Islamic authors but has different connotations depending on the subject. For Islamic astronomers, the dichotomy is generally between the ancient Greeks and themselves. For the contrast as used in literature between the pre- and early Islamic poets versus the later ones, see Geert Jan van Gelder, "Ancients and Moderns," *EI*3. Brill Online, 2014. Reference. McGill University. 19 January 2015 http://referenceworks.brillonline.com/entries/encyclopaedia-of-islam-3/ancients-and-moderns-SIM_0040. For a fuller version of this article, see also Geert Jan van Gelder, "Muḥdat̲h̲ūn," in *EI*2 (2004), 12:637–40.

[184] Maʾmūn sponsored two sets of observations (one in Baghdad in 828, by astronomers who included Yaḥyā ibn Abī Manṣūr, and another that lasted more than a year [between 831 and

centuries later, by the twelfth century, Jaghmīnī would have inherited a rather extensive corpus of *hay'a* works stemming from this period, many synthesized and transformed, to help him compose his elementary theoretical textbook on astronomy. What follows is a brief overview of some of the key astronomical textbooks within this tradition that Jaghmīnī might well have had at his disposal even though we cannot in every case prove influence. Such an overview will help us assess how the *Mulakhkhaṣ* fits into the *hay'a* genre.

It has been suggested that during the earliest stages of this formative period of science, the 'Abbāsid astronomer Ya'qūb ibn Ṭāriq (fl. late eighth-century Baghdad) composed one of the first *hay'a* works, based on the title being *Tarkīb al-aflāk* (On the Arrangement of the Orbs), and also that the work deals with planetary sizes and distances,[185] a common topic associated with most (though not all) *hay'a* works (for example, it is omitted in the *Mulakhkhaṣ*). This work, extant only in fragments, was composed circa 777-78 (so prior to the translations of most Greek scientific texts) and uses Indian techniques to compute the planetary distances. But as al-Bīrūnī aptly commented, the Hindu approach is markedly different than Ptolemy's "computation of the distances of the planets in the *Kitâb-almanshûrât*, and in which he has been followed both by the ancient and the modern astronomers."[186] Indeed, once Ptolemy's *Almagest* and *Planetary Hypotheses* were translated into Arabic in the ninth century, it is not an exaggeration to state that they become the formative works for the *hay'a* tradition. So it is not surprising that Bīrūnī, writing two centuries later, would view Ya'qūb ibn Ṭāriq's cosmology as unfamiliar and "based on a principle which is unknown to me in the present stage of my knowledge."[187]

833] near Damascus) with the intent of verifying Ptolemy's parameters in the *Almagest* and *Handy Tables*. One important improvement was determining new values for the obliquity of the ecliptic. See Lennart Berggren, "Ma'mūn," and Benno van Dalen, "Yaḥyā ibn Abī Manṣūr," in *The Biographical Encyclopedia of Astronomers* [hereafter cited as *BEA*], ed. Thomas Hockey et al., 2 vols. (New York: Springer-Verlag, 2007), 2:733 and 2:1249–50 [respectively]; and Ibn al-Nadīm, *Fihrist,* Dodge 2:653; Sayyid 2/1:237. In addition, Ma'mūn sent a group of scientists to the Plain of Sinjār in upper Mesopotamia in order to determine a more precise measurement for a meridian degree. See F. J. Ragep, "Islamic Reactions to Ptolemy's Imprecisions," in *Ptolemy in Perspective*, 124–25. See also Y. Tzvi Langermann, ("The Book of Bodies and Distances of Ḥabash al-Ḥāsib," *Centaurus* 28, no. 2 [1985]:108–28), who presents a portion of Ḥabash's Arabic text with English translation, which is a record of the scientific projects carried out by the Caliph al-Ma'mūn.

[185] See Kim Plofker, "Ya'qūb ibn Ṭāriq," in *BEA*, 2:1250–51; and David Pingree, "The Fragments of the Works of Ya'qūb ibn Ṭāriq," *Journal of Near Eastern Studies* 27, no. 2 (Apr., 1968): 98, 105–20. Pingree includes Bīrūnī's extracts from *Tarkīb al-aflāk* found in *Alberuni's India* (trans. Edward C. Sachau, 2 vols. [London: Trübner & Co., 1910], 2:67–68, 80) in addition to Bīrūnī's own comments. See also Ibn al-Nadīm, *Fihrist,* Dodge 2:659; Sayyid 2/1:245; Jamāl al-Dīn Abū al-Ḥasan 'Alī ibn Yūsuf al-Qifṭī (d. 1248), *Ta'rīkh al-ḥukamā'*, ed. J. Lippert (Leipzig: Dieterich'sche Verlagsbuchhandlung, 1903), 373; and Sezgin, *GAS*, 6:124–27.

[186] *Alberuni's India*, 2:69.

[187] See *Alberuni's India*, 2:70. Bīrūnī provides a brief synopsis of Hindu planetary theory and points out that it differs from the Ptolemaic system (69).

Bīrūnī was certainly no Hellenophile;[188] he simply was acknowledging the fact that by the eleventh century the impact of Hindu traditions for Islamic astronomers was overshadowed by the Ptolemaic one.

Yet one should not assume that embracing Ptolemaic astronomy sentenced Persian, Syriac, and other Greek sources into forced retirement. In support of this, we have the case of a recently identified tenth-century *hay'a* treatise indicating Sanskrit and Syriac influences. This is the discovery of the lost Arabic original of a work that had been incorrectly attributed in its Latin translation to Māshā'allāh (fl. Baghdad, 762–ca. 815), one of the early ʿAbbāsid astrologers associated with the courts of al-Manṣūr and al-Ma'mūn; according to Taro Mimura, it is probably a tenth-century composition.[189] The treatise, translated into Latin as *De scientia motus orbis* or *De elementis et orbibus coelestibus*, in twenty-seven chapters (with a longer version in forty chapters),[190] includes introductory chapters on *hay'a* dealing with phenomena in both the celestial and sublunar world, but with a focus on glorifying God and how the celestial orbs influence the sublunar region. Ptolemy is quoted; however, the treatise uses non-Ptolemaic planetary models, seemingly similar to Sanskrit ones and based on Syriac sources.[191]

Nevertheless, Ptolemaic astronomy gained a stronghold, and making Ptolemy's works more comprehensible became a high priority for Islamic astronomers. One of the earliest introductory accounts on various aspects of Ptolemaic spherical astronomy and planetary theory was compiled by Muḥammad ibn Kathīr al-Farghānī, a scholar affiliated with the ninth-century Baghdad ʿAbbāsid court. His thirty-chapter compendium on the science of the stars (*Jawāmiʿ ʿilm al-nujūm*),[192] composed between 833 (after Ma'mūn's death) and 857, has often been characterized

---

[188] See David Pingree, "Hellenophilia versus the History of Science," *Isis* 83, no. 4 (Dec., 1992): 554–55. Cf. Sabra, "Reply to Saliba," 342–43.

[189] See Taro Mimura, "The Arabic Original of (ps.) Māshā'allāh's *Liber de orbe*: Its Date and Authorship," *The British Journal for the History of Science* 48 (2015): 321–52.

[190] Ibn al-Nadīm lists the work as a "book known as The Twenty-Seven" (*Fihrist*, Dodge 2:650–51; Sayyid 2/1:242). So does al-Qifṭī, *Ta'rīkh al-ḥukamā'*, 327 and Rosenfeld and İhsanoğlu, *MAMS2*, 17 (no. 18), A2. For sources citing the Latin translation, see: Francis J. Carmody, *Arabic Astronomical and Astrological Sciences in Latin Translation: A Critical Bibliography* (Berkeley: University of California Press, 1956), 32–33 (no. 8: De motibus [De orbe]); David Pingree, "Mās͟hā' allāh," in *DSB* [1974], 9:162 [no. 25 in Pingree's list of 28]); Julio Samsó, "Mās͟hā' Allāh," in *EI2* (1991), 6:710–12; and Sezgin, *GAS*, 6:129 (no. 2).

[191] See David Pingree, "Māshā'Allāh: Some Sasanian and Syriac Sources," in *Essays on Islamic Philosophy and Science*, ed. George F. Hourani (Albany: State University of New York Press, 1975), 10–12.

[192] Farghānī's *Jawāmiʿ* is not really a summary, but more a compilation of selected parts of Ptolemy's *Almagest*; see Ibn al-Nadīm, *Fihrist*, Dodge 2:660; Sayyid 2/1:247. Many titles have been attributed to this work: Ḥājjī Khalīfa just gives *Kitāb al-Fuṣūl al-thalāthīn* (The Book of 30 chapters) in his *Kashf al-ẓunūn* (4:438–39); but al-Qifṭī refers to it as *Madkhal ilā ʿilm hay'at al-aflāk wa-ḥarakāt al-nujūm* (Introduction to the Science of the Structure (*hay'a*) of the Orbs and the Movements of the Stars) (*Ta'rīkh al-ḥukamā'*, 78), although Farghānī refers to his work as *ʿilm al-nujūm*, not *ʿilm al-hay'a*, and he restricts the use of the word *hay'a* to the title of a chapter discussing the "arrangement" of the nested orbs (see Ch.

as a popular summary of Ptolemy's *Almagest* due to the scope of his descriptive selections from it, which are replete with basic astronomical information, definitions, concepts, and parameters, many (but not all) "updated" Maʾmūn values (which Farghānī specifically references).[193] For anyone unacquainted with Ptolemaic astronomy or seeking a quick reference source, this single textbook introduces the reader to a wide range of topics that include *both* the celestial and terrestrial realms, although certain subjects mentioned in the *Almagest*, such as astronomical instruments, are noticeably absent.[194] It also lacks any illustrations. Aside from providing the names of the Islamic months (Ch. 1), Farghānī avoids matters directly applicable to religion and natural philosophy, thus making it more in line with what we have described as the "*hayʾah* tradition."[195] However, one might object to this categorization, finding fault with the work's (dis)organization and oversimplifications, and lack of attention to what one might call the "how-to's" of planetary motion, this including his presentation of Ptolemy's nested spheres (e.g., he constantly lumps the upper and lower planets together, is vague on positions of eccentrics, and so forth).[196] Nevertheless, given the time and place, one must credit Farghānī with providing a description of the "*hayʾa*" of the orbs for

12, *Jawāmiʿ*, 45–49 [e.g., Paris, BnF, ar. MS 2504, ff. 127b–128b]). For more on Farghānī, the astronomer-astrologer-engineer, plus an overview of the content of the *Jawāmiʿ*, see A. I. Sabra, "Al-Farghānī," in *DSB* (1972), 4:541–45 and Bahrom Abdukhalimov, "Aḥmad al-Farghānī and His Compendium of Astronomy," *Journal of Islamic Studies* 10, no. 2 (1999): 142–58. See also Fuat Sezgin's reprint of Golius's 1669 Arabic printed edition with Latin translation (Aḥmad ibn Muḥammad ibn Kathīr al-Farghānī [Alfraganus] (about 850 CE), *Jawāmiʿ ʿilm al-nujūm wa-uṣūl al-ḥarakāt al-samāwīya*, ed. Jacob Golius [Frankfurt am Main, 1986]) [hereafter cited as *Jawāmiʿ*]; Sezgin, *GAS*, 6:149–51; and Gregg DeYoung, "Farghānī," in *BEA*, 1:357.

[193] Farghānī sometimes provides both the old (Ptolemaic) and the new (Maʾmūnī) parameters, exemplified by his statement that a number of scholars give the Maʾmūn value of 23;35 for the ecliptic obliquity as an update to Ptolemy's 23;51 (see Ch. 5, *Jawāmiʿ*, 18; Paris, BnF, ar. MS 2504, f. 121a). Sometimes Farghānī gives only Maʾmūn's new information, such as his measurements for the Earth's circumference (20,400 miles) and the Earth's diameter (approx. 6,500 miles) (Ch. 8, *Jawāmiʿ*, 31; Paris, BnF, MS ar. 2504, f. 124a). However, sometimes he retains the old (perhaps unknowingly) rather than presenting the new, such as maintaining Ptolemy's precessional rate of 1°/100 (Ch. 13, *Jawāmiʿ*, 49–50; Paris, BnF, MS ar. 2504, f. 128b), versus replacing it with the updated 1°/66 value. For more specifics on Farghānī's parameters, and comparisons with other sources, see the Commentary below, including the charts.

[194] One would have thought Farghānī, being an engineer, would have included some discussion of instruments, especially since Ptolemy deals with instruments and their construction in the *Almagest*. Perhaps, he felt it was unnecessary since he had also composed a separate treatise on the astrolabe.

[195] This is Langermann's assessment (*Ibn al-Haytham's "On the Configuration of the World,"* 31).

[196] Langermann suggests that Ibn al-Haytham may have had Farghānī's *Jawāmiʿ* in mind when he criticized his predecessors for producing works that "fall short" in that they lack "an explicit enunciation of the way in which the motions of the stars take place on the various spheres" (*Ibn al-Haytham's "On the Configuration of the World,"* 26–28).

each of the planets and their distances from the Earth (Ch. 12), and further acknowledge his overall attempt to make more accessible a lot of information often deeply buried within the thirteen books of the *Almagest*, which includes Ptolemy's numerous parameters. One example is Farghānī's calculation of the distances of the planets from the Earth in miles (in Ch. 21), which seems to be based on an independent calculation using parameters from the *Almagest* rather than the *Planetary Hypotheses*.

It is not at all clear who Farghānī's target audience was (e.g., court officials or the general public); however, the *Jawāmiʿ* inspired a few Arabic commentaries—one by al-Qabīṣī (d. 967),[197] another by Ibn Sīnā's companion and literary secretary al-Jūzjānī (eleventh century),[198] and possibly a third by al-Bīrūnī[199]—which indicates the work was known, and taken seriously, by later scientists. Given that there are relatively few extant copies and commentaries of the *Jawāmiʿ*, certainly in comparison with other later *hayʾa* works and their commentaries, perhaps it should be considered more of a formative textbook within an Islamic context in contrast with those general astronomical works that came after it. On the other hand, its popularity and influence as an astronomical textbook in medieval Europe is undeniable, given the longevity of its wide circulation there.[200]

Another scholar active in promoting the exact sciences in ninth-century Baghdad was the Ṣābian Thābit ibn Qurra (221–88/836–901), who flourished during the reigns of several post-Maʾmūn ʿAbbāsid caliphs.[201] Renowned for his extensive number of translations and revisions of Greek works (which included the *Almagest*

---

[197] Al-Qabīṣī also wrote a work on sizes and distances. See Sezgin, *GAS*, 6:209 (nos. 1 and 2); and Rosenfeld and İhsanoğlu, *MAMS2*, 85 (no. 205), A3, A4.

[198] See F. Jamil Ragep, "The *Khilāṣ kayfiyyat tarkīb al-aflāk* of al-Jūzjānī: A Preliminary Description of Its Avicennian Themes," in *Avicenna and his Legacy: A Golden Age of Science and Philosophy,* ed. Y. Tzvi Langermann (Turnhout: Brepols, 2010), 303–8.

[199] The work is listed as corrections to Farghānī's chapters (*Tahdhīb fuṣūl al-Farghānī*) in D. J. Boilot, "L'oeuvre d'al-Beruni, Essai bibliographique," *Mélanges de l'Institut Dominicain d'Études Orientales* 2 (1955): 181 (no. 14); Sezgin, *GAS*, 6:274 (no. 13); and Rosenfeld and İhsanoğlu, *MAMS2*, 152 (no. 348), A26.

[200] The influence of al-Farghānī [Alfraganus] on medieval European astronomy is indicated by the number of Latin translations and printed editions of the *Jawāmiʿ* or *Elements*. There were two twelfth-century Latin translations and one into Hebrew by Jacob Anatoli (fl. thirteenth century) that served as a basis for a third sixteenth-century Latin translation. It is noteworthy that in the fifteenth century Regiomontanus lectured on Farghānī at the University of Padua; see Noel M. Swerdlow, "Science and Humanism in the Renaissance: Regiomontanus's Oration on the Dignity and Utility of the Mathematical Sciences," in *World Changes: Thomas Kuhn and the Nature of Science*, ed. Paul Horwich (Cambridge, MA: MIT Press, 1993), 131–68; and James S. Byrne, "A Humanist History of Mathematics? Regiomontanus's Padua Oration in Context," *Journal of the History of Ideas* 67, no. 1 (January 2006): 41, 43.

[201] The number of ʿAbbāsid caliphs that spanned Thābit's lifetime is rather impressive and include: al-Muʿtaṣim (r. 833–42); his son al-Wāthiq (842–47); his brother al-Mutawakkil (847–61) and his son al-Muntaṣir (861–62); al-Mustaʿīn (862–62); al-Muʿtazz (866–69) and al-Muʿtamid (870–92) [sons of al-Mutawakkil]; and al-Muʿtaḍid (892–902).

and *Planetary Hypotheses*), he also composed numerous astronomical compositions, several of them on Ptolemaic astronomy, among which a few can be classified as introductions.[202] It has been suggested that Thābit may have been familiar with, even influenced by, the work of his predecessor Farghānī.[203] However, Régis Morelon asserts that Farghānī's work was "more superficial" (without indicating in what sense) and cites their only commonalities as being that both had modified Ptolemy's parameters on the ecliptic obliquity (though each gives a slightly different value) and that both had agreed on the motion of the solar apogee (taken to be fixed by Ptolemy).[204] In fact there are some significant differences between the two regarding focus, scope, and content, at least as indicated by Thābit's two short extant works on Ptolemaic astronomy, his *Tashīl al-Majisṭī* and *Fī dhikr al-aflāk....*[205] But they also had some commonalties, and both differences and similarities are worth pointing out here for what they indicate about elementary astronomy texts during this period. Part of their differences lies in the fact that Thābit's introductions deal exclusively with the celestial realm, whereas Farghānī's

---

[202] Régis Morelon attributes to Thābit between thirty to forty astronomical works, with at least seven of these related to Ptolemaic astronomy; see *Thābit ibn Qurra: Œuvres d'astronomie* (Paris: Les Belles Lettres, 1987), XI–XXIII, XXV–XXVI. See also al-Qifṭī, *Taʾrīkh al-ḥukamāʾ*, 115–22 at 117; Ibn Abī Uṣaybiʿa, *ʿUyūn al-anbāʾ*, Beirut ed., 295–300 at 298; and Ibn al-Nadīm, *Fihrist* (Dodge 2:647–48; Sayyid 2/1:227–28); Sezgin, *GAS*, 6:163–70; and Morelon, "Ṯābit b. Qurra and Arab Astronomy in the 9th Century," *Arabic Sciences and Philosophy* 4 (1994): 111–12.

[203] Francis J. Carmody suggests a "possible" relationship between the two, and even that Thābit could have been influenced by Farghānī, based on connections within some minor works by Thābit. However, Carmody's speculation rests heavily on his use of the Latin translations for his analysis, and some of these may have been modified from the Arabic originals (*The Astronomical Works of Thabit b. Qurra* [Berkeley: University of California Press, 1960], 17, 117–18, 120).

[204] See Morelon, *Thābit ibn Qurra*, XLIV–XLV. Thābit gives 23;33 for the ecliptic obliquity (*Tashīl al-Majisṭī* [Morelon, *Thābit ibn Qurra,* Arabic: 8, line 7]), and Farghānī gives 23;35 compared with Ptolemy's value of 23;51,20 (*Almagest*, I.12). Each may have been relying on different Maʾmūn observations for their modifications; however, to somewhat muddy the waters, A. I. Sabra points out that Farghānī also gives the inclination of the ecliptic as 23;33 for the year 225 of Yazdigird (=857–58 CE) in his work on the astrolabe ("Farghānī," 543). See also Ragep, "Islamic Reactions to Ptolemy's Imprecisions," 129–30 (on "The Obliquity of the Ecliptic").

[205] Morelon refers to these two works by Thābit as introductions and presents an analysis and critical Arabic editions with French translations of both *Tashīl al-Majisṭī* (تسهيل المجسطي) ("L'Almageste simplifié") [Istanbul, Ayasofya MS 4832, ff. 52a–53b] and *Fī dhikr al-aflāk wa-khalaqihā* [correct to *ḥalaqihā*] *wa-ʿadad ḥarakātihā wa-miqdār masīrihā* (في ذكر الأفلاك وخلقها [!] وعدد حركاتها ومقدار مسيرها) ("Présentation des orbes des astres, de leur disposition [should be corrected to "rings"], du nombre de leurs mouvements et de la valeur de leur progression") [Istanbul, Ayasofya MS 4832, ff. 50a–51a (*Thābit ibn Qurra*, XIV, XIX, XX, XXIV–XXV, XXXVII–XLIII). Also see Carmody for a brief overview of the content of the Latin translation of Thābit's *Tashīl al-Majisṭī* [=*De Hiis que indigent antequam legatur Almagesti*] (*The Astronomical Works of Thabit b. Qurra*, 21, 117–18).

*Jawāmi*ʿ was far more ambitious in the range of subject matter he covered. So given Thābit's focus on the celestial bodies and their movements, he excludes certain terrestrial-related subjects such as determining the sphericity and the centrality of the Earth (found in Farghānī [Chs. 3 and 4]), and the discussion of the inhabited world (contained in Farghānī [Ch. 8]).[206] Interestingly, the word *hay'a* never enters Thābit's picture; it at least makes a cameo appearance in Farghānī [Ch. 23]). Thābit's presentation of material is pedagogically far better organized and structured, certainly when compared to Farghānī's tendency to conflate topics and his piecemeal approach,[207] but then again Thābit confined his subject to the arrangement of the celestial orbs (*tarkīb al-aflāk*), and did not discuss the divisions of the sublunar realm. On the other hand, both have several things in common: they both provide basic astronomical definitions that underscore the fact that a technical terminology was well-established by the ninth century (though in both cases certain terms are in need of refinement[208]); and both felt no compulsion to provide illustrations for their readers. Both also seem content to present parameters as rough approximations, which I find a bit puzzling (e.g., both give the Sun's daily motion simply as 59 minutes of arc even though Ptolemy's value is 0;59,8,17,13,12,31). Now one can obviously attribute this to the context of introductory works (but then again Jaghmīnī gives 0;59,8,17), but as mentioned above, this was a period when the Caliph Ma'mūn was sponsoring astronomical observations that produced more precise parameters. In any event, both are certainly keen on incorporating Ma'mūn's new results into their works; however, here we find some differences (minor and more significant ones) between the two scholars. Whereas Farghānī maintains the Ptolemaic value for precession, Thābit acknowledges modifications have been made, though he provides no specific parameters;[209] and Farghānī presents Ma'mūn's updated values for the Earth's diameter and circumference, and Thābit omits these. On the other hand, Thābit presents the values for the planetary distances, and furthermore he uses the values contained in the *Planetary Hypotheses* (but without citing his source),[210] whereas Farghānī calculates parameters for determining the nearest distances of the planets from the Earth based on the *Almagest*.

---

[206] Generally speaking, absent from Thābit are topics associated with *hay'at al-arḍ*, such as the seven climes (discussed in Ptolemy's *Almagest* and *Geography* [both works that Thābit translated] and also in Farghānī's *Jawāmi*ʿ). On the other hand, Thābit does include items peripherally associated with terrestrial localities, such as definitions for the meridian and horizon circles, zenith, and so forth.

[207] In fact, Thābit's organization of definitions in the *Tashīl al-Majisṭī* is strikingly similar to Jaghmīnī's, whose definitions are found in his separate chapters on circles and on arcs (cf. *Mulakhkhaṣ*, I.3 and I.4); the similarity is such that it is worth considering the former as a model for Jaghmīnī, especially since most other *hay'a* works did not have separate chapters on circles and arcs.

[208] For example, both Farghānī and Thābit seem unconcerned about distinguishing when to use sphere (*kura*) versus orb (*falak*) to mean a constituent part of the general configuration of the world; but this is also an issue that continues well into the thirteenth century.

[209] See Morelon, *Thābit ibn Qurra, Tashīl al-Majisṭī* (Arabic: 16, lines 10–11).

[210] See Morelon, *Thābit ibn Qurra, Tashīl al-Majisṭī* (Arabic: 14, line 5–15, line 3). Since Thābit doesn't reference the values, Carmody is apparently unaware that the values contained

All of this should remind us that scientists working on the same subject matter, in roughly the same time and location, do not necessarily have access to the same information or are knowledgeable of all available extant sources or are even aware of all new developments. Nor might they have the same views on what the scope and content of their subject entailed.

Furthermore, one should not assume that new information will be assimilated immediately. Here we have the example of al-Kindī (fl. ninth century),[211] who demonstrates a clear "unreadiness to discard all the vestiges" of the ancient heritage even in light of new developments.[212] It is well known that al-Kindī considered it his "personal task to serve as an Arab transmitter and interpreter" of difficult Greek philosophical and scientific works, and to popularize them for "the curious student or interested layperson."[213] However, if his intention (as he claimed) was to elucidate texts such as the *Almagest* for beginners, his choice to "faithfully follow" Theon's extremely technical *Commentary on the Almagest* as his model for his *Almagest* commentary (*Kitāb fī al-ṣināʿa al-ʿuẓmā*) was odd, as was his decision to discuss the first eight chapters of Book I, which focus on isoperimetric problems related to the Earth's sphericity.[214] One would assume this subject would have had a limited appeal for inclusion in an elementary astronomical textbook, even al-Kindī's simplified rendition of it. However, al-Kindī's decision to use Theon as his source, and also to examine specific issues in great detail (such as determining the Earth's diameter) from within a Greek context, completely ignoring any new astronomical developments, indicates his strong commitment to ancient sources. It did not go unnoticed that "No mention is made by al-Kindī of the measurement of the meridian under al-Maʾmūn"..."which is inconceivable that he should not have known about it."[215] Rosenthal suggests (halfheartedly) that al-Kindī's rivalry with the Banū Mūsā, who were active in establishing the new measurements during this period, may have been a contributing factor in al-Kindī's decision.

---

within his *De Hiis* are from the *Planetary Hypotheses*, since he claims that "there is no evidence that Thâbit knew this work" (*The Astronomical Works of Thabit b. Qurra*, 19; cf. 130, 137). Neugebauer makes the point that Thābit ibn Qurra "fully confirms the numbers from the Planetary Hypotheses" (*HAMA*, 2:920 and n23).

[211] Al-Kindī also flourished during the reigns of several caliphs who included al-Maʾmūn, al-Muʿtaṣim, and al-Mutawakkil. A seminal article is Rosenthal's "Al-Kindī and Ptolemy"; and for more on his writings on astronomy and mathematics, see also Ibn al-Nadīm, *Fihrist* (Dodge 2:618–20; Sayyid 2/1:187–89); Sezgin, *GAS*, 6:151–55 at 153 (no. 1).

[212] Rosenthal, "Al-Kindī and Ptolemy," 455.

[213] Rosenthal, "Al-Kindī and Ptolemy," 440, 444–45, 455. See also A. I. Sabra, "Some Remarks on Al-kindi as a Founder of Arabic Science and Philosophy," in *Dr. Mohammad Abdulhadi Abu Ridah Festschrift*, ed. Abdullah O. Al-Omar (Kuwait: Kuwait University Press, 1993), 601–7.

[214] For more on Theon's *Commentary of the Almagest*, see above *I.3.2c: The Ptolemaic Aftermath.* For specifics regarding the content of al-Kindī's *Kitāb fī al-ṣināʿa al-ʿuẓmā,* which includes evidence that his source was Theon's *Commentary*, see Rosenthal, "Al-Kindī and Ptolemy," 436–53, esp. 446.

[215] Rosenthal, "Al-Kindī and Ptolemy," 454.

An account of the Palmyra-Raqqa scientific expedition in Syria is reported by the Banū Mūsā in a treatise entitled *Ḥarakat al-aflāk* (Motion of the Orbs).[216] And the exact same passage (in fact the entire extant fragment) is contained in another more extensive anonymous treatise attributed to Qusṭā ibn Lūqā, another ninth-century scholar, of Greek Christian origin who composed and translated numerous scientific works. Either attribution makes this theoretical astronomical treatise, which cites Ptolemy and the *Almagest*, an example of an early *hayʾa* work.[217] This text strikingly contains some forty-eight two-dimensional mathematical illustrations that complement extensive descriptions of various aspects of celestial motions and terrestrial phenomena (such as the lunar and solar eclipses, retrogradation, and so forth). The planets are treated individually, i.e., they are not generically lumped together (a characteristic of other astronomical treatises that led to criticism); however, no attempt has been made to provide a coherent physical picture of the universe. Noticeably absent (from the extensive figures) is an illustration of the configuration of the world; and also the word *hayʾa* (as far as I could tell) does not appear throughout the entire treatise (it is written only in the codex's table of contents, which, as mentioned, is not in the copyist's hand). Nevertheless, the word *falak* (not *kura*) is systematically used throughout the treatise, which may be indicative of physical underpinnings at work.[218] Clearly, a more careful analysis of this text and its parameters is needed for the future. For our present purposes, we can say that this treatise, insofar as it was ever meant as a "teaching" textbook, is

[216] F. J. Ragep provides the passage on the expedition from the *Ḥarakat al-aflāk* (both the Arabic and an English trans.), and situates the expedition within the broader context of the complexity of introducing new parameters, and balancing tradition and innovation in Islamic science (*Tadhkira*, 2:502–10). See also F. J. Ragep, "Islamic Reactions to Ptolemy's Imprecisions," 122–25. For listings of this work by the Banū Mūsā, see Sezgin, *GAS*, 6:147 (no. 3); and Rosenfeld and İhsanoğlu, *MAMS2*, 35–36 (no. 74), A3. The passage is contained in an extant fragment of Damascus, Ẓāhiriyya 4489, f. 12a–b; and the treatise begins "qāla Banū Mūsā" (f. 1b).

[217] See Oxford, Bodleian Library, Seld. 11, ff. 38b–85b (the passage is on ff. 47b–48a). This text describes both the celestial and terrestrial realms (unlike the Damascus fragment which only deals with the terrestrial realm); and a codex table of contents lists it as *Hayʾat al-aflāk* by Qusṭā ibn Lūqā, but this is clearly in a different hand than the witness itself. Though the text itself is anonymous, George Saliba has consistently attributed this early *hayʾa* work to Qusṭā; see Saliba, "Early Arabic Critique of Ptolemaic Cosmology: A Ninth-Century Text on the Motion of the Celestial Spheres," *Journal for the History of Astronomy* 25, no. 2 (1994): 119; "Arabic versus Greek Astronomy," 328, 330; *A History of Arabic Astronomy*, 17; and *Islamic Science and the Making of the European Renaissance* (Cambridge, MA: MIT Press, 2007), 18, 262. For more on Qusṭā and his works, see Ibn al-Nadīm, *Fihrist* (Dodge 2:694–95; Sayyid 2/1:292–94); Elaheh Kheirandish, "Qusṭā ibn Lūqā al-Baʿlabakkī," in *BEA*, 2:948–9; al-Qifṭī, *Taʾrīkh al-ḥukamāʾ*, 262–63; and Rosenfeld and İhsanoğlu, *MAMS2*, 59–60 (no. 118).

[218] The illustrations in this treatise are more mathematical than physical depictions of the orbs, along the lines one finds in the *Almagest*; however, many diagrams in astronomical works, even *hayʾa* works, use "mathematical" simplifications rather than the full, spherical versions.

clearly not well-organized (the subject matter is in one continuous stream distinguished only by subtitles). It does, though, present the reader with many parameters (again as approximations[219]), including the latest values gleaned from the scientific expeditions.

Al-Kindī's unwillingness to abandon Greek traditions in light of new developments is somewhat ironic given his advocacy of upholding Ptolemy's imperative of scientific advancement, which demanded "the necessity of [building on] the consecutive labors of scholars and thinkers."[220] As Rosenthal noted, the novelty of a subject may need assimilation time during its pioneering stages,[221] but al-Battānī (d. 317/929) provides us with a prominent counter-example. In the preface to his great astronomical *Zīj,* Battānī explicitly informs us that his work is also in accordance with Ptolemy's imperative for scientific advancement;[222] but in contrast to al-Kindī, he presents, within fifty-seven chapters, new and more precise astronomical parameters (many beyond sexagesimal seconds) based on his observational activities that spanned over forty years (264–306/877–918).[223] The focus of Battānī's *al-Zīj al-Ṣābi'*,[224] as its title indicates, is concerned predominantly with

[219] E.g., the values given for the climes are rounded Ptolemaic ones (see Oxford, Bodleian Library, Seld. 11, f. 44b and Damascus, Ẓāhiriyya 4489, f. 11b).

[220] Rosenthal, "Al-Kindī and Ptolemy," 445, 447.

[221] Rosenthal, "Al-Kindī and Ptolemy," 455.

[222] Carlo A. Nallino, *Al-Battānī Sive Albatenii Opus Astronomicum,* 3 vols. (Milan, 1899–1907), 3:7 [hereafter cited as *Zīj*]. As noted by Willy Hartner: "Al-Battānī tells us that errors and discrepancies found in the works of his predecessors had forced him to compose this work in accordance with Ptolemy's admonition to later generations to improve his theories and inferences on the basis of new observations, as he himself had done to those made by Hipparchus and others" ("Al-Battānī," in *DSB* [1970], 1:508).

[223] Much has been written on Abū ʿAbd Allāh Muḥammad ibn Sinān ibn Jābir al-Battānī al-Ḥarrānī al-Ṣābi' (known as Albatenius in the Latin West). In addition to Hartner, "Al-Battānī," brief summations of his new parameters are contained within Carlo A. Nallino, "Battānī," in *EI2* (1960), 1:1104–5; Julio Samsó, "Battani, Al-," in *Medieval Science, Technology, and Medicine: An Encyclopedia*, ed. Thomas F. Glick, Steven J. Livesey, Faith Wallis (New York: Routledge, 2005), 79–80; and Samsó, "Al-Battānī," in *Encyclopaedia of the History of Science, Technology, and Medicine in Non-Western Cultures*, 91. F. J. Ragep summarizes trepidation in Islam before Battānī, and Battānī's criticism and alternatives, in "Al-Battānī, Cosmology, and the Early History of Trepidation in Islam," in *From Baghdad to Barcelona. Essays on the History of the Exact Sciences in Honour of Prof. Juan Vernet*, ed. Josep Casulleras and Julio Samsó (Barcelona, 1996), 283–90. Additional sources include: Ḥājjī Khalīfa, *Kashf al-ẓunūn*, 2: col. 970 [Flügel, *Lexicon,* 3:564 (no. 6946)]; Ibn al-Nadīm, *Fihrist* (Dodge 2:661–62; Sayyid 2/1:249); Qifṭī, *Ta'rīkh al-ḥukamā'*, 280; E. S. Kennedy, "A Survey of Islamic Astronomical Tables," 132–33, 154–56 (Battānī [no. 55] plus other *zīj*es influenced by al-Battānī); Sezgin, *GAS*, 6:182–87; and Benno van Dalen, "Battānī," in *BEA*, 1:101–3. Also see *Mulakhkhaṣ*, Commentary, II.3[9] and charts that contain his parameters.

[224] The title *al-Zīj al-Ṣābi'* or "The Ṣābian *zīj*" is a reference to Battānī's Ṣābian ancestral roots, and probable links with Ḥarrān in southern Anatolia. This also suggests connections with Battānī's older contemporary, the Ṣābian Thābit ibn Qurra, despite Carmody's statement

practical rather than theoretical aspects of astronomy. However, Battānī, like Farghānī, includes a general description of the nested orbs, and also deals with many overlapping topics that make it relevant for *hayʾa* works, though the word itself is rarely used (and then to signify a general structure or configuration of the universe, not a scientific discipline).[225] However, Battānī's objective was not to provide a teaching text; rather, his priorities were acquiring and presenting more accurate parameters, so he was less concerned with couching them in a coherent physical cosmology.[226] This is evident from the fact that Battānī often oversimplifies his descriptions of the celestial realm (something he shares with Farghānī and others). He also is not beyond getting the modeling wrong; for example, in the case of Mercury he presents the deferent center as the center of mean motion (whereas it should be the equant point).[227] In short, Battānī's claim to fame was due to his level of competency in providing various improved parameters for planetary motion; and it is not an exaggeration to state that his *Zīj* became one of the main reference sources for generations of Islamic astronomers, including those scholars working on *hayʾa* textbooks. Battānī is the only "authority" outside of Ptolemy that Jaghmīnī specifically cites in the *Mulakhkhaṣ* (II.3[9]), an indication that Battānī's influence was pervasive some three centuries after he flourished.

I.3.3b The Post-Moderns

Although Jaghmīnī only specifically cites Ptolemy and his *Almagest* and Battānī, he clearly relied on a variety of other unnamed authorities and reference sources.[228] Given that he flourished some three centuries after the so-called "Moderns," and in Khwārizm (a region somewhat distant, but not isolated, from Mesopotamian scientific activities), it is not surprising that he would inherit the ensuing work of other scholars whose writings had altered the understanding of theoretical astronomy. For

---

that Battānī's *Zīj* shows "no influence of [Thābit's] work or methods" (stated without explanation) (*The Astronomical Works of Thabit b. Qurra,* 18–19).

225 Langermann, however, wants to situate al-Battānī, along with Farghānī, within the "*hayʾa* literature"; he views Battānī's general description of the nested orbs (Ch. 31) and presentation of parameters for planetary distances and sizes (Ch. 50) as indications of physical concerns associated with *hayʾa* works (*Ibn al-Haytham's "On the Configuration of the World,"* 25–29). Cf. Carlo Nallino, who in stating that the solidity of the spheres was held by almost all Muslim writers, gives Battānī as the one counter-example who left the question uncertain ("Sun, Moon, and Stars [Muhammadan]," 99, col. 2, n4).

226 Battānī's extremely lengthy explanations (throughout the *Zīj*) that accompany his parameters can be deadly for pedagogical purposes. On the other hand, E. S. Kennedy viewed his detailed contextualizing of parameters as rewarding gateways into understanding the underlying mathematics behind the numbers ("A Survey of Islamic Astronomical Tables," 123).

227 Hartner states that anyone familiar with Ptolemy would be struck by Battānī's insufficient and inaccurate explanations, and he points out some of Battānī's "particularly bewildering features" contained in his *Zīj* ("Al-Battānī," 509). See also Samsó, "Battani, Al-," 79.

228 For example, Jaghmīnī alludes to Ptolemy's *Geography* (*Mulakhkhaṣ*, II.1[2]).

example, we mentioned that Jaghmīnī omits astrological topics in the *Mulakhkhaṣ*; however, this weeding out of astrology from *hay'a* works only began in earnest in the eleventh century, as we see with Ibn Sīnā's categorization. It was after all a subject sanctioned by Ptolemy (who also includes astrological terms and concepts in his *Almagest*[229]). Another example is from tenth-century Basra, where the Ikhwān al-Ṣafā' included both astronomy and astrology in Epistle 3 (entitled "On *Astronomia*") of their encyclopedic work, the astronomy evidently a handmaiden to astrological applications. It would seem that their purpose was not to present a summary account of Ptolemaic astronomy, but rather to provide moral guidance through astronomical knowledge, i.e., the orbs being stairways to Heaven. In sum, the work (an introduction and thirty-one chapters) gives an overview of the stars, planets, and zodiacal signs employing basic Ptolemaic principles, but without explanations of planetary models (e.g., the terms epicycles and eccentrics never appear); and Ptolemy's *Almagest* is cited only once [Ch. 26] and in the context of its application for salvation.[230]

The introduction of various aspects of theoretical astronomy for application to astrology was not unique. For example, the *Kitāb al-Tafhīm*, the astrological primer of al-Bīrūnī (d. ca. 442/1050), which he composed in both Arabic and Persian, certainly cannot be overlooked as providing a valuable user-friendly reference of astronomical terms, concepts, and explanations, even though it is ostensibly an astrological text.[231] Bīrūnī's "true" attitude towards astrology has been questioned, and it has been stated that he really believed that the basic tenets of astrology were spurious and that its practitioners were unscrupulous. But given his twenty-three or

---

[229] For example, see Ptolemy, *Almagest*, Book VIII.4 (Toomer, 407–8, nn185, 187, 190).

[230] The exact date of the 52 epistles of the Ikhwān al-Ṣafā' (the Brethren of Purity) remains as mysterious as the authors themselves. The corpus has four general divisions (Mathematics, Natural Philosophy, Sciences of the Soul and Intellect, and Theology), and Epistle 3 is contained in Mathematics, one of fourteen parts dealing with the mathematical sciences. Their citations range from the Qur'ān and ḥadīths to Pythagoras and Aristotle; and some of their possible sources include: Ptolemy's *Tetrabiblos*, Farghānī's *Jawāmi'*, Abū Ma'shar's *Introduction to Astrology*, Battānī's *Zīj*, and Qabīṣī's *Introduction to Astrology*. For most of the information on the Ikhwān presented here, see F. Jamil Ragep and Taro Mimura (eds. and trans.), *Epistles of the Brethren of Purity. On* Astronomia. *An Arabic Critical Edition and English Translation of Epistle 3* (Oxford: Oxford University Press in association with The Institute of Ismaili Studies, 2015). Also see: Yves Marquet, "Ikhwān al-Ṣafā'," in *EI2* (1971), 3:1071–76; Rosenfeld and İhsanoğlu, *MAMS2*, 90–91 (no. 226), E1, A1; Sezgin, *GAS*, 6:234–39; and Živa Vesel, "Ikhwān al-Ṣafā'," commissioned for *BEA*: http://islamsci.mcgill.ca/RASI/BEA/Ikhwan_al-Safa%27_BEA.htm.

[231] Bīrūnī is explicit in stating that the purpose of the *Kitāb al-Tafhīm* is to provide definitions of astronomical terms, by way of questions and answers, to help facilitate their further application elsewhere; and he informs us: "I have begun with Geometry and proceeded to Arithmetic and the Science of Numbers, then to the structure of the Universe, and finally to Judicial Astrology, for no one is worthy of the style and title of Astrologer who is not thoroughly conversant with these four sciences" (*Tafhīm*, 1).

so compositions on the subject, we cannot deny both the demand for works on astrology and its multifaceted role within Islamic society and among scholars themselves.[232]

That Bīrūnī was a fellow Khwārizmian of Jaghmīnī, albeit two centuries earlier, also underscores the fact that greater Central Asia was known for being a locus of scientific activity and creativity, and knowledge from this region disseminated throughout Islamic lands.[233] It is undeniable that many prominent scholars hailed from this area, one renowned example being Bīrūnī's contemporary Abū ʿAlī Ibn Sīnā.[234] However, there were many others (some recognizable by their *nisbas*), such as Muḥammad ibn Mūsā al-Khwārizmī (d. ca. 830), Abū Jaʿfar al-Khāzin al-Khurāsānī (d. ca. 971), Abū al-Wafāʾ al-Būzjānī (d. 997 or 998), and Abū Saʿīd al-Sijzī (d. ca. 1020). Bīrūnī mentions the observational improvements he found in al-Khāzin's *Tafsīr al-Majisṭī*, another commentary concerned with isoperimetric problems of the *Almagest*, Book One.[235] He also exchanged astronomical data and measurements with Būzjānī (a Baghdad transplant from Khurāsān), who also composed a work entitled *al-Majisṭī*;[236] and befriended the prolific mathemati-

---

[232] See Edward S. Kennedy, "Al-Bīrūnī," in *DSB* (1970), 2:152, 155–57.

[233] İhsan Fazlıoğlu refers to the regions of Transoxiana, Khurāsān, and Iran as "philosophical and scientific *granaries* of Islamic civilization" for Ottoman lands [italics in the original] ("The Samarqand Mathematical-Astronomical School: A Basis for Ottoman Philosophy and Science," *Journal for the History of Arabic Science* 14 [2008]: 8n13).

[234] For a preliminary overview analyzing Ibn Sīnā's astronomical works, see F. Jamil Ragep and Sally P. Ragep, "The Astronomical and Cosmological Works of Ibn Sīnā: Some Preliminary Remarks," in *Sciences, techniques et instruments dans le monde iranien (Xe–XIXe siècle),* études réunies et présentées par N. Pourjavady et Ž. Vesel (Tehran, 2004), 3–15.

[235] See Bīrūnī, *Al-Qānūn al-Masʿūdī*, 3 vols. (Hyderabad, 1954–56), 2:653 [hereafter cited as *Qānūn*]. Recall that the subject of isoperimetrics (contained in the fragment attributed to Abū Jaʿfar al-Khāzin [Paris, BnF, MS ar. 4821, ff. 47–68]) was the focus of Theon of Alexandria's Ptolemaic commentary on the *Almagest*, Book I, and other scholars (see above *I.3.2c: The Ptolemaic Aftermath*). In the *Qānūn,* Bīrūnī reports al-Khāzin's improved observations, along with those of the ninth-century Maʾmūn astronomers Khālid al-Marwarrūdhī, ʿAlī ibn ʿĪsā, and Sind ibn ʿAlī; but apparently he was critical of al-Khāzin in other works (Yvonne Dold-Samplonius, "Al-Khāzin," in *DSB* [1973], 7:334–35). See also Emilia Calvo, "Khāzin," in *BEA*, 1:628–29; Roshdi Rashed, *Founding Figures and Commentators in Arabic Mathematics: A History of Arabic Science and Mathematics Volume 1,* ed. Nader El-Bizri (London: Routledge; Beirut: Center for Arab Unity Studies, 2012), Ch. IV: Abū Jaʿfar al-Khāzin: Isoperimetrics and Isepiphanics; Rosenfeld and İhsanoğlu, *MAMS*2, 82 (no. 194), A3; and Sezgin, *GAS*, 6:190 (no. 1). For more on the other scholars, see Marvin Bolt, "Marwarrūdhī," (2:740) and Bolt, "ʿAlī ibn ʿĪsā al-Asṭurlābī," (1:34), both in *BEA*; Rosenfeld and İhsanoğlu, *MAMS*2, 26 (no. 42: Marwarrūdhī), 28 (no. 47: ʿAlī al-Asṭurlābī), 28–29 (no. 48: Sanad ibn ʿAlī); and Sezgin, *GAS*, 6:159, 143–44, and 138 [respectively]).

[236] According to Behnaz Hashemipour, Būzjānī's *al-Majisṭī* presents new observational data and trigonometric applications for astronomy, but he was not known for introducing any "theoretical novelties" ("Būzjānī," in *BEA*, 1:188–89). See also the Arabic edition by ʿAlī Mūsā, *Majisṭī Abī al-Wafāʾ al-Būzjānī* (Beirut: Markaz Dirāsāt al-Waḥda al-ʿArabiyya, 2010).

cian/astronomer al-Sijzī, who flourished in Khurāsān for some period and composed a work enticingly entitled *Kitāb al-aflāk*.[237] Bīrūnī's various relationships highlight the fact that many scholars known for their compositions on more "practical" aspects of astronomy (such as instruments, observations, compiling *zīj*es, and so forth) were also writing on theoretical issues—though many of these works are either no longer extant or have yet to be carefully examined. Bīrūnī also showcases the vibrant exchange of information among scholars irrespective of place during the eleventh century, a phenomenon certainly not confined to this scholar or period alone.

It is also in the eleventh century that Ibn al-Haytham (who flourished in the more westerly regions of Basra and Cairo) composes his *al-Maqāla fī hay'at al-'ālam* (Treatise on the Configuration of the World),[238] a *hay'a* work often championed as the first attempt to physicalize the mathematical constructs of Ptolemaic astronomy. Putting aside the veracity of this claim for the moment, it is certainly undeniable that his fifteen-chapter work influenced generations of scholars throughout Islamic lands and also had a major impact on astronomical planetary theory in the Latin West.[239] Ibn al-Haytham assessed (I believe correctly) that no previous work on

[237] This work is listed with this title in both Rosenfeld and İhsanoğlu (*MAMS*2, 113 [no. 296], A2) and Sezgin, *GAS*, 6:225 [no. 1]). But this title is not mentioned in Glen van Brummelen, "Sijzī," in *BEA*, 2:1059 nor in Yvonne Dold-Samplonius, "Al-Sijzī," in *Encyclopaedia of the History of Science, Technology, and Medicine in Non-Western Culture*, 159–60. The witness I checked (Tehran, Majlis-i Shūrā MS 174, which was kindly provided to me by Mr. Sajjad Nikfahm-Khubravan) is referred to on the library's frontispiece label as a work on judicial astrology. The content (240 pages) seems to deal with detailed parameters for sizes and distances and various celestial motions based on Ptolemy, and it also contains extensive tables. Al-Sijzī was apparently known for his astrological compilations and commentaries, which included at least forty-five geometrical and fourteen astronomical works.

[238] Langermann provides an edition of the text, along with an English translation and notes, in *Ibn al-Haytham's "On the Configuration of the World."* See also Langermann, "Ibn al-Haytham," in *BEA*, 1:556–57.

[239] Ibn al-Haytham [Latinized as Alhacen or Alhazen] became known in Europe in the thirteenth century, and his *Configuration* was translated into Spanish, Hebrew, and Latin (see A. I. Sabra, "Ibn al-Haytham," in *DSB* [1972], 6:197–98, 210). Its influence on Renaissance scholars, particularly Peurbach's *Theoricae* (a work on planetary theory composed in 1460), was noted by E. J. Aiton, who concluded that "Peurbach evidently drew upon Ibn al-Haytham's (Alhazen's) *On the Configuration of the World* or some later work based on this" ("Peurbach's *Theoricae novae planetarum*: A Translation with Commentary," *Osiris* 3 [1987]: 7–8). Some thirty years earlier than Aiton, Willy Hartner had also discussed Peurbach's dependency on Islamic astronomers and compared his Mercury model with that of Ibn al-Haytham's. Hartner also recognized Jaghmīnī's interest in the physical reality of the orbs (albeit misplaced to the fourteenth-century [124n89]), and lumped Jaghmīnī together with Ibn al-Haytham in asserting, "The dependency of early Renaissance astronomers on ALHAZEN and AL-JAGHMĪNĪ is beyond doubt. Yet I am unable to tell at the moment from which of the two (possibly from both), and through which channels, they drew their information" ("The Mercury Horoscope of Marcantonio Michiel of Venice: A Study in the History of Renaissance Astrology and Astronomy," *Vistas in Astronomy*, ed. Arthur Beer [London: Pergamon Press, 1955], 1:124–35).

theoretical astronomy had actually explained to the reader how the various components of the Ptolemaic models operated and ultimately fit together to form a coherent whole, certainly not with a straightforward and non-technical depiction. Based on available evidence, theoretical astronomical textbooks prior to Ibn al-Haytham tended to be general overviews, summaries, and/or (overly) technical, selective discussions. In comparison, *On the Configuration of the World* attempts to match the mathematical models of the *Almagest* with physical structures to account for the various motions of the celestial bodies. However, Ibn al-Haytham does not provide parameters, proofs, a discussion of sizes and distances, or even illustrations.[240] Furthermore, since his focus is presumably explaining the hows of the celestial components, terrestrial topics are kept to a minimum; and philosophical and astrological topics are omitted altogether.

However, it was not Ibn al-Haytham's aim in the *Configuration* to question Ptolemaic theory; this was reserved for criticisms found in his other works such as his *al-Shukūk ʿalā Baṭlamyūs* (Doubts About Ptolemy), which addressed irregularities and violations by Ptolemy of his own principles in three of his works: the *Almagest, the Planetary Hypotheses,* and the *Optics*. Ibn al-Haytham was truly remarkable in being both prolific and creative.[241] But he was also exceptional in his ability to articulate underlying ideas and sentiments upheld by many Islamic scholars, as exemplified by his *Configuration*, and he also conveyed in his statements the duties of the scholar to question scientific authorities in the quest for truth, which is contained in his Introduction to *al-Shukūk*.[242]

Ibn al-Haytham's work was certainly remarkable and undoubtedly inspirational for future scholars; nevertheless, to claim that he "single-handedly established

---

240 However, there are indications that Ibn al-Haytham may have wanted to include illustrations. The closing statements at the end of the chapters on the orbs of the Sun, Moon, Mercury, Venus, the Upper Planets, the Fixed Stars, and the Highest Orbs seem to indicate that figures should follow (Langermann, *Ibn al-Haytham's "On the Configuration of the World,"* 131 [209], 150 [272], 177 [321], 196 [337], 206 [359], 215 [374], 223–24 [382] (English); and 37, 46, 54, 57, 60, 63, 65 (Arabic)).

241 A daunting combination, which probably contributed to speculation that one man alone could not have written all the works attributed to him, and that there were two Ibn al-Haythams, one mathematically inclined, another philosophically inclined. For evidence supporting the one Ibn al-Haytham position (which I endorse), see A. I. Sabra's two articles: "One Ibn al-Haytham or Two? An Exercise in Reading the Bio-bibliographical Sources," *Zeitschrift für Geschichte der Arabisch-Islamischen Wissenschaften* 12 (1998): 1–50; and Sabra, "One Ibn al-Haytham or Two? Conclusion," *Zeitschrift für Geschichte der Arabisch-Islamischen Wissenschaften* 15 (2002/2003): 95–108. Cf. the work of Roshdi Rashed, who denies that the Ibn al-Haytham who composed *On the Configuration of the World* also authored *Doubts on Ptolemy*; see "The Configuration of the Universe: A Book by al-Ḥasan ibn al-Haytham?" *Revue d'histoire des sciences* 60 (2007/1): 47–63, where Rashed summarizes and reaffirms his position in light of criticism.

242 A. I. Sabra provides a marvelous translation in "Ibn al-Haytham: Brief Life of an Arab Mathematician: Died circa 1040," *Harvard Magazine*, September-October 2003, 54.

physical cosmography in Islam" overshadows the fact that he was often making explicit what was already implicit in previous theoretical works.[243]

However, Ibn al-Haytham should be duly recognized as a "pioneering inspiration rather than a prototype to be emulated."[244] Naṣīr al-Dīn al-Ṭūsī devotes an entire chapter to the configuration of the epicycle orbs of the planets according to Abū ʿAlī ibn al-Haytham in a Persian appendix to his *Risālah-yi Muʿīniyya* (written in 1235);[245] and ʿAbd al-Jabbār al-Kharaqī (477–553/1084–1158)[246] explicitly credits Ibn al-Haytham as being an important influence for motivating him to consider solid spheres as opposed to imaginary circles in astronomy, and inspiring his attempts to reconcile physics with mathematical models.[247] In turn, Kharaqī's compositions on theoretical astronomy, especially his *Muntahā al-idrāk fī taqāsīm al-aflāk,* in which he explicitly stated that *ʿilm al-hayʾa* follows theology in standing and nobility in showing God's wisdom, and his shorter, more popular *al-Tabṣira fī ʿilm al-hayʾa* (both written in Arabic and composed ca. 526-27/1132-33),[248] would play critical roles in the development of *hayʾa.*[249] In fact, the *Tabṣira*

[243] F. J. Ragep points out that Ibn al-Haytham "seems to go out of his way to indicate that previous [astronomical] work has assumed the existence of solid spheres" (*Tadhkira*, 1:30–33 at 31).

[244] See Ragep, *Tadhkira*, 1:33.

[245] See F. Jamil Ragep, "Ibn al-Haytham and Eudoxus: The Revival of Homocentric Modeling in Islam," in *Studies in the History of the Exact Sciences in Honour of David Pingree*, ed. Charles Burnett et al. (Leiden: E. J. Brill, 2004), 786–809.

[246] ʿAbd al-Jabbār al-Kharaqī has often been confused with an older contemporary, a Shams al-Dīn Abū Bakr Kharaqī (both sharing the same *nisba*); however, we can confidently date our Kharaqī (i.e., the one who authored *hayʾa* works) based on information from contemporary primary sources. See Hanif Ghalandari, "A Survey of the Works of '*Hayʾa*' in the Islamic Period with a Critical Edition, Translation and Commentary of the Treatise *Muntahā al-Idrāk fī Taqāsīm al-Aflāk* written by Bahāʾ al-Dīn al-Kharaqī (d. 553 AH/1158 AD)," Ph.D. diss., Tehran, Oct. 2012, 4–5 [in Persian with an Arabic edition]. See also Cemil Akpınar, "Harakī," in *İslam Ansiklopedisi* (Istanbul: TDV, 1997), 16:94–96 [Turkish]. For a German translation of the introductions to both the *Muntahā* and *Tabṣira*, see Eilhard Wiedemann and Karl Kohl, "Einleitung zu Werken von al-Charaqi," *Sitzungsberichte der Physikalisch-Medizinischen Sozietät zu Erlangen* 58 and 59 (1926–27): 203–18; repr. in E. Wiedemann, *Aufsätze zur arabischen Wissenschafts-geschichte* (Hildesheim: Georg Olms, 1970), 2:628–43.

[247] Kharaqī cites both Ibn al-Haytham and Abū Jaʿfar al-Khāzin in the *Muntahā* (e.g., Berlin, Staatsbibliothek, Landberg MS 33, f. 2a), but he omits al-Khāzin in the *Tabṣira* (e.g., Istanbul, Ayasofya MS 2581, f. 2b). See also Ghalandari, "A Survey of the Works of '*Hayʾa*' in the Islamic Period," 149–50.

[248] Ghalandari, "A Survey of the Works of '*Hayʾa*' in the Islamic Period," Abstract and 7.

[249] Some indications are: Muʾayyid al-Dīn al-ʿUrḍī's thirteenth-century *Tabṣira* commentary and quotes from Kharaqī's works (see Saliba, "The First Non-Ptolemaic Astronomy at the Maraghah School," *Isis* 70, no. 4 (Dec., 1979): 572; repr. in *A History of Arabic Astronomy*, 114); Ṭūsī referencing Kharaqī simply as "the author [*ṣāḥib*] of the *Muntahā al-idrāk*" (صاحب منتهى الإدراك) in his *Ḥall-i mushkilāt-i Muʿīniyya* (Ragep, "Ibn al-Haytham and Eudoxus," 797 [Persian], 805 [Eng. trans.]); Quṭb al-Dīn al-Shīrāzī including three of his

was an unnamed source used extensively by Jaghmīnī throughout the *Mulakhkhaṣ*.[250] Kharaqī's works would codify the basic structure of subsequent *hay'a* works: an introduction and (most importantly) the two-part division of regions into the arrangement (*tarkīb*) of the celestial bodies and their motions and the configuration (*hay'a*) of the Earth. Other *hay'a* works might also devote a chapter or section to the subject of sizes and distances; and a discussion of chronology, included as a separate section in the *Muntahā*, tended to be downplayed in later works.[251] The credit for this new delineation of astronomy associated with *hay'a* works, i.e., presenting the cosmos as a coherent whole with subject matter divided into two basic arenas—i.e., the upper bodies of the celestial region ("cosmography") and the lower bodies of the terrestrial realm ("geo-graphy")—has usually been attributed to Kharaqī's two Arabic treatises.[252] However, Kharaqī also wrote another *hay'a* work, in Persian, entitled *al-'Umda* dedicated to the Khwārizm Shāh Atsiz (r. 521–51/1127–56); and it is among the list of "unattributed" *hay'a* works mentioned by Shīrāzī in his explicit to the *Nihāya*.[253] Kharaqī's Persian treatise, as do its two Arabic counterparts, contains the signature two-part division of the cosmos.

Although I have yet to examine Kharaqī's Persian *hay'a* treatise carefully, its mere existence raises interesting questions that challenge our views regarding the role of Persian compositions of theoretical astronomy, especially their relationship

---

works in his list of well-known *hay'a* books in the explicit to his *Nihāya* (Ragep, "Shīrāzī's *Nihāyat al-Idrāk*: Introduction and Conclusion," 51 [Arabic], 55 [Eng. trans.]); and Shīrāzī's student 'Ubaydī (d. 751/1350) playfully incorporating the use of *tabṣira* into his commentary title to Ṭūsī's *Tadhkira* (*Bayān al-Tadhkira wa-tibyān al-tabṣira*) (Ragep, *Tadhkira*, 1:61). See also Petra G. Schmidl, "'Urḍī," in *BEA*, 2:1161–62.

[250] See the Commentary for references to Kharaqī, and especially the examples provided in which Jaghmīnī paraphrases sections from Kharaqī's *Tabṣira* (II.1[10]: his description of latitudes from 63 to 66 degrees; and II.3[5]: regarding the astrolabe/gnomon exercise). Rather than charging Jaghmīnī with plagiarism, one should keep in mind that Kharaqī's works were probably common knowledge, and so Jaghmīnī may have felt no compunction to paraphrase him.

[251] As mentioned earlier, both Kharaqī and Jaghmīnī do not include sizes and distances in the *Tabṣira* and *Mulakhkhaṣ* (respectively), most likely considering it inappropriate subject matter for a *hay'a basīṭa* work; however, Kharaqī's *Muntahā* contains a chapter on the subject [*Maqāla* II, *bāb* 17, in 2 parts (*faṣlān*)]. Some later *hay'a* works would devote an entire section to sizes and distances (e.g., Ṭūsī's *Tadhkira*). Regarding subjects related to chronology (such as years, months, hours), Jaghmīnī in the *Mulakhkhaṣ* lumps the various topics together in his final chapter of Part Two (on an explanation of the Earth and what pertains to it) under the umbrella title of "Miscellaneous Items" (*Mulakhkhaṣ*, II.3). In comparison, Ṭūsī informs us that subjects related to chronology have no place in a *hay'a* work, and he buries these topics in a section of a chapter in Book III (see Ragep, *Tadhkira*, 1:36, 36n12, 300–3 (III.10[3], lines 10–12)).

[252] See Ragep, *Tadhkira*, 1:36.

[253] See Ragep, "Shīrāzī's *Nihāyat al-Idrāk*: Introduction and Conclusion," 51 [Arabic], 55 [Eng. trans.]. Only recently have we become aware that the identity of Shīrāzī's unnamed author of *al-'Umda li-ūlī al-albāb* was Kharaqī, and also that the work was composed in Persian.

to Arabic treatises during this formative period. The standard narrative has been that "from the fifth/eleventh century onwards [at least]...the language par excellence of science" was "almost exclusively in Arabic."[254] Indeed, the assumption was that any Persian treatises on scientific topics are *later* translations of their Arabic counterparts, perhaps attempts to convey scientific information for court members or a lay audience, people more comfortable with the vernacular Persian and less familiar with Arabic. The possibility that a Persian treatise gave rise to an Arabic one is typically downplayed (or dismissed). A good example highlighting this point is the general assumption that Bīrūnī's Arabic version of his *Kitāb al-Tafhīm* preceded the Persian, although there is no evidence to support any priority. Furthermore, it has also been suggested that the Persian rendition was not necessarily done by Bīrūnī himself (who I might add was trilingual in Persian, Arabic, and [his mother tongue] Khwārizmian), again based on unfounded speculation.[255]

So a prevalent narrative is that it would be difficult to point to "scientific texts and say that there was an indigenous Persian scientific production which was independent of the contemporary Arabic production."[256] However, even a preliminary comparison of Kharaqī's three works indicates striking similarities between them, and especially between the Arabic *Muntahā* and the Persian *ʿUmda*.[257] This clearly

---

[254] See C. Edmund Bosworth, "The Persian Contribution to Islamic Historiography in the Pre-Mongol Period," in *The Persian Presence in the Islamic World,* ed. Richard G. Hovannisian and Georges Sabagh (Cambridge: Cambridge University Press, 1998), 231.

[255] Although Gilbert Lazard points out that the priority between the two versions has never been explicitly established, he confidently concludes Arabic seniority based on his impression that the Arabic is written in a concise and elegant style, whereas the Persian is written in a more clumsy and belabored way ("Souvent l'arabe est concis, net élégant là où le persan paraphrase plus ou moins et donne l'impression d'une certaine gaucherie et d'une certaine lourdeur" [60]). He also notes that the Persian is highly dependent on Arabic scientific terms, but it is not clear why this is an argument in favor of the priority of the Arabic version. Furthermore, he cites Bīrūnī's objection to composing scientific works in Persian (as expressed in his work on *Pharmacology* [*Kitāb al-Ṣaydala*]). Lazard is not alone in referring to this latter work and citing Bīrūnī's statement that Persian was a language "fit only for the recital of bedtime stories and legends of kings" (E. S. Kennedy, "Al-Bīrūnī," 155), and also that Persian was a language far "less precise and less lexically rich for scientific purposes" [than Arabic] (Bosworth, "The Persian Contribution to Islamic Historiography in the Pre-Mongol Period," 231–32). George Saliba also reiterates this sentiment in "Persian Scientists in the Islamic World: Astronomy from Maragha to Samarqand," in *The Persian Presence in the Islamic World*, 126, 146. Nevertheless, unlike Lazard, Kennedy concluded that Bīrūnī alone prepared both versions of the *Tafhīm* (154). See G. Lazard, *La langue des plus anciens monuments de la prose persane* (Paris: C. Klincksieck, 1963), 58–62 (no 10: *Al-Tafhīm*).

[256] Saliba, "Persian Scientists in the Islamic World," 127. A major counterexample is the Ṭūsī-couple, which was first presented in Persian in Ṭūsī's appendix to his *Risālah-yi Muʿīniyya*; see F. Jamil Ragep, "The Origins of the Ṭūsī-Couple Revisited," forthcoming.

[257] For example, for the *Tabṣira*, Kharaqī has removed the entire section on chronology that is included in both the *Muntahā* and the *ʿUmda*; the following compares the divisions of these three works: *Muntahā*: Part I: 20 chs; Part II: 17 chs; Part III: 11 chs; *ʿUmda*: Part I: 25 chs; Part II: 15 chs; Part III: 12 chs; *Tabṣira*: Part I: 22 chs; Part II: 14 chs. With an admittedly

supports the notion that the two "productions" are at least somehow interrelated; and it would be rash to conclude from this simply that the Persian is a direct translation or derivative from the "original" Arabic, certainly not without a more careful examination of the content of these treatises. Fortunately, here we have the example of another extant Persian *hay'a* work dating from this period, one that (as far as I know) has no Arabic counterpart. The *Gayhān Shinākht* (Knowledge of the Cosmos) was composed ca. 498/1104 by Kharaqī's contemporary Qaṭṭān al-Marwazī (465–548/1072–1153), known for being a physician and also a polymath, who stemmed from a family of scholars among the learned circles of Merv.[258] This three-part treatise contains what we have been calling the "classical" two-part division of the universe plus a section on chronology, and the early date indicates that this may very well have been the inspiration for Kharaqī's structure (with an alternative possibility being another unknown source that influenced both).[259] In any event, the mere existence of two Persian *hay'a* treatises, especially at this formative stage, is a step in problematizing the view that Persian astronomy was concerned more with "astrological computations and less with theoretical astronomical issues."[260]

The wide range of subject matter covered in Qaṭṭān al-Marwazī's *Gayhān Shinākht* on theoretical astronomical issues is impressive, and so it would be reasonable to assume Jaghmīnī would have known of this work, especially given the place and date of composition.[261] Nevertheless, Jaghmīnī's parameters are clearly gleaned from the "authorities" of Ptolemy and Battānī; and Kharaqī's *Tabṣira* (more than the *Muntahā*) is another main source of structure and material for him, given

brief skim of the *ʿUmda*, I noted that the *Muntahā* and *ʿUmda* cite Abū ʿAlī ibn al-Haytham and Abū Jaʿfar al-Khāzin, whereas the latter is omitted in the *Tabṣira*.

[258] See Behnaz Hashemipour, "Qaṭṭān al-Marwazī," in *BEA*, 2:943–44 and "Gayhānshenākht: A Cosmological Treatise," in *Sciences, techniques et instruments dans le monde iranien (Xe–XIXe siècle)*, 77–84 [in Persian]; Storey, *Persian Literature,* 1:45–46 (no. 82: ʿAin al-Zamān Abū ʿAlī al-Ḥasan b. ʿAlī b. M. al-Qaṭṭān al-Marwazī); and Ghalandari, "A Survey of the Works of '*Hay'a*' in the Islamic Period," 26, 147 (English Abstract). A facsimile of this treatise has recently been published along with an introduction [in Persian]; see al-Ḥasan ibn ʿAlī Qaṭṭān, *Gayhān Shinākht*, Chāp-i 1 (Tehran, 2012).

[259] It is certainly possible that contemporary scholars residing in the same locale may not be aware of each other; however, the Khwārizm Shāh Atsiz provides a common link since Qaṭṭān al-Marwazī is known to have corresponded with Rashīd al-Dīn Waṭwāṭ, who was the Shāh's chief secretary (see Hashemipour, "Qaṭṭān al-Marwazī"; and Natalia Chalisova, "Waṭwāṭ, Rašid-al-Din," in *Encyclopædia Iranica* (New York, 1996–), available online at http://www.iranicaonline.org/articles/watwat-rasid-al-din.

[260] Saliba, "Persian Scientists in the Islamic World," 144.

[261] The date of copying given in one colophon is Tuesday, 21 Ramaḍān 586 H (=Tuesday, 29 Oct. 1190), which indicates it was in circulation during the time Jaghmīnī was composing the *Mulakhkhaṣ*. Also, the facsimile is not the only witness. The published *Gayhān Shinākht* lists several other witnesses (67–69), and I was able to check no. 2 on the list (Tehran, Majlis-i Shūrā MS 202), which contains the exact same colophon; however, in this case, one was probably copied from the other.

that the *Mulakhkhaṣ* shares its signature two-part division of the celestial and terrestrial regions, and, more significantly, Jaghmīnī directly lifts several parts from the *Tabṣira* and incorporates them into the *Mulakhkhaṣ*.[262]

This would certainly make Badr al-Dīn al-Qalānisī's request for yet another elementary astronomical textbook quite puzzling without a closer examination of the differences between the *Tabṣira* and the *Mulakhkhaṣ* in both subject matter[263] and organization.[264] However, for our purposes here, even a simple comparison of the number of chapters contained in both works shows us immediately how the *Mulakhkhaṣ* is by far a less complex work than Kharaqī's introductory alternative:

	Jaghmīnī's *Mulakhkhaṣ*	Kharaqī's *Tabṣira*
Introduction	Introduction	Introduction (includes an extensive discussion of mathematical terms)
Part One (*hay'at al-samā'*)	5 chapters	22 chapters (5 chapters have extensive subdivisions)
Part Two (*hay'at al-arḍ*)	3 chapters	14 chapters

262 Extant copies of Kharaqī's *Tabṣira* far outnumber those of his *Muntahā*. In fact, I am currently unaware of any commentaries on the *Muntahā*. The *Tabṣira* commentaries (all seemingly composed in the thirteenth and fourteenth centuries), indicate that the treatise disseminated widely, was studied in Yemen and in eighteenth-century Egypt, and was translated into Hebrew. It is noteworthy that one commentator, Ibn al-Turkmānī (fl. Cairo), was the father of Kamāl al-Dīn al-Turkmānī, who authored a *Mulakhkhaṣ* commentary (see Appendix II, no. 5). See Fazlıoğlu, "Kamāl al-Dīn al-Turkmānī," in *BEA*, 1:609. See also Langermann, "Kharaqī," in *BEA*, 1:627; Ghalandari, "A Survey of the Works of '*Hay'a*' in the Islamic Period," 8; and İzgi, *Riyazî ilimler*, 1:405–6 (no. 7).

263 Here are some additional features and differences between the contents of the two works: (1) in discussing the equinox points Jaghmīnī references the two holidays of Nayrūz and Mihrjān (see II.2[2]), and this may be an indication of Persian influence; (2) Jaghmīnī cites al-Shāfiʿī and Abū Ḥanīfa (see II.3[2]); and, most significantly, (3) Jaghmīnī has considerably condensed the *Tabṣira*'s introductory sections, essentially eliminating Kharaqī's section dealing with mathematical definitions (such as point, line, straight line, and so forth) and giving only the briefest account of the general properties of bodies. In fact, Jaghmīnī has drastically abbreviated Kharaqī's section (contained in Chapter One of the *Tabṣira*) that deals with bodies from the perspective of natural philosophy. The explanation that Jaghmīnī provides in the introduction to the *Mulakhkhaṣ* (Intr.[1]) regarding the simple and composite bodies barely hints at its connection with Aristotelian natural philosophy.

264 An example of this is that Jaghmīnī gives an explanation of *all* the orbs in one chapter (I.1) and *all* the motions of the orbs in another separate one (I.2), whereas Kharaqī, similar to Ṭūsī in his *Tadhkira*, combines the descriptions and motions for a planet together (usually in a separate, self-contained chapter).

Clearly, Kharaqī's popular "abridged" version of the *Muntahā* was still fairly extensive, with quite a bit of technical detail for an elementary treatise. One could well conclude that there must have been a growing need for a more simplified "user-friendly" textbook on theoretical astronomy.

Hence, Jaghmīnī's *Mulakhkhaṣ* enters the picture, both literally and figuratively. So what now follows is an overview summarizing what Jaghmīnī does—and also does not do—in comparison with some of these earlier works on theoretical astronomy. This allows us to assess how the *Mulakhkhaṣ* fits into the transformations that began to occur within the discipline of *hayʾa* that transformed the way it was being taught, with one very important result being the emergence of a new kind of *hayʾa* textbook that was conducive for a more general audience.

§ I.4 Jaghmīnī's *Mulakhkhaṣ*: A Beginners Text, but Not for the Untutored

Jaghmīnī's plain or simplified (*basīṭa*) epitome of *hayʾa* is in fact anything but simple-minded.[265] Unlike other introductory astronomical textbooks that on the one hand present wide-ranging but non-coherent overviews or on the other hand target specific astronomical problems, Jaghmīnī provides fundamental information to comprehend the broad picture of the universe (from top to bottom) that is conceptually packaged. He gives basic definitions and rules along with parameters in easily accessible lists to account for various planetary motions; many of the latter have been updated from Ptolemy's *Almagest* for a twelfth-century Khwārizmian audience, the fruits of the work of Battānī and the Maʾmūn observations (the so-called "Moderns") and subsequent scholars. He omits mathematical proofs and the topic of sizes and distances. He also eliminates information that the student could (or should) seek elsewhere, such as within practical handbooks with their astronomical tables (*zīj*es) or the *anwāʾ* literature.[266] The general subject of astrology never enters the picture, at least as a science that interprets celestial signs and makes predictions, even though he is undoubtedly aware of its popularity (at least in some circles). However, Jaghmīnī assumes that the student is familiar with certain components of astrology, such as the signs of the zodiac and the constellations, inasmuch as he incorporates the movement of the zodiacal signs and constellations into various discussions, in particular when he discusses their appearances for the various climes.[267]

[265] Cf. David A. King, who viewed the ubiquitous *Mulakhkhaṣ* as no more than consisting "mainly of a nontechnical digest of Ptolemaic astronomy" ("The Astronomy of the Mamluks," *Isis* 74, no. 4 [Dec., 1983]: 552).

[266] The *anwāʾ* literature was the discipline that consisted of a corpus of folklore material that developed from astronomical mapping and weather prognostication, and was modified to conform with the 28 lunar mansions (see Ch. Pellat, "Anwāʾ," in *EI2* [1960], 1:523–24).

[267] As far as I know, no *hayʾa* work devoted an independent section to the constellations; more common is to direct the reader elsewhere. For example Ṭūsī states that that the "knowledge of the fixed stars and that which concerns them [is] a separate discipline" (Ragep, *Tadhkira,* II.4[9–12], 1:37, 128–29). Interestingly, Rudloff and Hochheim duly point

In fact, Jaghmīnī has the clear expectation that the student already has had some previous astronomical training, as evidenced by the following few examples: a student must be familiar with how to use an astrolabe for the "hands-on" exercise in II.3[5], since Jaghmīnī provides no definitions of its parts or operating procedures; the student should be able to perform computations using sexagesimal notation, often beyond seconds (see I.5[37] for exercises); and the student should be familiar with astronomical dating, in particular the Alexandrian calendar (Dhū al-Qarnayn) as in I.5[26].

Jaghmīnī's challenge pedagogically was to simplify difficult material, while ensuring that the information presented was both detailed and accurate, unlike the case, as we have seen, with Roman sources. Many of the astronomical textbooks Jaghmīnī inherited contained complex explanations (Proclus's *Hypotyposis*), long-winded discussions (Kharaqī's *Muntahā*), oversimplifications (Farghānī's *Jawāmiʿ*), additional literary references (Geminus's *Introduction to the Phenomena*), and/or incorrect statements and depictions (recall that Battānī incorrectly depicted the Mercury model in his *Zīj*). Indeed, the success of the *Mulakhkhaṣ* was Jaghmīnī's ability to meet this challenge and make the complex look simple. He presented basic astronomical information with an objective style that exuded authority while also providing expansive asides meant to aid and reassure the student. The reinforcing, pedagogical style is witnessed throughout the textbook in statements such as "as you will come to know…" and "as you already learned…" (statements he interjects at least twenty-one times!). In addition, the *Mulakhkhaṣ* contains several diagrams, which, as we have seen, are not often found in earlier introductions such as Farghānī's *Jawāmiʿ*. These diagrams are not lavish or elaborate but simply functional with pedagogical value. Perhaps this explains why the original text is not as extensively illustrated as other *hayʾa* works, such as those of Kharaqī or Ṭūsī, or as later commentaries on the *Mulakhkhaṣ*. We should also note that this is not a "passive" treatise but one replete with pedagogical exercises.

But pedagogy here has its limits. Jaghmīnī does not seek to provide moral guidance to the reader by using examples from astronomy, unlike what one finds in Epistle 3 ("On *Astronomia*") of the Ikhwān al-Ṣafāʾ. In fact, it bears mentioning that the *Mulakhkhaṣ* only touches on religious needs when it relates to determining the direction of Mecca and determining the prayer times, with distinctions noted between the Ḥanafī and Shāfiʿī schools. God seemingly remains a silent partner.

All of this brings us to the pressing questions of what inspired the commissioning of this treatise, and who was the target audience? The new delineation of astronomy that dealt with both the upper and lower bodies connected the unchanging celestial realm (*hayʾat al-samāʾ*) with the ever-changing sublunar one of man (*hayʾat al-arḍ*). (Recall Ibn al-Haytham kept terrestrial topics to a minimum in his *On the Configuration of the World*.) The study of *hayʾa* then could provide a

---

out that the *Mulakhkhaṣ* lacks a star catalogue; however, they assume it was because Jaghmīnī would have found the catalogue of Ṣūfī to be sufficient ("Die Astronomie," 214–15). They are referring to ʿAbd al-Raḥmān al-Ṣūfī, the tenth-century author of the lavishly-illustrated *Book of Constellations (Kitāb ṣuwar al-kawākib)*, who describes 48 Ptolemaic constellations based on the *Almagest* (see Paul Kunitzsch, "Ṣūfī," in *BEA*, 2:1110).

picture of God's entire creation, both that of the perfect and that of the corruptible, and offer another approach to serve God.[268] The assertion made by Quṭb al-Dīn al-Shīrāzī in his *Nihāya* introduction that the discipline of *ʿilm al-hayʾa* was "the most noble of the sciences" (with his support being a citation from the Qurʾān [269]), indicates that within Islamic society there was an ever-growing segment of the population that had begun to view the study of *hayʾa* as a way to glorify the Creator. Kharaqī explicitly tells us in his introduction to his *Muntahā* that the study of *ʿilm al-hayʾa* is a rational and noble approach for attaining a better understanding of God through His creation; presumably he still believed this when he composed his *Tabṣira*, but felt there was no need to explicitly state it there, nor did Jaghmīnī in his *Mulakhkhaṣ*.[270] The *Mulakhkhaṣ* then provided the essential keys to unlocking knowledge of His created universe (without attempting to address "why" the celestial and terrestrial realms operate the way they do), which potentially made it an ideal addition to the madrasa curriculum.[271]

[268] In other words, astronomy "in the service of Islam" (a term coined by David King) is valued for its theoretical applications to achieve a better understanding of the physical world of God's creation, not only for its practical applications for religious ritual and more utilitarian needs. For the latter, see D. A. King, *Astronomy in the Service of Islam* (Aldershot: Ashgate Variorum Reprints, 1993).

[269] Quṭb al-Dīn al-Shīrāzī cites Quʾrān III.191 to link the heaven and the Earth in the "Introduction" to his *Nihāya*: "Whoever—standing, sitting or reclining—recalls God and reflects on the creation of the heavens and the Earth [will say]: Our Lord! Thou hast not created this in vain" (Ragep, "Shīrāzī's *Nihāyat al-Idrāk*: Introduction and Conclusion," 49 [Arabic], 54 [Eng. trans.]).

[270] It is, though, pointed out in some commentaries; for example, Qāḍīzāde states that the discipline of *hayʾa* is one by which one learns about the Creator, namely, from substances and accidents (*Sharḥ al-Mulakhkhaṣ*, Istanbul, Ayasofya MS 2662, f. 2b). See also, F. J. Ragep, "Freeing Astronomy," 51, 64.

[271] I develop the case that at least part of the readership of Jaghmīnī's *Mulakhkhaṣ* were students studying in madrasas in Sally P. Ragep, "Maḥmūd ibn Muḥammad ibn ʿUmar al-Jaghmīnī's *al-Mulakhkhaṣ fī al-hayʾa al-basīṭa*: An Edition, Translation, and Study," Ph.D. diss., McGill University, 2015, chapter 3. It is my contention that, beginning in the twelfth century, interconnected conceptual and textual transformations began to occur within the discipline of *hayʾa* that transformed the way it was taught. One very important result is the emergence of a new kind of *hayʾa* textbook that was conducive for a more general audience. My current research further examines evidence establishing that the mathematical sciences (with a focus on theoretical astronomy) were being studied in Islamic institutions, especially the madrasas; see Sally P. Ragep, "The Teaching of Theoretical Astronomy in Pre-modern Islam: Looking Beyond Individual Initiatives," in *Schüler und Meister*, ed. Andreas Speer and Thomas Jeschke, *Miscellanea Mediaevalia* 39 (Berlin: De Gruyter, 2016), 557–68.

Editorial Procedures

§ II.1 Editorial Procedures

II.1a Establishing the Text

Although there are a large number of extant manuscript witnesses of the *Mulakhkhaṣ*, either standing as independent texts or incorporated into a commentary or supercommentary, it was possible to establish an edition that, I claim, is close to the author's original. Fortunately, there was a relatively simple way to eliminate the vast majority of extant manuscripts as candidates for the author's original. These witnesses contain modifications that, as I explain in Commentary, II.1[4] ("the second clime"), could have only occurred after the publication of the *Tadhkira* by Naṣīr al-Dīn al-Ṭūsī in 659/1261, i.e., well after Jaghmīnī's lifetime.[1] Next, I identified a dedication and poem to Badr al-Dīn al-Qalānisī that occurs in only a very few manuscripts. Thus I chose MSS B, F, and S, which also contain the pre-*Tadhkira* parameters, for the edition. There were two additional manuscripts containing the original parameters that I also used: one, MS K, has the dedication but not the poem; and MS L, which lacks both but has the earliest known copy date (644/1246-47). One could then distinguish these five manuscripts based on their prefaces: three have the poem and dedication (MSS B, F, and S); one has only the dedication (MS K); and one has neither (MS L). These divergent prefaces come together in the preface (Pref.[2]) and continue to the end of the treatise with relatively minor variants; these are listed in the text apparatus. The five principal manuscripts used for the edition are described in detail in the next section §II.

There is no autograph copy, and no single manuscript establishes the "original" version. Each has some deficiency. For example, the oldest one (MS L) lacks the original preface; MS F has one folio missing; MS S has many grammatical mistakes; and MSS B and K contain various mistakes and are further contaminated by one or more commentaries. Nevertheless, using MSS F, L, and S, I claim that the edited text is very close to the author's original, given the remarkably few variants between these three unaffiliated manuscripts and the plausible explanations for divergences

[1] See also Ragep, "On Dating Jaghmīnī and His *Mulakhkhaṣ*," 462–64.

S.P. Ragep, *Jaghmīnī's Mulakhkhaṣ*, Sources and Studies in the History of Mathematics and Physical Sciences, DOI 10.1007/978-3-319-31993-3_2

in MSS B and K (usually due to misreadings or misunderstandings by the copyists, or additional material from one or more commentaries). My occasional use of the commentaries usually confirmed my readings. An early commentary by Yūsuf ibn Mubārak al-Alānī (ca. 735/1334) [Istanbul, Topkapı Sarayı Müzesi, Ahmet III MS 3308] had the original values for the climes, while ʿAbd al-Wājid (d. 838/1435) [Istanbul, Süleymaniye Library, Laleli MS 2127] clearly struggled (as I did) with the range of numbers and gave both the original and the post-Ṭūsī parameters for the climes. Other commentators gave the post-Ṭūsī values.

II.1b Establishing the Figures

The figures in the manuscripts displayed different degrees of meticulousness; but generally speaking, MS L had the best diagrams. There was also a range of labeling the figures: some copyists being quite detailed, others sparse. Specific figures are occasionally missing; these are noted in the Figure Apparatus. My procedure was to follow the basic structure of the figures (which was usually similar in all manuscripts) and then use the text and context to decide on which labels to include. In a number of cases, I checked commentaries to confirm or clarify, but never used them to supplement or modify my five core manuscripts. Variants to my edited figures are noted in the Figure Apparatus, which follows the edited text.

II.1c Variants and Orthography

Since I used only five manuscripts to establish the text, I describe each along with its incipit, preface, and colophon in § II.2, and I note all variants in the text apparatus with the exception of minor orthographic differences. As noted below in § II.3: Explanation of Signs and Conventions, I have generally modernized the orthography for writing *hamza*s, numbers, and numerals; divergences are not noted except where there could be alternative readings (such as between *thulth* and *thalāth*). When giving variants, I have written these as they occur in the text, providing or leaving out the dots, vowels, and *hamza*s as given.

II.1d Parameters

Four out of five of my main manuscripts used the alphanumeric system for numbering parameters. The exception was MS B; here parameters were often omitted altogether, but it is noteworthy that when included in the text, the copyist wrote them in unit fractional form, an indication of a late Ottoman style (Berlin, Staatsbibliothek, MS or. oct. 1511, p. 33):

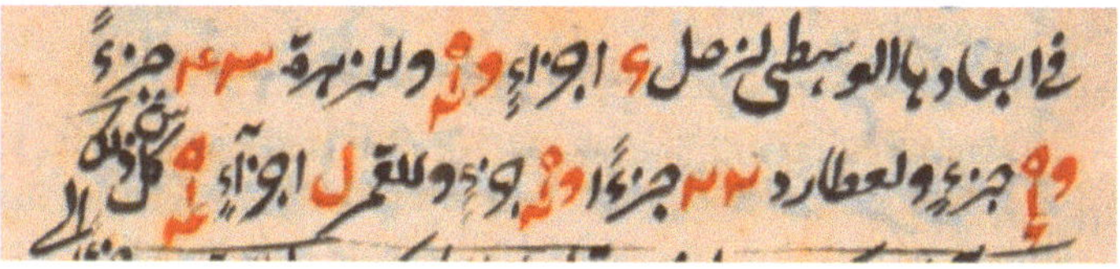

Since the alphanumeric system lends itself to ambiguity, and inattentive copyists could often introduce mistakes—for example by omitting a dot which would lead one to read a ج (3) as a ح (8) or by forgetting to add a stroke to ك (20) causing it to be read as a ل (30)—I relied on the context to confirm a value, either in the main text or as a variant. In general, values given by Ptolemy and Battānī allowed me to control the text. When this was not possible, or when further confirmation was needed, commentaries in which the parameters were written out in words proved valuable; however, cautious judgment had to be applied in recognition that parameters were often "updated" by commentators (for example, by changing Jaghmīnī's Ptolemaic ones to those found in Naṣīr al-Dīn al-Ṭūsī's *Tadhkira*). Alānī's commentary (one of the older ones) alone seemed to contain non-contaminated values, so it was particularly valuable for establishing/confirming some of the parameters. A significant example of this occurs in fixing the date that Jaghmīnī gives for the position of the planetary apogees; misreading a single letter ش (300) instead of ث (500) can make a 200-year difference, but fortunately both context and Alānī's commentary provide us with the correct Alexandrian date of 1517, which also gives added confirmation of Jaghmīnī's dates (see I.5[26], and *I.2.3b: Dating the* Mulakhkhaṣ).

§ II.2 Description of the Manuscripts

The following list contains the five principal manuscripts that have been used to establish the edition.

<table>
<tr><th></th><th>Sigla</th><th>Description of Manuscript</th></tr>
<tr><td rowspan="3">1.</td><td rowspan="3">ب
[=B]</td><td>Berlin, Staatsbibliothek, MS or. oct. 1511, pp. 6–64. The codex contains several treatises, with a total of 667 pages, all in the same hand. On p. 667, a date of 1275/1858 is given. A more expansive colophon is on p. 623, where we learn that the copyist is a certain ʿAbd al-Karīm Bulghārī (عبد الكريم بلغاري) who finished copying that particular work on Wednesday, 24 Jumādā I 1275 H [28-29 Dec. 1858 CE] in Tashkent (al-Tāshkand) in the Kallah Khānah quarter.
MS B is contaminated with commentary comments; but despite the late date, it includes Jaghmīnī's dedicatory poem. Its use of unit fractions is discussed above.</td></tr>
<tr><td>Incipit and Preface: pp. 6–7

بسم الله الرحمن الرحيم الحمد لله كِفاءِ اِفضالِه والصّلوة على رسوله محمدٍ وآله يقول الشيخُ الامام الهمام الفاضل الكامل المتبحّر شرف الدّين محمود بن محمّد بن عُمر الجَغمنىّ رحمه الله انه نُقل الى اعزّة الاحباب وخالصة الاصحاب انّ مولاى الامامَ الاجلَّ البارعَ المنعم بدْرَ الملّة والدّين عزيز الملوك والسلاطين شفآء الارواح خاتمة الحكما محمّد بن بهْرامَ القلانسىُّ اشار الىَّ انْ اجمع فى علْم الهيئة كتابا نَقرُبُ بن الاحتصار والبيان وبجمعُ انجار اللّفْظ الى بسط المعانى فعددْتُ ذلك من نعمه المتوالية وبادرْتُ الى امثال(!) اشارته العاليةِ وقُلتُ ✽ يا لَها من اشارةٍ صدَرتْ لى رفعت رُتبى واعلَت محلّى صدرتْ لى من الكرم المروحىّ ✽ بدْرِدْن الهُدى الامامُ الأَجلِّ قد رآنى اهلاً لامْرٍ حطيْر ✽ ليس مثلى بمثل ذاك باهل غير آنّى بذلتُ فى ذاك جُهْدِىْ ✽ امثالاً(!) لامْرِهِ اىَّ بذْل قد دعانى لذاك لطفا وبرّا ✽ لا افىقاراً لى بضاعة مثلى فالفت هذا الكتاب فى هيئة العالم تدكرةً منّى بَعْدى لكلّ عالمٍ مُتحَرّيِا فيه التّلحيْص مع البيان وايجازَ الالفاظ الى بسط المعانى على حَسب الامكان وسميته الملخِّصَ فى الهيئة ليكون اِسمُه دالّاً على معناهُ وظاهرهُ مُحبراً عن معنا فحواه وجعلتَه(!) ىشتمل على مقدّمة ومقالتىن</td></tr>
<tr><td>In the name of God, the Beneficent, the Merciful. Praise be to God as much as His bestowal of bounty, and may a benediction be upon His Messenger Muḥammad and his family. The magnanimous, worthy, perfect, erudite Shaykh Imām Sharaf al-Dīn Maḥmūd ibn</td></tr>
</table>

	Sigla	Description of Manuscript
		Muḥammad ibn ʿUmar al-Jaghmīnī, may God have mercy upon him, states: the dearest of friends and the sincerest of companions conveyed to me that my master, the highly esteemed, proficient, refined Imām Badr al-Milla wa-l-Dīn, cherished by kings and sultans, the healer of spirits, the seal of the sages, Muḥammad ibn Bahrām al-Qalānisī, proposed that I compile a book on [the subject of] *ʿilm al-hayʾa,* being both an abridgement and an exposition, and combining a succinctness of words with an elucidation of meanings. I considered this a delightful entrustment and I hastened to comply with his lofty proposal, and I said: | Oh what a proposal came my way; it raised my rank and it advanced my standing. It came to me from the noble one who (?); | the highly esteemed Imām, the full moon [Badr] of the true religion. He considered me worthy for a momentous task; | [but] the likes of me is not worthy of such a thing as that. Nevertheless, I expended every effort for that; | complying with his command whatever sacrifice. He called upon me for that in kindness and piety; | not requiring the offerings of such as myself. I composed this book on *hayʾat al-ʿālam* [Configuration of the World] as a memento from me for every scholar after me seeking an epitome on [*hayʾa*] with an exposition, and a succinctness of words with an elucidation of meanings, according to [my] ability. And I have called it "The Epitome on Theoretical Astronomy," so that its name will indicate its connotation and its literal sense will inform its signification; and I arranged it so as to comprise an introduction and two parts.
		Colophon: p. 64 والله اعلم بالصواب واليه المرجع والمآب And God is most knowing of truth, and to Him are the refuge and the final return.
2.	ف [=F]	**Philadelphia, University of Pennsylvania, Rare Book & Manuscript Library, LJS MS 388, ff. 2b–19b**. The codex of 19 folios contains only this one witness. It is written in a *nastaʿlīq* script. Formerly owned by Muḥammad ibn al-Dawla, 1246 [1830-31], it bears Qajar seal imprints on ff. 1a and 19b. It was sold by Sam Fogg Ltd., cat. 22 (July 2000), no. 60 to Lawrence J. Schoenberg in 2011. (See Crofton Black, *Transformation of Knowledge: Early Manuscripts from the Collection of Lawrence J. Schoenberg* [London: Paul Holberton, 2006], 55 (LJS 388); and http://dla.library.upenn.edu/dla/medren/record.html?id=MEDREN_5068122).The witness was completed day 2 [i.e., Monday] 29

	Sigla	Description of Manuscript
		Rabīʿ I 786 H [probably, Sunday-Monday, 22-23 May 1384 CE]. MS F and MS S are closely aligned. The folios in MS F (2–19), however, are bound in the wrong order; the correct order should be 2–7, 10, 13, 11, 12, 8, 9, 14–19. In addition, a folio is missing between 17b and 18a, and this includes Figure 8 along with text. I have indicated the beginning and the end of the missing passage (contained in II.3[1–4]) in both the Arabic edition and English translation between two asterisks [*]…[*].
		Incipit and Preface: f. 2b بسم اللـه الرحمن الرحم وبه نسـعں الحمد للـه كِفاء افضاله والصلوة على ىىه محمد وآله ~~محمد~~ قال الامام البارع الاجل العلامه استاد الورى شرف الافاصل عدىم الاماثل ملك الفضلا حاتم الحكمآء محمود ں محمد ں عمر ~~الفقتهى~~ (؟) الجعمينى الخوارزمى رحمه له إىه نقل الىّ اعزّه الاحباب وحالصه الاصحاب اں مولاى الامام الاجل البارع المنعم بدر المله والدں فخر الاسلام والمسلمين عرىر الملوك والسلاطين سفآء الارواح خاىمه الحكمآء محمد بں بهرام القلانسىّ رحمه اللـه اشار الىّ ان احمع فى علم الهيئة كىاىا يُقرِنُ ىں الاختصار والبيان ويحمع احار اللفط الى بسط المعانى فعَدَدتُ ذلك من نعمة المتوالية وبادَرتُ الى امتثال اشارته العاليه وقلت يا لها من اشارة صدرت لى ❁ رفعَتْ رتبتى واعلَتْ محلّ ❁ صدرت لى من الكرىم المرجّى ❁ بدردِىن الهدى الامام الاجلّ ❁ قد رانى اهلا لامرٍ خطير ❁ ليس مثلى لمثل ذاك باهل ❁ عىر انّى بذلت فى ذاك حهدى ❁ امتىالا لامره اىّ بذل ❁ قد دَعانى لذاك لطفا وَبرّا ❁ لا افىقارا الى بضاعة مثلى ❁ والّفتُ هذا الكتابَ على حسب الامكان قاصدا للتلخيص فيه ىں («مع» في الهامش مع رمز «ح») البيان وسميته الملخّص فى الهيئة ليكون اسمه مُخبراً عن معناه وطاهره دالا على فحواه وجعلته مشىملا على مقدمة ومقالتىں
		In the name of God, the Beneficent, the Merciful, and from Whom we seek assistance. Praise be to God as much as His bestowal of bounty, and may a benediction be upon His Prophet Muḥammad

	Sigla	Description of Manuscript
		and his family. The proficient, highly esteemed, most learned Imām, teacher of mankind, most noble of the worthies, he without peer, king of the eminent ones, seal of the sages, Maḥmūd ibn Muḥammad ibn ʿUmar (al-Faqīhī: [may have been crossed out]) al-Jaghmīnī al-Khwārizmī, mercy upon him, has said: the dearest of friends and the sincerest of companions conveyed to me that my master, the highly esteemed, proficient, refined Imām Badr al-Milla wa-l-Dīn, the pride of Islam and Muslims, cherished by kings and sultans, the healer of spirits, the seal of the sages, Muḥammad ibn Bahrām al-Qalānisī, may God have mercy upon him, proposed that I compile a book on [the subject of] *ʿilm al-hayʾa*, being both an abridgement and an exposition, and combining a succinctness of words with an elucidation of meanings. I considered this a delightful entrustment and I hastened to comply with his lofty proposal, and I said: Oh what a proposal came my way; | it raised my rank and it advanced my standing. | It came to me from the noble one who inspires hope; | the highly esteemed Imām, the full moon [Badr] of the true religion. | He considered me worthy for a momentous task; | [but] the likes of me is not worthy of such a thing as that. | Nevertheless, I expended every effort for that; | complying with his command whatever sacrifice. | He called upon me for that in kindness and piety; | not requiring the offerings of such as myself. | I composed this book according to [my] ability, aiming for an epitome on [*hayʾa*] that is also an exposition. And I have called it "The Epitome on Theoretical Astronomy," so that its name will inform its connotation and its literal sense will indicate its signification; and I arranged it so as to comprise an introduction and two parts.
		Colophon: f. 19b والله الموفق والمستعان وعليه الاعتماد والتكلان اتفق الفراغ من كتابه يوم ٢ ٢٩ شهر المبارك ربيع الاول سنه ٧٨٦ هجريه And God is the One who bestows success and from Whom one seeks assistance, and in Whom is the greatest support and trust. The completion of its copying occurred on day 2 [i.e., Monday], the 29th of the blessed month of Rabīʿ I in the year 786 hijra [probably, Sunday-Monday, 22-23 May 1384 CE].

	Sigla	Description of Manuscript
3.	ك **[=K]**	**Cambridge UK, Cambridge University Library, MS Or. 593, ff. 1b–38b** [=Trinity, R. 13.21]; the codex contains 109 folios written in a Persian *naskh* script. According to Edward G. Browne, it is dated 764 [1362–63] and the codex was bought from Élias Géjou on October 30, 1905. (See *A Supplementary Hand-List of the Muḥammadan Manuscripts, in the Libraries of the University and Colleges of Cambridge* [Cambridge, 1922], 205). E. H. Palmer gives the date incorrectly as 1582-83 (*A Descriptive Catalogue of the Arabic, Persian and Turkish Manuscripts in the Library of Trinity College, Cambridge* [Cambridge: Deighton Bell and Co., 1870], 50–52.) For an online description, see http://www.fihrist.org.uk/profile/manuscript/abef3293-10e8-4e05-8142-f15e28786ae9. The title page states it was owned by a Muṣṭafā ibn Ḥasan al-Farḍī in the year 1180 [1766-67].
		Incipit and Preface: f. 1b بِسمِ ٱللهِ ٱلرَّحْمَنِ ٱلرَّحِيمْ الحمد لله («كفاء» غير مقروء) افضاله والصلاة علي رسوله محمد وآلهِ قال قال(؟) الشيخ الامام الاجل البارع العلامه استاد الوري شرف الاماثل ملك الفضلا خاتمة الحكماءِ محمود بن محمد بن عمر الجغميني الخوارزمي تغمدهُ الله تعالي برحمته ان اعزة الاحباب وخالصة الاصحاب بدر الملة والدين فخر الاسلام والمسلمين عزيز الملوك والسلاطين راحة الاشياخ وشفا الارواح محمد بن بهرام القلانسِي احمد الله عواقبه اشار ان اجمع في علم الهيية كتابا يقرن بين الاختصار والبيان ويجمع ايجاز اللفظ الي بسط المعاني فعددت ذلك من نعمه المتواليه وبادرت الي امتثال اشارته العاليه والفت هذا الكتاب علي حسب الامكان قاصداً للتلخيص فيه مع البيان وسميتُه الملخص («في» غير مقروء) الهيئة اسمه مخبرا عن معناه وظاهره دالا علي فحواه وجعلته يشتمل علي مقدمة ومقالتين
		In the name of God, the Beneficent, the Merciful. Praise be to God as much as His bestowal of bounty, and may a benediction be upon His Messenger Muḥammad and his family. The highly esteemed, proficient, most learned Shaykh Imām, teacher of mankind, most noble of peers, king of the eminences, seal of the sages, Maḥmūd ibn Muḥammad ibn ʿUmar al-Jaghmīnī al-Khwārizmī, may God

	Sigla	*Description of Manuscript*
		Almighty protect him with His grace, has said: the dearest of friends and the sincerest of companions, Badr al-Milla wa-l-Dīn, the pride of Islam and Muslims, cherished by kings and sultans, comforter of Shaykhs, the healer of spirits, Muḥammad ibn Bahrām al-Qalānisī, may God find his outcomes praiseworthy, proposed that I compile a book on [the subject of] *ʿilm al-hayʾa,* being both an abridgement and an exposition, and combining a succinctness of words with an elucidation of meanings. I considered this a delightful entrustment and I hastened to comply with his lofty proposal. I composed this book according to [my] ability, aiming for an epitome on [*hayʾa*] that is also an exposition. And I called it "The Epitome on Theoretical Astronomy," so that its name will inform its connotation and its literal sense will indicate its signification; and I arranged it so as to comprise an introduction and two parts.
		Colophon: f. 38b والله الموفق للصواب والحمد لله وحده وصلى الله علي سيّدنا محمد واله وصحبه وسلم في تارخ سنه ٧٦٤ احسن الله عاقبتها بمه وكرمه And God is the One who bestows truth, praise be to God alone. May God bless our master Muḥammad and his family and companions and grant them salvation on the date of the year 764 [1362-63 CE], may God make its outcome favorable by His grace and munificence.
4.	ل **[=L]**	**Istanbul, Süleymaniye Library, Laleli MS 2141, ff. 61b–81a**; the codex contains 94 folios. This witness was copied in 644 H [1246-47 CE], making it the oldest extant *Mulakhkhaṣ* to date. The title page and f. 94a both contain an endowment stamp: Sulṭān Salīm Khān [i.e., Selīm III] ibn Sulṭān Muṣṭafā Khān 1217 [1802-3]. (See Günay Kut and Nimet Bayraktar, *Yazma Eserlerde Vakıf Mühürleri Waqif* [Ankara, 1984], 41 [no. 15].)
		Incipit and Preface: f. 61b بسم الله الرَّحمن الرَّحيم الحمدُ للهِ كِفآ اِفضاله والصلوة على نبيه محمد وآله نقول عبد الله الفقر الى رحمته محمودُ ىن محمدِ ىن عمر الجغمينى رحمه الله انى الفتُ هذا الكتابَ فى هيئه العالم تدكرةً منّى نعدِى لكل عالم مُتحرياً فىه اللخيصَ مع

	Sigla	Description of Manuscript
		البياں وانجازَ الالفاط الى نَسط المعانى على حسب الامكان وسميتُه الملخّص في الهيئه لىكون اسمه دالا على معناه وظاهره مُخْبراً عن فحواهُ وحعلتُه يَشتمل على مقدمة ومقالتين
		In the name of God, the Beneficent, the Merciful. Praise be to God as much as His bestowal of bounty, and may a benediction be upon His Prophet Muḥammad and his family. The servant of God (ʿAbd Allāh) in need of His compassion, Maḥmūd ibn Muḥammad ibn ʿUmar al-Jaghmīnī, may God have mercy upon him, states: I composed this book on *hayʾat al-ʿālam* [Configuration of the World] as a memento from me for every scholar after me seeking an epitome on [*hayʾa*] with an exposition, and a succinctness of words with an elucidation of meanings, according to [my] ability. And I have called it "The Epitome on Theoretical Astronomy," so that its name will indicate its connotation and its literal sense will inform its signification; and I arranged it so as to comprise an introduction and two parts.
		Colophon: f. 81a وبالله الىوفىق تم الكتاب تم الكتاب فى شهور خمد («٦٤٤» في الهامش) هحريه With God is success. The book was completed, the book was completed [*sic*] in the months of 644 hijra [1246-47 CE].
5.	س [=S]	**Paris, Bibliothèque nationale de France, MS arabe 2330, ff. 48b–82b**; the codex contains a total of 116 folios. Written in a *naskh* script, the codex contains at least 14 witnesses. (See Baron William MacGuckin de Slane, *Catalogue des manuscrits arabes / par le baron de Slane* [Paris: Imprimerie nationale, 1883–95], 408–9.) This witness was completed the night of Friday, 19 Dhū al-qaʿda 787 [Thursday evening/Friday morning, 21-22 Dec. 1385 CE].
		Incipit and Preface: ff. 48b–49a بسم الله الرحمن الرحيم رب الهم بالصواب الحمدُ لله كفأ افضاله والصلوةُ على نبيّه محمدٍ وعلى اله قال الامامُ البارعُ الاجلُ العلّامةُ استادُ الوري شرفُ الافاضِل عدمُ الاماثِل مَلكُ الفضلأ خاتمُ الحكماءِ محمودُ بن محمد بن عمرَ الفقيهِى الجَغْمينِىّ

	Sigla	*Description of Manuscript*
		الخوارِزمِیّ رحمه الله اِنه نَقَلَ الیّ اعزّةُ الاحباب وخالصةُ الاصحاب آن مولانا الامامُ الاجلُ البارِعُ المنعمُ بَدْرُ الملّةِ والدّین فخرُ الاسلام والمسلمین عزیز الملوكِ والسلاطین شفاءُ الارواح خاتمةُ الحکماء محمدُ بن بَهرام القلانِسیّ رحمه الله آشارَ الیّ ان اجمعَ فی علم الهیئَةِ کتابا یقرن بین الاختصار والبیان وتَجمع ایجاز اللفظ الی بسط المعان فعددتُ ذلك من نِعَمِه المتوالیه وبادرت الی امتثال اشارتِه العالیةِ ۞ وقلتُ شعرا ۞ یا لَها من اِشارةٍ صَدرت لي ۞ رفعَتْ رُتبتی واَعلَتْ مَحلّ صدَرَتْ لی من الکریم المرَجّی ۞ بدرِدین الهدی الامامُ الاجلّ قد رآنی آهلا لامرٍ خطیر («عطم» تحت السطر) ۞ لیس مثلی لمثل ذاك باَهْلِ غیر آنی بذلتُ فی ذاك جهدی ۞ امتثالاً لامرِه ايَّ بَذْلِ قد دعانی لذاك لُطْفاً وبِرّاً ۞ لا افتقاراً الی بضاعة مِثلی وَالَّفْتُ هذا الکتابَ علی حَسب الاِمکان قاصداً للتلخیص فیه مع البیان وسمیتُه الملخصَ فی الهیئةِ لیکون اسمه مخبرا («دالا» تحت السطر مع رمز «ح») عن معناه وظاهرُه دالّا علی فحواه وجعلتُه تشتمل علی مقدمة ومقالتین
		In the name of God, the Beneficent, the Merciful, Lord may You inspire Truth. Praise be to God as much as His bestowal of bounty, and may a benediction be upon His Prophet Muḥammad and his family. The proficient, highly esteemed, most learned Imām, teacher of mankind, most noble of the worthies, he without peer, king of the eminent ones, seal of the sages, Maḥmūd ibn Muḥammad ibn ʿUmar al-Faqīhī al-Jaghmīnī al-Khwārizmī, may God have mercy upon him, has said: the dearest of friends and the sincerest of companions conveyed to me that our master, the highly esteemed, proficient, refined Imām Badr al-Milla wa-l-Dīn, the pride of Islam and Muslims, cherished by kings and sultans, the healer of spirits, the seal of the sages, Muḥammad ibn Bahrām al-Qalānisī, may God have mercy upon him, proposed that I compile a book on [the subject of] *ʿilm al-hayʾa,* being both an abridgement and an exposition, and combining a succinctness of words with an elucidation of meanings. I considered this a delightful entrustment

	Sigla	*Description of Manuscript*
		and I hastened to comply with his lofty proposal, | and I said in verse |: Oh what a proposal came my way; | it raised my rank and it advanced my standing. It came to me from the noble one who inspires hope; | the highly esteemed Imām, the full moon [Badr] of the true religion. He considered me worthy for a momentous task; | [but] the likes of me is not worthy of such a thing as that. Nevertheless, I expended every effort for that; | complying with his command whatever sacrifice. He called upon me for that in kindness and piety; | not requiring the offerings of such as myself. I composed this book according to [my] ability, aiming for an epitome on [*hay'a*] that is also an exposition. And I have called it "The Epitome on Theoretical Astronomy," so that its name will inform its connotation, and its literal sense will indicate its signification; and I arranged it so as to comprise an introduction and two parts.
		Colophon: f. 82b والله الموفق والمستعان وعليه التكلان اتفق الفراغ عن كتابه لىله الجمعه التاسع عسر من سهر دى القعده من سىه سبع وثمانىن وسبعمائه الحمد لله وحده وصلى الله على سىدنا محمد واله وسلم And God is the One who bestows success and from Whom one seeks assistance, and in Whom is the greatest trust. The completion of its copying occurred during the night of Friday, the nineteenth of the month of Dhū al-qaʿda of the year 787 [Thursday evening-Friday morning, 21-22 December 1385 CE]. Praise be to God alone, and may God bless our master Muḥammad and his family and grant them salvation.
		[ملاحظة بيد آخر] وفرغ عن قرائه فى اول سهر ربيع الاخر من سنه ثمان وثمانين وسبعما<به> على الشيخ علاى الدين الموقت بحلب فسح الله في مدته **[A note in another hand]:** The reading of this under the Shaykh ʿAlāʾ al-Dīn the Timekeeper was completed at the beginning of the month of Rabīʿ II of the year 788 [early May 1386 CE] in Aleppo, may God extend its duration.

§ II.3 Explanation of Signs and Conventions Used in the Arabic Critical Edition and Apparatus

For the Arabic edition, the following conventions have been used:

1. The orthography and rules for *hamza* follow modern conventions.
2. The dotting of ي follows the rules used by printers in Syria and Lebanon.
3. *Tanwīn* is generally added (but not on feminine *tāʾ* endings).
4. *Shaddas* have been supplied (except for sun letters and *nisbas*).
5. Short vowels have been provided sparingly as aids to the reader and/or to avoid ambiguity.

Critical Apparatus

[	Separates reading in edition from any variant
:	Separates variant and manuscript *sigla*
+	Added in
–	Missing from
=	Indicates another variant
(…)	Editor's comments
ب (B)	Berlin, Staatsbibliothek, MS or. oct. 1511, pp. 6–64
س (S)	Paris, Bibliothèque nationale de France, MS arabe 2330, ff. 48b–82b
ف (F)	Philadelphia, University of Pennsylvania, LJS MS 388, ff. 2b–19b
ك (K)	Cambridge UK, Cambridge University Library, MS Or. 593, ff. 1b–38b
ل (L)	Istanbul, Süleymaniye Library, Laleli MS 2141, ff. 61b–81a
با	بياض (blank)
تا	تحت السطر في (under the line in)
شا	مشطوب في (crossed out in)
طا	مطموس، غير مقروء، إلخ (smudged, unreadable, etc.)
فا	فوق السطر في (above the line in)
ها	في الهامش في (in the margin in)

Edition, Translation, and Apparatuses

S.P. Ragep, *Jaghmīnī's Mulakhkhaṣ*, Sources and Studies in the History of Mathematics and Physical Sciences, DOI 10.1007/978-3-319-31993-3_3

[Preface]

In the Name of God, the Beneficent, the Merciful

[1] Praise be to God as much as His bestowal of bounty, and may a benediction be upon His Prophet Muḥammad and his family. The proficient, highly esteemed, most learned Imām, teacher of mankind, most noble of the worthies, he without peer, king of the eminent ones, seal of the sages, Maḥmūd ibn Muḥammad ibn ʿUmar al-Faqīhī al-Jaghmīnī al-Khwārizmī, may God have mercy upon him, has said: the dearest of friends and the sincerest of companions conveyed to me that our master, the highly esteemed, proficient, refined Imām Badr al-Milla wa-l-Dīn, the pride of Islam and Muslims, cherished by kings and sultans, the healer of spirits, the seal of the sages, Muḥammad ibn Bahrām al-Qalānisī, may God have mercy upon him, proposed that I compile a book on [the subject of] *ʿilm al-hayʾa,* being both an abridgement and an exposition, and combining a succinctness of words with an elucidation of meanings. I considered this a delightful entrustment and I hastened to comply with his lofty proposal, and I said in verse:

Oh what a proposal came my way;	it raised my rank and it advanced my standing.
It came to me from the noble one who inspires hope;	the highly esteemed Imām, the full moon [Badr] of the true religion.
He considered me worthy for a momentous task;	[but] the likes of me is not worthy of such a thing as that.
Nevertheless, I expended every effort for that;	complying with his command whatever sacrifice.
He called upon me for that in kindness and piety;	not requiring the offerings of such as myself.

[2] I composed this book according to [my] ability, aiming for an epitome on [*hayʾa*] that is also an exposition. And I have called it "The Epitome on Theoretical Astronomy," so that its name will inform its connotation, and its literal sense will indicate its signification; and I arranged it so as to comprise an introduction and two parts.[1] The Introduction is about an explanation of the divisions of the bodies in general terms. The First Part concerns an explanation of the orbs and what pertains to them, and there are five chapters: (1) On the configurations of the orbs; (2) On an explanation of the motions of the orbs; (3) On an explanation of the circles; (4) On an explanation of the arcs; (5) On what occurs to the planets in their motions and what is connected with this. The Second Part concerns an explanation of the configurations of the Earth and what pertains to it, and there are three chapters: (1) On the inhabited part of the Earth and its latitude, its longitude,

[1] Up to this point, there is considerable variation among the manuscripts. For the different versions, see *§II.2: Description of the Manuscripts*. For the edition and translation, I have mainly followed MSS F and S.

[١]بسم الله الرحمٰن الرحيم[٢]

[١] الحمد لله كِفاء[٣] إفضاله والصلوة على نبيّه[٤] محمّد وآله[٥]. قال الإمام البارع الأجلّ العلّامة أستاذ الورى شرف الأفاضل عديم الأماثل ملك الفضلاء خاتم الحكماء محمود بن محمّد بن عمر الفقيهي الجغميني الخوارزمي رحمه الله[٦]: إنّه نَقَلَ إليّ[٧] أعزّة الأحباب وخالصة الأصحاب أنّ مولانا[٨] الإمام الأجلّ البارع المنعم بدر الملّة والدين فخر الإسلام والمسلمين[٩] عزيز الملوك والسلاطين شفاء الأرواح خاتمة الحكماء محمّد بن بهرام القلانسي رحمه الله[١٠] أشار إليّ أن أجمع في علم الهيئة كتاباً يقرن[١١] بين الاختصار والبيان ويجمع إيجاز اللفظ إلى بسط المعاني[١٢] فعَدَدتُ ذلك من نعمة المتوالية وبادرت إلى امتثال[١٣] إشارته العالية وقلت شعراً[١٤]:

يا لها من إشارة صدرت لي / رفعتْ رتبتي وأعلتْ محلّ[١٥]
[١٦]صدرت لي[١٧] من الكريم المرجّي[١٨] / بدرِ دين الهدى الإمام الأجلّ
قد رآني أهلاً لأمر خطير[١٩] / ليس مثلي لمثل[٢٠] ذاك بأهل
غير أنّي بذلت في ذاك جهدي / امتثالاً[٢١] لأمره أيّ بذل
قد دعاني لذاك لطفاً وبرّاً / لا افتقاراً إلى[٢٢] بضاعة مثلي

[٢] وألّفتُ هذا الكتاب على حسب الإمكان قاصداً للتلخيص فيه مع[٢٣] البيان وسمّيته الملخّص في الهيئة ليكون اسمه مخبراً[٢٤] عن معناه وظاهره دالّاً على فحواه وجعلته يشتمل[٢٥] على مقدّمة ومقالتيْن[٢٦]. المقدّمة [٢٧]في[٢٨]بيان أقسام الأجسام على الإجمال. المقالة الأولى[٢٩] في بيان الأفلاك وما يتعلّق بها وهي خمسة أبواب: آ [٣٠] في هيئات الأفلاك؛ ب[٣١] في بيان[٣٢] حركات الأفلاك؛ حـ[٣٣] في بيان الدوائر؛ د[٣٤] في بيان القسي؛ ﻫ[٣٥] فيما يعرض للكواكب في حركاتها وما يتّصل بذلك. [٣٦]المقالة الثانية في بيان هيئات[٣٧] الأرض وما يتعلّق بها وهي ثلاثة أبواب: [٣٨]آ[٣٩] في المعمور من الأرض[٤٠] وعرضه وطوله

and its division into the climes; (2) On the characteristics of the equator and locations having latitude; (3) On miscellaneous items.

Introduction

On an Explanation of the Divisions of the Bodies in General Terms

[1] The bodies are two kinds: simple, which are those that cannot be [further] broken down into bodies of different natures; and composite, which are those that can be [further] broken down into bodies of different natures, such as minerals, plants, and animals. There are two kinds of simple bodies: elements, namely, earth, water, air, fire, and the aethereal bodies, which are the orbs with what is in them. Every simple body, when left unimpeded in its natural state, is—as has been shown in another science—spherical in form. Hence, the elements, in their totality, and the aethereal bodies have spherical shapes. However, on the Earth, inasmuch as it admits of [geological] formations, there are undulations that occur on its surface due to reasons external to it, such as we observe by way of valleys, hills, and so forth. But these undulations do not detract from its being spherical in shape as a whole, like with an egg: if kernels of barleycorn were stuck on it, this would not detract from its overall shape. Similarly, the water is spherical, despite the fact that it is not completely round, since emerging from its surface are elevations from the Earth. Likewise, the air is spherical yet its concave surface is irregular as well due to undulations in it from the water and the Earth. The fire is a spherical shape that is truly round [both] convexly and concavely according to the most correct opinion.

[2] All the orbs are spherical in shape and these spheres enclose one another. The Earth is in the middle, then the water that encloses it, then the air, then the fire, then the orb of the Moon, then the orb of Mercury, then the orb of Venus, then the orb of the Sun, then the orb of Mars, then the orb of Jupiter, then the orb of Saturn, then the orb of the Fixed Stars, and then the Orb of Orbs, which is called the Greatest Orb; it is the orb that encloses all the bodies, nothing being beyond it, neither vacuum nor plenum. Every enclosing [orb] is contiguous with that enclosed by it, which is adjacent to it according to the aforementioned arrangement. To the totality of these bodies—the elements, the orbs, and what is within them—is extended the name "The World." This is its illustration:

وقسمته إلى الأقاليم؛ ب[٤١] في خواصّ خطّ الاستواء والمواضع التي لها عرض؛ حـ[٤٢] في أشياء منفردة[٤٣].

المقدّمة

في بيان أقسام الأجسام على الإجمال

[١] الأجسام قسمان: بسائط وهي التي لا تنقسم إلى أجسام مختلفة الطبائع؛ ومركّبات وهي التي تنقسم إلى أجسام مختلفة الطبائع[٤٤] [٤٥]كالمعدنيات والنبات والحيوان. فالبسائط قسمان: عناصر وهي الأرض [٤٦]والماء والهواء والنار وأجرام أثيرية وهي الأفلاك بما فيها. وكلّ جسم بسيط[٤٧] إذا خُلِّي وطبعه[٤٨] فهو على ما بيّن في غير هذا العلم كريّ الشكل. فالعناصر بجملتها والأجرام الأثيرية كريّة الأشكال. إلّا أنّ الأرض لقبولها التشكيلات[٤٩] وقعت في سطحها تضاريس لأسباب[٥٠] خارجة عنها كما نشاهد[٥١] من الوهاد والتلال[٥٢] [٥٣]ونحوهما. لكنّ هذه[٥٤] التضاريس لا تقدح[٥٥] في كونها كريّة الشكل[٥٦] بجملتها كالبيضة لو[٥٧] أُلزِقت بها حبّاتُ[٥٨] [٥٩]شعير[٦٠] لم يقدح[٦١] ذلك في شكل[٦٢] جملتها. وكذا الماء كريّ[٦٣] إلّا أنّه ليس بتامّ الاستدارة لأنّه خرج عن سطحه[٦٤] ما ارتفع من الأرض. وكذا الهواء كريّ إلّا أنّ سطحه المقعّر[٦٥] مضرّس أيضاً[٦٦] بحسب تضاريس ما فيه من الماء والأرض. والنار كريّة الشكل صحيحة الاستدارة تحديباً وتقعيراً بالرأي الأصحّ[٦٧].

[٢] والأفلاك كلّها كريّة الأشكال وهذه الكرات يحيط[٦٨] بعضها ببعض. والأرض[٦٩] في الوسط ثمّ الماء فهو[٧٠] محيط [٧١]بها[٧٢] ثمّ الهواء ثمّ النار ثمّ فلك القمر ثمّ فلك عطارد[٧٣] ثمّ فلك الزهرة ثمّ فلك الشمس ثمّ فلك المرّيخ ثمّ فلك المشتري ثمّ فلك زحل ثمّ فلك الثوابت ثمّ فلك الأفلاك ويسمّى [٧٤]الفلك الأعظم[٧٥] وهو الفلك المحيط بجميع الأجسام ليس ورائه شيء لا خلاء ولا[٧٦] ملاء[٧٧]. وكلّ محيط يماسّ المحاط به الذي يليه في الترتيب المذكور[٧٨]. وبجملة[٧٩] هذه الأجرام[٨٠] من العناصر والأفلاك وما فيها يُطْلق عليها[٨١] اسم العالم. وصورتها هذه[٨٢]:

Illustration of the Orbs

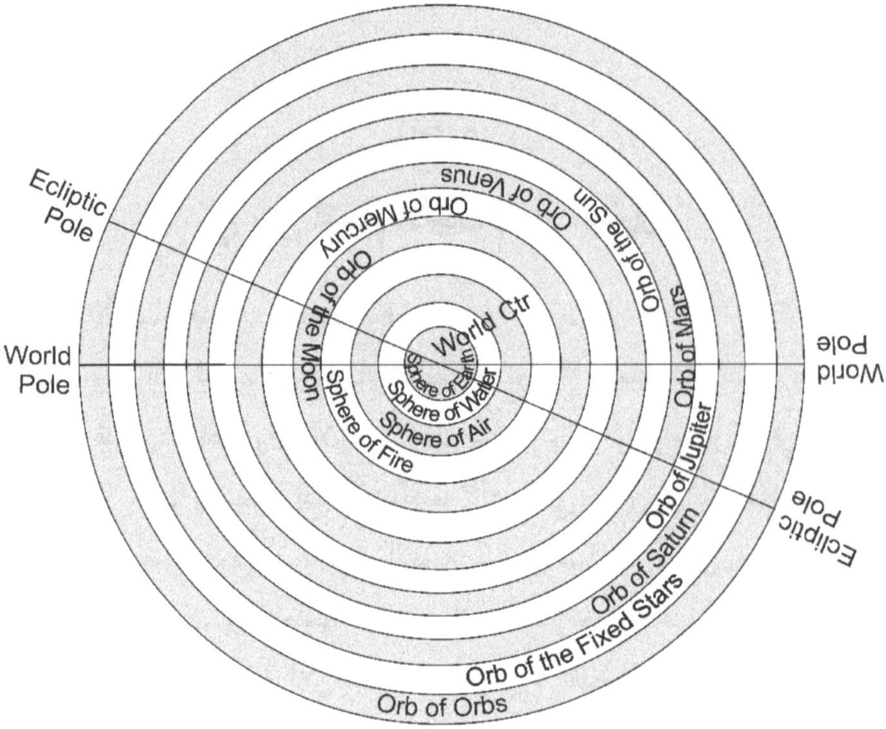

[Figure 1]

صورة الأفلاك

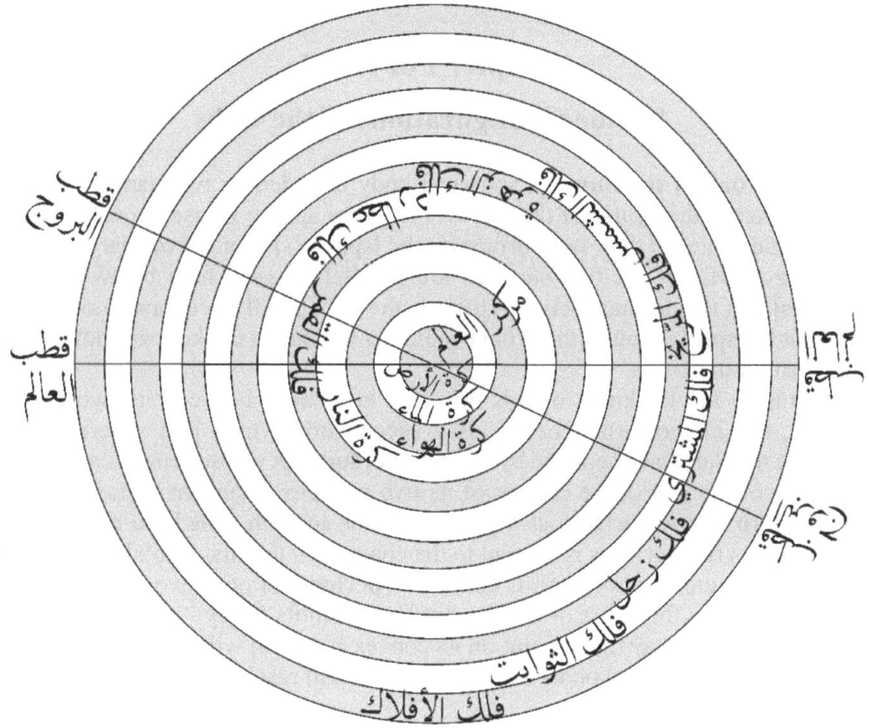

[شكل ١]

The First Part

On an Explanation of the Orbs and What Pertains to Them

Chapter 1 of Part I

On the Configurations of the Orbs

[1] **The orb of the Sun** is a spherical body bounded by two parallel surfaces whose center is the center of the world. For every sphere whose two surfaces are parallel, the center of their two surfaces is the [sphere's] center. For every solid orb enclosing the Earth, its two surfaces are parallel. I mean here by two parallel [surfaces] that the distance between them is the same in all directions—not varying such that the sphere would [then] have a thinner part and a thicker part, but rather it is uniform in thickness.

[2] Inside the thickness of this orb, i.e., in what is between its two parallel surfaces, not in the [orb's] cavity, is a second orb, which is a spherical body enclosing the Earth and bounded by two parallel surfaces whose center is away from the center of the world; the convex of its two surfaces is tangent to the convex of the first [orb's] two surfaces at a point common to both called the **apogee**. The concave of its two surfaces is tangent to the concave of the first [orb's] two surfaces at a point common to both and is called the **perigee**. In other words, this second [orb] is inside the thickness of the first [orb]—not in its cavity—and shifted to one side of it in such a way that a point on its convex [surface] will reach the convex of the first [orb], and a point on its concave [surface will reach] the concave of the first [orb].

[3] Thus necessarily the first [orb] becomes by [the second orb] two spheres [i.e., two spherical bodies] whose surfaces are not parallel but rather of variable thickness, one of the two encloses [the second orb], and the other is enclosed in it. The thinner part of the enclosing [spherical body] is that which is adjacent to the apogee; and the thicker part is that which is adjacent to the perigee. And the thinner part of the enclosed [spherical body] and its thicker part are in reverse. Each one of them is a **complementary [body]**. This second orb is called the **eccentric**, and the first is called the **parecliptic orb** because on its circumference is the circle that is also called the parecliptic orb, which you will learn about in the chapter on circles.

[٨٣] المقالة الأولى
في بيان الأفلاك وما يتعلّق بها[٨٤]

الباب الأوّل من المقالة[٨٥] الأولى
في هيئات[٨٦] الأفلاك

[١] **فلك الشمس** جرم كريّ [٨٧] يحيط به سطحان متوازيان مركزهما مركز العالم. وكلّ كرة متوازية السطحيْن فمركز سطحيْها[٨٨] هو مركزها[٨٩]. وكلّ فلك مجسّم [٩٠] شامل للأرض فهو متوازي السطحيْن. وأعني بالمتوازييْن هنا[٩١] أنّ البعد بينهما واحد من جميع الجهات[٩٢] لا يختلف حتّى[٩٣] يكون للكرة جزء[٩٤] أرقّ وجزء أغلظ بل هي متشابهة الثخن.

[٢] وفي داخل ثخن هذا الفلك، [٩٥] أيْ فيما بين سطحيْه المتوازييْن لا في جوفه، فلك ثانٍ هو جرم[٩٦] كريّ شامل للأرض يحيط به سطحان متوازيان مركزهما[٩٧] خارج عن مركز العالم، محدّب سطحيْه مماسّ لمحدّب سطحيْ الأوّل على نقطة مشتركة بينهما تسمّى[٩٨] **الأوج** ومقعّر سطحيْه مماسّ لمقعّر سطحيْ الأوّل على نقطة مشتركة بينهما وتسمّى[٩٩] **الحضيض**[١٠٠] [١٠١] أيْ يكون هذا الثاني في داخل ثخن الأوّل لا في جوفه مائلاً إلى جانب منه بحيث[١٠٢] تصل[١٠٣] نقطة من[١٠٤] محدّبه إلى محدّب الأوّل ونقطة من مقعّره[١٠٥] [١٠٦] إلى مقعّر الأوّل.

[٣] فبالضرورة[١٠٧] يصير[١٠٨] به الأوّل كرتيْن غير متوازيتيْ[١٠٩] السطوح بل[١١٠] مختلفتيْ الثخن، إحداهما[١١١] حاوية له، والأخرى محوية[١١٢] فيه[١١٣]. ورقّة الحاوية[١١٤] ممّا يلي الأوج؛ وغلظها ممّا يلي الحضيض. ورقّة المحوية وغلظها بالخلاف[١١٥]. وتسمّى[١١٦] كلّ واحدة[١١٧] منهما **متمّماً**. وهذا الفلك الثاني يسمّى[١١٨] **الخارج المركز** والأول يسمّى[١١٩] [١٢٠] **الفلك الممثّل** لأنّ على محيطه[١٢١] الدائرة المسمّاة أيضاً بالفلك الممثّل [١٢٢] وستعرفها في باب الدوائر.

[4] **The Sun** is a solid, spherical body fixed in the body of the eccentric orb, embedded in it in such a way that [the Sun's] diameter is equal to the thickness of the orb and its surface is tangent to [the orb's] two surfaces.

[5] **As for the orbs of the upper planets and Venus,** they are exactly the same as the orb of the Sun, there being no difference at all between them and it except that they have small orbs that do not enclose the Earth. Rather, they are fixed [and] embedded in the bodies of their eccentric orbs in such a way that the surface of each one of them is contiguous with the two surfaces of its deferent, in the manner of the body of the Sun in its eccentric orb. These small orbs are called **epicycle orbs**.

[6] **A planet** in [the epicycle orbs] is a solid, spherical body fixed in the body of the epicycle orb, embedded in it in such a way that its surface is tangent to the surface of the epicycle at a common point between them. The eccentric orbs, with the exception of the Sun, are called **deferents** [sing: *ḥāmil*] on account of their carrying [*ḥaml*] the centers of the epicycles, because they, I mean the centers, are like parts of them [i.e., the deferents].

Illustration of the Sun's Orb

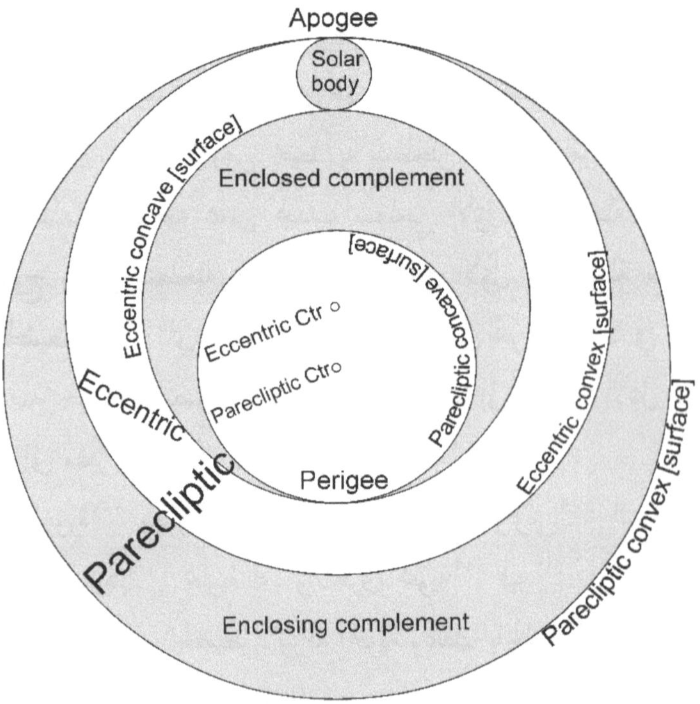

[Figure 2]

[٤] **والشمس** جرم كريّ مُصْمَت مركوز[١٢٣] في جرم الفلك الخارج المركز مغرق فيه بحيث يساوي قطرها ثخن الفلك ويماسّ[١٢٤] سطحها سطحيْه.

[٥] **وأمّا أفلاك الكواكب العلويّة والزهرة**، فهي بِعينها كفلك الشمس، لا فرق بينها[١٢٥] وبينه البتّة إلّا أنّ لها أفلاكاً صغاراً غير شاملة للأرض [١٢٦]بل هي مركوزة مغرقة في أجرام أفلاكها الخارجة[١٢٧] المراكز[١٢٨] بحيث يماسّ[١٢٩] سطح كلّ واحد منها سطحيْ حامله[١٣٠]، بمنزلة جرم الشمس في فلكها الخارج المركز. وتسمّى هذه الأفلاك الصغار[١٣١] **أفلاك التداوير**.

[٦] **والكوكب** فيها[١٣٢] جرم كريّ مصمت مركوز[١٣٣] في [١٣٤]جرم فلك التدوير مغرق فيه بحيث يماسّ سطحه سطح التدوير على نقطة مشتركة بينهما. والأفلاك الخارجة المراكز، لغير الشمس، تسمّى **حوامل** لحملها مراكز التداوير[١٣٥] لأنّها أعني المراكز كأجزاء منها[١٣٦].

صورة فلك الشمس

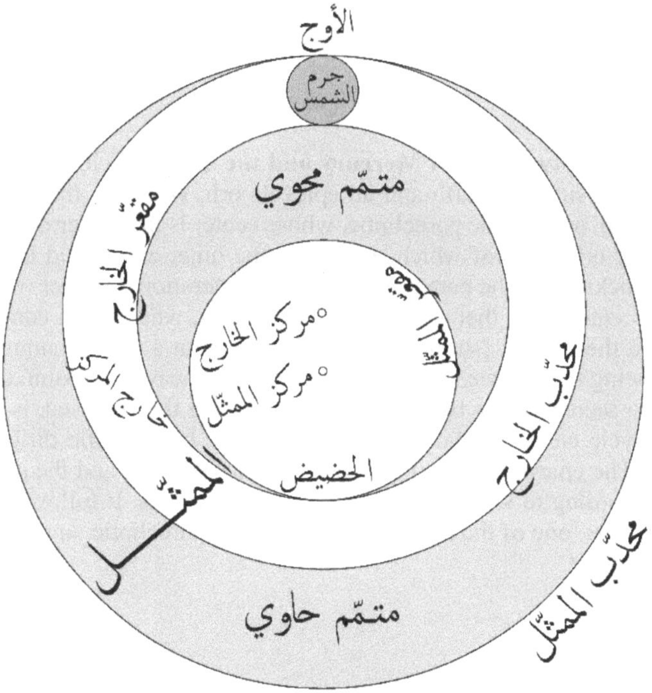

[شكل ٢]

Illustration of the Orbs of the Upper Planets and Venus

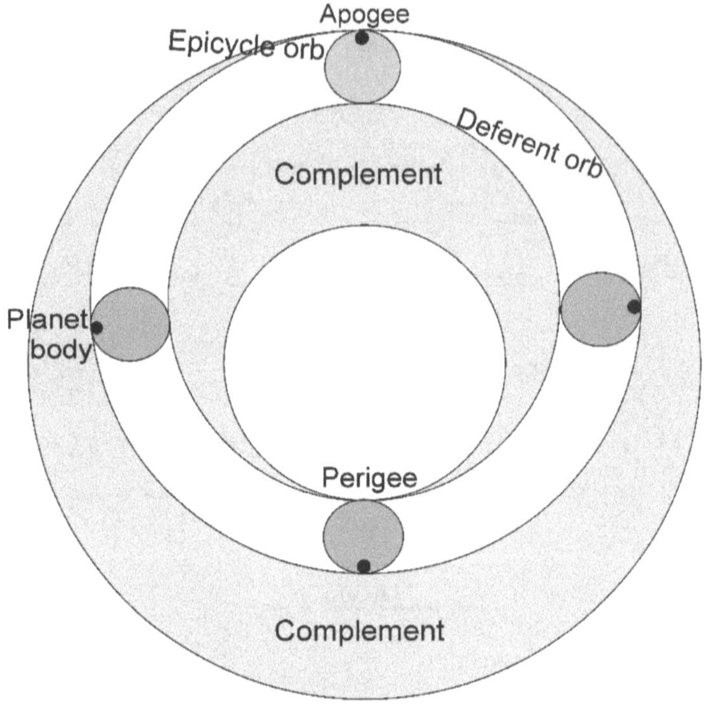

[Figure 3]

[7] **As for the two orbs of Mercury and the Moon**, each of them consists of three orbs enclosing the Earth and an epicycle orb; however, the orb of Mercury includes an orb, namely the parecliptic, whose center is the center of the world, and two eccentric orbs, one of which, enclosing the other and called the **dirigent**, is within the thickness of the parecliptic as in the illustration. In other words, it is like the other eccentric orbs that are in their parecliptics, whereby its convex [surface] is tangent to the convex [surface] of the parecliptic at a point common to both of them, this being the apogee, and its concave to its concave at a point, this being the perigee. The second of the two eccentrics, this being the enclosed, is the **deferent** for the epicycle center within the thickness of the body of the dirigent as in the illustration. The epicycle orb is in the body of the deferent, and the planet is in the epicycle, according to what we stated for other epicycles. It follows that Mercury has two apogees, one of them being as a part of its parecliptic, and the second as a part of its dirigent.

صورة أفلاك الكواكب العلوية والزهرة

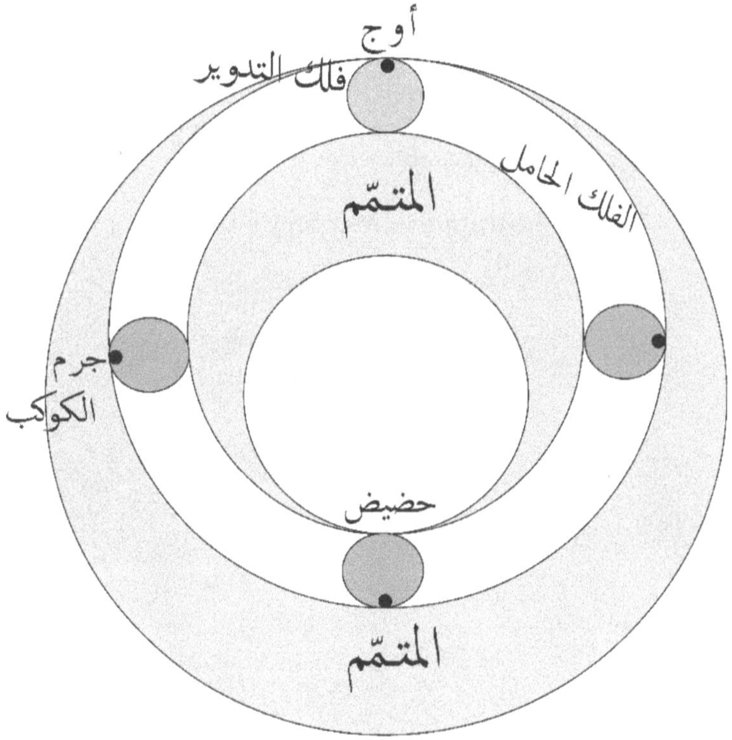

[شكل ٣]

[٧] [١٣٧] **وأمّا فلكا عطارد والقمر** فكلاهما مشتمل[١٣٨] على ثلاثة أفلاك شاملة للأرض وعلى فلك تدوير إلّا أنّ فلك عطارد مشتمل على فلك هو الممثّل مركزه مركز العالم، وعلى فلكيْن خارجيْ المركز[١٣٩]، أحدهما، وهو الحاوي للآخر ويسمّى **المدير**، في داخل ثخن الممثّل، على الرسم، أيْ[١٤٠] كسائر الأفلاك الخارجة المراكز في ممثّلاتها [١٤١]بحيث يماسّ محدّبه[١٤٢] محدّب الممثّل[١٤٣] على نقطة مشتركة بينهما، وهي[١٤٤] الأوج، ومقعّره[١٤٥] مقعّره [١٤٦]على نقطة[١٤٧]، وهي الحضيض. والثاني من الخارجي المركز[١٤٨] وهو المحوي[١٤٩] وهو **الحامل** لمركز التدوير[١٥٠] في داخل ثخن جرم[١٥١] المدير على الرسم[١٥٢]. وفلك التدوير في جرم الحامل والكوكب[١٥٣] في التدوير على ما ذكرنا في سائر التداوير[١٥٤]. ويلزم أنْ يكون لعطارد أوجان، أحدهما كالجزء من ممثّله، والثاني كالجزء من مديره.

[8] **The Moon's orb** includes two orbs, their center being the center of the world, and a deferent orb. One of the first two, which encloses the second, is called the ***jawzaharī*** and the parecliptic. The second, called the **inclined,** is in the cavity of the *jawzaharī*, not in its thickness. The deferent is in the thickness of the inclined as in the illustration. The epicycle is in the deferent, and the Moon is in the epicycle, according to what we have stated.

[9] From these circles, one may conceive the manner of what we have stated regarding the configurations of the orbs.

Illustration of Mercury's Orb

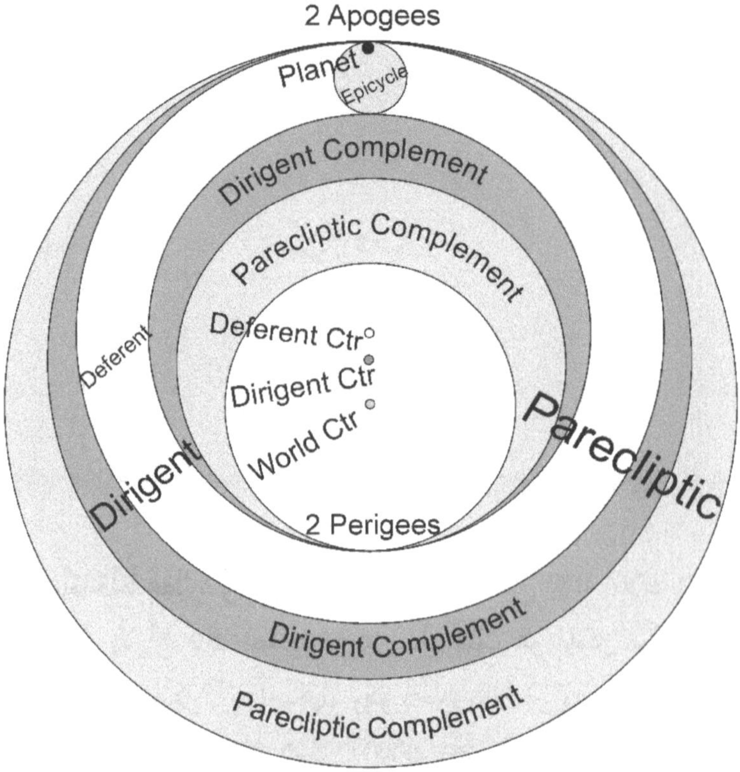

[Figure 4]

[٨] **وفلك القمر** مشتمل على فلكيْن،[155] مركزهما[156] مركز العالم، وفلك حامل. أحد الأوّليْن وهو المحيط بالثاني يسمّى **الجَوْزَهرِي** والممثّل. والثاني يسمّى[157] **المائل**[158] في جوف الجَوْزَهرِي لا في ثخنه. والحامل في ثخن المائل على الرسم. والتدوير في الحامل والقمر في التدوير على نحو ما ذكرنا[159].

[٩] ومن هذه[160] الدوائر يتصوّر كيفيّة ما ذكرنا من هيئات الأفلاك.

صورة فلك عطارد

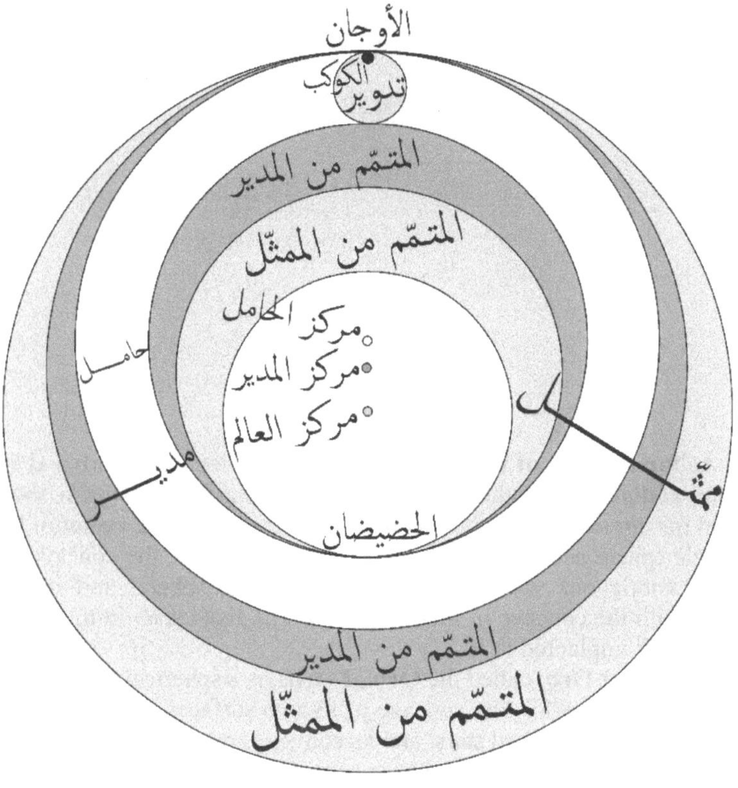

[شكل ٤]

Illustration of the Moon's Orb

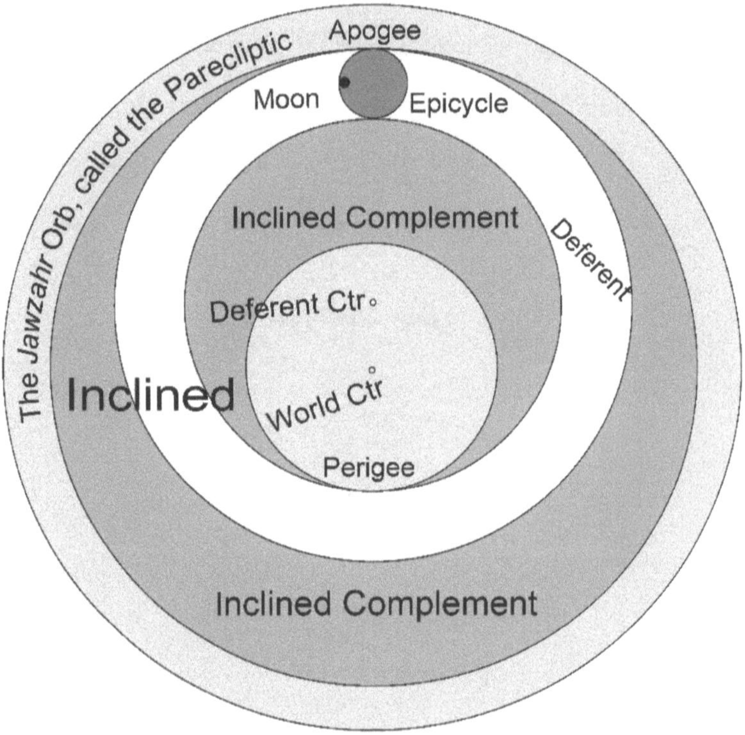

[Figure 5]

[10] **As for the Orb of the Fixed Stars**, it being the eighth orb and called the **orb of the zodiac** [lit., of the signs] whose meaning you will learn about in the chapter on the circles, it is a spherical body whose center is the center of the world. It is a single sphere according to the most correct opinion; the concave of its two surfaces is contiguous with the convex of Saturn's sphere, and its convex is contiguous with the concave of the Great Orb. The fixed stars in their entirety are embedded [and] implanted in it.

[11] **The Great Orb**, called the **Orb of Orbs**, is a spherical body whose center is the center of the world. The concave of its two surfaces is contiguous with the convex of the orb of the fixed stars, and its convex is not contiguous with anything, since there is nothing beyond it, neither vacuum nor plenum.

صورة فلك القمر

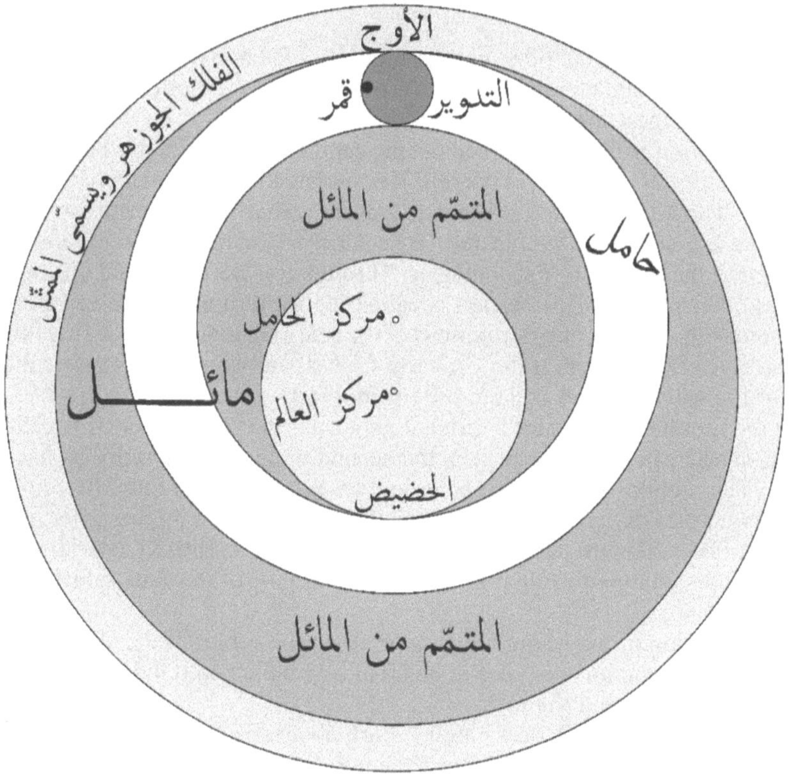

[شكل ٥]

[١٠] [161]**وأمّا فلك الثوابت**[162] وهو الفلك الثامن ويسمّى[163] **فلك البروج** وستعرف معنى[164] هذا في باب الدوائر فجرم[165] كريّ مركزه[166] مركز العالم. وهو[167] كرة واحدة على الرأي الأصحّ؛ مقعّر سطحيْه[168] يماسّ محدّب [169]كرة زحل، ومحدّبها[170] يماسّ مقعّر الفلك الأعظم. والكواكب الثابتة بأجمعها مركوزة مغرقة فيه.

[١١] **والفلك الأعظم**[171] ويسمّى **فلك الأفلاك** جرم كريّ مركزه[172] مركز العالم. مقعّر سطحيْه يماسّ[173] محدّب فلك الثوابت ومحدّبها[174] لا يماسّ شيئاً إذ ليس ورائه [175]شيء لا خلاء ولا ملاء[176].

Chapter 2 of Part I

On The Motions of the Orbs

[1] The motions of the orbs, in their multiplicity, have two divisions: a motion from East to West and a motion from West to East. As for motion that is from East to West, among these are:

[2] the motion of the Great Orb about the center of the world, this being the rapid motion whereby the rotation [of the orb] is completed in approximately one day and one night. The motion of the remaining orbs and what is in them follows from it, since they are contained within the Great Orb, following as the motion of that contained to the motion of the container. Through it is the rising and setting of the Sun and other planets. This motion is called the **motion of the universe** and **the prime motion**, since among the motions of the celestial bodies it is the first [motion] one perceives and through it the Universe moves; and its two poles are called the **poles of the world**, and its equator the **equinoctial**;

[3] the motion of Mercury's dirigent about its eccentric center. It is called the **motion of the apogee** since in it is the second apogee of Mercury as has come before. [The motion] is upon two poles and an equator that are not the equinoctial and the two poles of the World, nor the zodiacal equator and its two poles; you will learn about both of them later. It is in each nychthemeron [lit., a day with its night] **0;59,8,20**, this being equivalent to the mean [motion] of the Sun, which you will learn about later;

[4] the motion of the Moon's *jawzahar* about the center of the World upon the zodiacal equator and its two poles, it being in a nychthemeron **0;3,10,37**; this is the **motion of the head and the tail**;

[5] the motion of the Moon's inclined orb about the center of the World, upon an equator and two poles that are not the equinoctial nor the zodiacal equator, and not their poles, it being in a nychthemeron **11;9,7,43**; this is the **motion of the apogee of the Moon.**

[6] As for motion that is from West to East, among these are: the motion of the Orb of the Fixed Stars, it being a slow motion about the center of the World that traverses, according to the opinion of most Moderns, **one degree in**

الباب الثاني[١٧٧] من المقالة الأولى
في حركات الأفلاك

[١] حركات الأفلاك[١٧٨] على كثرتها قسمان: حركة من المشرق إلى المغرب وحركة من المغرب إلى المشرق[١٧٩]. فأمّا[١٨٠] الحركة التي هي[١٨١] من المشرق إلى المغرب فمنها:

[٢] حركة الفلك الأعظم حول مركز العالم وهي الحركة السريعة التي بها تتمّ[١٨٢] دَوْرته في قريب من يوم وليلة[١٨٣]. وتلزمها حركة سائر الأفلاك وما فيها إذ هي في ضمن الفلك [١٨٤]الأعظم لزوم حركة المظروف لحركة الظرف. وبها طلوع الشمس وسائر الكواكب وغروبها. وتسمّى[١٨٥] هذه الحركة **حركة الكلّ والحركة الأولى** لأنّها أوّل ما[١٨٦] يُعرف من حركات الأجرام السماوية وبها [١٨٧]يتحرّك الكلّ؛ ويسمّى قطباها **قطبي[١٨٨] العالم** [١٨٩]ومنطقتها[١٩٠] **معدّل النهار**؛

[٣] ومنها حركة مدير[١٩١] عطارد حول مركزه الخارج. وتسمّى[١٩٢] **حركة الأوج** إذ فيه الأوج الثاني لعطارد[١٩٣] كما سلف[١٩٤]. [١٩٥]وهي على قطبين ومنطقة غير معدّل النهار وقطبي العالم وغير منطقة البروج وقطبيها وستعرفها[١٩٦]. وهي[١٩٧] في[١٩٨] كل يوم بليلته ٠ نط ح ك[١٩٩] وهو[٢٠٠] مثل وسط الشمس[٢٠١] وستعرفه[٢٠٢]؛

[٤] ومنها حركة جوزهر القمر حول مركز العالم[٢٠٣] على منطقة البروج وقطبيها في اليوم بليلته ٠ جـ ي لز[٢٠٤] وهي **حركة الرأس والذنب**؛

[٥] ومنها حركة [٢٠٥]الفلك المائل للقمر حول مركز العالم على منطقة وقطبين غير معدّل النهار ومنطقة البروج[٢٠٦] وغير أقطابها في اليوم بليلته يا ط ز مجـ[٢٠٧] وهي **حركة أوج القمر**.

[٦] وأمّا الحركة التي[٢٠٨] من المغرب إلى المشرق فمنها[٢٠٩]: حركة فلك الثوابت وهي حركة بطيئة حول مركز العالم، تقطع[٢١٠] على رأي أكثر المتأخّرين **جزءاً واحداً**[٢١١] **في**

sixty-six solar years or sixty-eight lunar [years]—and you will learn about both later—upon an equator, also called the **zodiacal orb**, and the **zodiacal equator** [lit., orb of the zodiacal signs and equator of the zodiacal signs], and upon two poles, which are not the World poles, called the **zodiacal poles**. It follows that its equator intersects the equinoctial; this discussion will be completed in the chapter on circles;

[7] the motions of the parecliptic orbs about the center of the World, equivalent to the motion of the Orb of the Fixed Stars and upon its equator and two poles, as if they move with it. These are the motions of the apogees and nodes [*jawzahar*s], except for one of Mercury's two apogees, namely that in the dirigent, and except for the Moon's apogee, its parecliptic, and its nodes;

[8] the motion of the Sun's eccentric orb upon an equator in alignment with the zodiacal equator, two poles that are not its two poles, and an axis that is parallel to the axis of the zodiacal orb, it being in a nychthemeron **0;59,8,20**;

[9] the motions of the deferent orbs about their eccentric centers, upon equators and poles that are not the two equators of the Great Orb nor the zodiacal orb and their [respective] poles, they being in each nychthemeron:

for Saturn:	**0;2,0,35**
for Jupiter:	**0;4,59,16**
for Mars:	**0;31,26,40**
for Venus:	**0;59,8,20**
for Mercury:	**1;58,16,40**
for the Moon:	**24;22,53,22**.

This motion is called the **mean [motion] of the planet**; it is also called the **motion of latitude**, which is just the **motion of longitude** when taken with respect to the zodiacal orb; we will make clearer the explanation of this in the chapter on circles. It is also called the **motion of the center**.

ست وستين سنة شمسية أو ثمان وستين[٢١٢] **قمرية** وستعرفها[٢١٣]، على[٢١٤] منطقة تسمّى أيضاً [٢١٥]**فلك البروج** و**منطقة البروج**[٢١٦] وعلى قطبين غير قطبي العالم يسمّيان[٢١٧] [٢١٨]**قطبي البروج**. ويلزم أن تقاطع[٢١٩] منطقتها معدّل النهار وسيتمّ [٢٢٠]هذا[٢٢١] الكلام في باب الدوائر[٢٢٢]؛

[٧] ومنها حركات الأفلاك الممثلة حول مركز العالم مثل حركة فلك الثوابت وعلى منطقتها وقطبيها كأنّها تتحرّك[٢٢٣] بها؛ وهي حركات الأوجات والجوزهرات سوى أحد أوجي عطارد أي الذي هو في المدير وسوى أوج القمر وممثله وجوزهره؛

[٨] ومنها حركة الفلك الخارج المركز[٢٢٤] للشمس على منطقة[٢٢٥] مسامتة[٢٢٦] لمنطقة البروج وقطبين[٢٢٧] غير قطبيها ومحور مواز لمحور فلك[٢٢٨] البروج[٢٢٩] [٢٣٠] وهي في اليوم بليلته ٠ نط ح ك[٢٣١]؛

[٩] ومنها حركات[٢٣٢] الأفلاك[٢٣٣] [٢٣٤]الحاملة حول مراكزها الخارجة على مناطق[٢٣٥] وأقطاب غير منطقتي الفلك الأعظم وفلك البروج وأقطابهما[٢٣٦]. وهي في كل يوم بليلته:

لزحل	٠ ب ٠ له [٢٣٧]
للمشتري[٢٣٨]	٠ د نط يو [٢٣٩]
للمرّيخ[٢٤٠]	٠ لا كو م [٢٤١]
[٢٤٢]للزهرة[٢٤٣]	٠ نط ح ك [٢٤٤]
لعطارد[٢٤٥]	ا نح يو م [٢٤٦]
للقمر[٢٤٧]	[٢٤٨]كد كب نج كب [٢٤٩].

وتسمّى[٢٥٠] هذه **الحركة وسط الكوكب**[٢٥١]، تسمّى[٢٥٢] أيضاً **حركة العرض**، وهي بعينها[٢٥٣] **حركة الطول** إذا أضيفت إلى فلك البروج. وسنزيد[٢٥٤] وضوح بيان هذا في باب الدوائر؛ وتسمّى[٢٥٥] أيضاً[٢٥٦] **حركة المركز**.

[10] As for the motions of the orbs of the epicycles about their centers, they deviate from what we mentioned, namely the two divisions of motions, because the motions in their upper parts obviously differ in direction from motions in their lower parts since they do not enclose the Earth. In other words, if the upper motion is from West to East, then the lower motion is from East to West; this is the case for the epicycles of the five vacillating planets.[2] If the upper motion is from East to West, then the lower motion is in reverse; and this is the case for the Moon's epicycle. Nevertheless, what is stated and accepted concerning the course of the epicycles with respect to the zodiac, this being established in the astronomical handbooks [*zīj*es], is that which is in the sequence of the signs, whether it is for the upper motion as in the case of the vacillating planets or the lower motion as for the Moon. The motions of the epicycles in each nychthemeron are:

for Saturn:	**0;57,7,44**
for Jupiter:	**0;54,9,3**
for Mars:	**0;27,41,40**
for Venus:	**0;36,59,29**
for Mercury:	**3;6,24,7**
for the Moon:	**13;3,53,56**.

This motion is called the **motion of anomaly** and the **proper motion** of the planet; and God knows best.

2 Jaghmīnī makes a distinction between the term *mutaḥayyira* [vacillating planets], which designates the five retrograding planets, i.e., Saturn, Jupiter, Mars, Mercury and Venus, and the more general term *al-sayyāra*, which designates all seven planets, including the Sun and Moon. For a fuller discussion, see Commentary I.2[10].

[١٠] أمّا[٢٥٧] حركات أفلاك[٢٥٨] التداوير على مراكزها فهي خارجة عمّا ذكرنا من قسمي الحركات لأنّ حركات أعاليها لا محالة مخالفة[٢٥٩] [٢٦٠]في الجهة لحركات أسافلها لكونها غير شاملة للأرض، أعني إن كانت حركة الأعلى[٢٦١]من المغرب إلى المشرق فحركة الأسفل من المشرق إلى المغرب[٢٦٢] وذلك لتداوير[٢٦٣] الخمسة المتحيّرة، وإن كانت حركة الأعلى[٢٦٤] من المشرق إلى المغرب فحركة الأسفل بالخلاف وذلك لتدوير[٢٦٥] القمر. لكن المذكور[٢٦٦] المعتبر[٢٦٧] من[٢٦٨] مسير التداوير بالنسبة إلى البروج، وهو[٢٦٩] المثبت في الزيجات، هو ما كان على توالي البروج [٢٧٠]سواء كان حركة الأعلى كما في المتحيّرة أو حركة الأسفل كما في القمر. وحركات التداوير في كل يوم بليلته:

[٢٧١]لزحل[٢٧٢]	٠ نز ز مد [٢٧٣]
للمشتري[٢٧٤]	٠ ند ط ج [٢٧٥]
للمرّيخ[٢٧٦]	٠ كز ما م [٢٧٧]
للزهرة[٢٧٨]	٠ لو نط كط [٢٧٩]
لعطارد[٢٨٠]	ج و كد ز [٢٨١]
للقمر	يج ج نج نو [٢٨٢].

وهذه الحركة تسمّى **حركة الاختلاف**[٢٨٣] **والحركة**[٢٨٤] **الخاصّة** للكوكب[٢٨٥] والله اعلم[٢٨٦].

Chapter 3 of Part I

On the Circles

[1] The circle is either a **great [circle]**, which bisects the World and its center is obviously the center of the World, or it is not a great [circle], which does not bisect it, and let it be called a **small [circle].**

[2] **The equinoctial [circle]** [lit., that which balances the day], called the **right orb**, you already know. In fact, it is called the equinoctial because when the Sun is on line with it, day and night are "balanced" in all regions, i.e., are equal. The circle located in its plane upon the face of the Earth is called the **equator**, I mean the circle that occurs upon the Earth's surface as we imagine the equinoctial intersecting the World. The circles parallel to [the equinoctial circle] are called **day-circuits**; they are small, imagined [circles] that are traced by the rotation of the Great Orb by every point assumed on it.

[3] **The zodiacal circle**, also called the zodiacal orb and the zodiacal equator, you have already learned of. The circles that are in its plane, I mean the circles that occur upon the surfaces of the parecliptic orbs when we imagine the zodiacal circle intersecting with the World, are also called parecliptic orbs. With reference to these circles one determines the quantity of **longitude** for the motions of the planets and the Sun; since when we imagine a line extending from the center of the World to the plane of the zodiacal orb passing through the centers of the planets, then if it happens that the endpoint of that line falls on the zodiacal equator, then its point of incidence will be the [projected] position of the planet along the zodiacal orb, and thereupon the planet will have no latitude. [But] if the [endpoint] falls away from the zodiacal equator, we imagine a circle passing through the zodiac's two poles and the endpoint of that line that intersects the zodiacal equator; then the intersection point between that circle and the zodiacal equator is the position of the [projected] planet along the zodiacal orb, and thereupon the planet will have latitude. So the position of the [projected] planet is one of these indicated points. Then as the planet moves forth, the [projected] point moves along the zodiacal orb; this is the meaning of the motion of the planet in longitude.

الباب الثالث من المقالة الأولى
في الدوائر

[١] **الدائرة** إمّا **عظيمة**، وهي التي تنصّف[٢٨٧] العالم ومركزها لا محالة مركز العالم، وإمّا غير عظيمة وهي التي لا[٢٨٨] تنصّفه[٢٨٩] ولتسمَّ[٢٩٠] **الصغيرة**.

[٢] **معدّل النهار**[٢٩١]، وتسمّى[٢٩٢] **الفلك المستقيم**[٢٩٣] وقد عرفتها. وإنّما سمّيت معدّل النهار لأنّ الشمس إذا سامتتها اعتدل[٢٩٤] الليل والنهار في جميع النواحي أي استويا. والدائرة التي في سطحها على [٢٩٥]وجه الأرض تسمّى **خطّ الاستواء**، أعني الدائرة التي تحدث على سطح [٢٩٦]الأرض[٢٩٧] [٢٩٨]عند توهّمنا[٢٩٩] معدّل النهار قاطعاً للعالم[٣٠٠]. والدوائر الموازية[٣٠١] لها تسمّى **المدارات اليوميّة**[٣٠٢]، وهي صغار موهومة ترتسم بدور الفلك الأعظم من كلّ نقطة تفرض[٣٠٣] عليه.

[٣] **دائرة البروج**[٣٠٤]، وتسمّى[٣٠٥] فلك البروج ومنطقة البروج، وقد عرفتها. [٣٠٦]والدوائر[٣٠٧] التي في سطحها، أعني الدوائر التي تحدث على سطوح الأفلاك الممثّلة عند توهّمنا دائرة البروج قاطعة للعالم تسمّى[٣٠٨] أيضاً بالأفلاك الممثّلة. وبالنسبة إلى هذه الدائرة[٣٠٩] تقدّر[٣١٠] كمّية **طول**[٣١١] حركات الكواكب والشمس[٣١٢]، لأنّا إذا توهّمنا خطّاً يخرج [٣١٣]من مركز العالم إلى سطح فلك البروج مارّاً بمراكز الكواكب[٣١٤]، فإن اتّفق أن وقع طرف ذلك الخطّ في منطقة البروج، فموقعه هو مكان الكوكب من فلك البروج، وحينئذ[٣١٥] لا يكون للكوكب[٣١٦] عرض. وإن وقع خارجاً عن منطقة البروج، توهّمنا دائرة مارّة بقطبي البروج[٣١٧] وطرف[٣١٨] ذلك الخطّ مقاطعة[٣١٩] [٣٢٠]لمنطقة البروج، فيكون نقطة التقاطع بين تلك الدائرة وبين منطقة البروج مكان الكوكب من فلك البروج، ويكون للكوكب عرض حينئذ[٣٢١] فمكان الكوكب إحدى هاتين النقطتين المذكورتين. [٣٢٢]فكلّما تحرّك الكوكب تحرّكت النقطة[٣٢٣] على فلك البروج، وهذا هو[٣٢٤] المعنى بحركة الكوكب في الطول.

[4] The circles parallel to [the zodiacal equator] are called **parallels of latitude**. They are imagined small [circles] that are traced with the rotation of the Eighth Orb by each point assumed on it.

[5] Since the two poles of the zodiac are not the two poles of the World, it follows that the zodiacal circle will intersect the equinoctial at two opposite points. One of the two, from which the zodiacal orb sets out northward in the sequence [of the signs], is called the **vernal equinox point**; the other [is called] the **autumnal equinox point**. Its maximum distance from it, I mean the distance of the zodiacal circle from the equinoctial, will be at two points: one of the two is toward the north and is called the **summer solstice point**; the other is toward the south and is called the **winter solstice point**. Thereby, then, are designated four points on the zodiacal circle by which it becomes four parts. The period of time the Sun traverses each fourth of [the zodiac] is the period of one of the four seasons of the year. Then we imagine for each one of two adjoining quarters of the [zodiac] two points, the distance of each one of them from the other is the same as the distance of the other from the nearer of the two endpoints of the quarter to it. Then we imagine six great circles, all intersecting one another at two opposite points, namely the two poles of the zodiac: one of them passes through the two poles of the World, the two poles of the zodiac, and the two solstice points; this is called the **solstitial colure** (lit., the **[great] circle passing through the four poles)**, and its two poles are the two equinox points. The other passes through the two equinox points and its two poles are the two solstice points. The remaining [great circles] pass through the four imaginary points lying in the two designated quarters, and through four other points, being opposite the designated ones, that are on the two remaining quarters facing the designated ones. So the Eighth Orb is thus divided by these six circles into 12 divisions, each division called a **zodiacal sign.** The arc that is between every two of these circles along the zodiacal equator is also called a zodiacal sign; for this reason, it is called the zodiacal orb. The parecliptic orbs and also the Great Orb are likewise divided by the imagined planes of these circles into 12 zodiacal signs.

[6] **The horizon circle** is a great circle that separates what is seen of the [celestial] orb from what is not seen; and with reference to it one determines rising and setting. Its two poles are the **zenith** and the **nadir**, and it bisects

[٤] والدوائر [٣٢٥]الموازية لها تسمّى[٣٢٦] **مدارات العرض**. وهي صغار موهومة ترسّم بدور الفلك الثامن من[٣٢٧] كلّ نقطة تفرض عليه.

[٥] ولمّا كان قطبا البروج غير قطبي العالم لزم أن[٣٢٨] تقاطع دائرة البروج معدّل النهار عند نقطتين متقابلتين. إحداهما وهي التي يأخذ[٣٢٩] منها فلك البروج على التوالي إلى الشمال تسمّى **بنقطة الاعتدال الربيعيّ**، والأخرى **بنقطة الاعتدال الخريفيّ**. وتكون غاية بُعدها عنه[٣٣٠]، أعني بُعد دائرة البروج عن معدّل النهار، عند نقطتين إحداهما ممّا يلي الشمال وتسمّى [٣٣١]**نقطة**[٣٣٢] **الانقلاب الصيفيّ**، والأخرى ممّا يلي الجنوب وتسمّى **نقطة**[٣٣٣] **الانقلاب الشتويّ**. فتُعيّن[٣٣٤] بذلك لدائرة البروج أربع نقط[٣٣٥] تصير بها أرباعاً. ومدّة[٣٣٦] قطع الشمس [٣٣٧]كلّ ربع منها هي مدّة فصل من أربعة[٣٣٨] فصول السنة. ثمّ نتوهّم[٣٣٩] على ربعين متلاصقين منها على كلّ[٣٤٠] واحد منهما نقطتين بُعد كل واحدة[٣٤١] منهما عن الأخرى مثل بُعد الأخرى عن[٣٤٢] أقرب طرفي الربع إليها. [٣٤٣]ثمّ نتوهّم[٣٤٤] ست دوائر عظام تتقاطع[٣٤٥] بأجمعها على نقطتين متقابلتين هما قطبا البروج[٣٤٦]: [٣٤٧]إحداهما تمرّ[٣٤٨] بقطبي العالم وبقطبي [٣٤٩]البروج[٣٥٠] وبنقطتي[٣٥١] الانقلابين وهذه تسمّى **بالدائرة المارّة بالأقطاب الاربعة**، وقطباها نقطتا[٣٥٢] الاعتدالين. والأخرى تمرّ[٣٥٣] بنقطتي[٣٥٤] الاعتدالين وقطباها نقطتا[٣٥٥] الانقلابين[٣٥٦]. والأربع الباقية تمرّ بالنقط[٣٥٧] الأربع المتوهّمة على الربعين[٣٥٨] المفروضين، وبأربع نقط[٣٥٩] أخر[٣٦٠] مقابلة للمفروضة هي على[٣٦١] [٣٦٢]الربعين الباقيين المقابلين[٣٦٣] للمفروضين. فينقسم الفلك الثامن بهذه الدوائر الست اثنى عشر قسماً، كلّ قسم منها يسمّى[٣٦٤] **بُرجاً**. والقوس التي بين كلّ دائرتين منها[٣٦٥] من منطقة البروج تسمّى[٣٦٦] أيضاً برجاً، ولهذا تسمّى[٣٦٧] بفلك البروج. وبالسطوح الموهومة[٣٦٨] لهذه الدوائر تنقسم[٣٦٩] الأفلاك الممثّلة والفلك الأعظم أيضاً[٣٧٠] باثني عشر بُرجاً.

[٦] **دائرة الأفق** دائرة عظيمة تفصل[٣٧١] بين ما يُرى من الفلك وبين ما لا يُرى، [٣٧٢]وبالنسبة إليها يُعرف الطلوع والغروب. وقطباها **سَمْتا**[٣٧٣] **الرأس والقدم** وتنصّف

the equinoctial at two points: one is called the **east point** and [place of] rising of the equinox; the other the **west point** and [place of] setting of the equinox. The line connecting them is called the **east-west line** and the equinox line. The circles parallel to it are called **almucantars**.

[7] **The meridian circle** is a great circle that passes through the two poles of the World, and the zenith and nadir. Its two poles are the east and west points. It bisects the horizon circle at two points, one of them called the **south point** and the other **the north point**; and the line connecting them is called the **meridian line**. This line and the east-west line are etched on the surfaces of sundials [*rukhāmāt*].

[8] **The altitude circle**, also called **the azimuth circle**, is a great circle that passes through the zenith and nadir, and through the endpoint of a line extending from the World center to the surface of the highest orb, having passed through the center of a star or the Sun. It intersects the horizon circle at right angles at two points that are not fixed but rather shift along the horizon circle commensurate with the shifting of the star or the Sun. Each one of them is called an **azimuth point**. The arc along the horizon circle that is between [one of] them and either the east or west point is called the **arc of the azimuth**; what is between [one of] them and either the north or south point is called the **complement of the azimuth.** This circle coincides with the meridian circle twice in a nychthemeron.

[9] **The circle of the initial azimuth [prime vertical]** is a great circle that passes through the zenith and nadir, and through the east and west points. Its two poles are the north and south points. It intersects the meridian circle at the zenith and nadir points; in fact, it is called by that [name] because when an altitude circle coincides with it, it does not have an azimuth arc. The circuit that is tangent to it is called the circuit of that locality, this [passing through] the zenith for its residents.

[10] **The declination circle** is a great circle passing through the two poles of the equinoctial. With it one determines the distance of a star from the equinoctial and the inclination of the zodiacal orb from the equinoctial, in other words the **first declination,** which you will learn about later.

معدّل النهار بنقطتين: يُقال لإحداهما[٣٧٤] **نقطة المشرق** ومطلع الاعتدال وللأخرى **نقطة المغرب** ومغرب الاعتدال. ويقال للخطّ الواصل بينهما **خطّ المشرق والمغرب** وخطّ الاعتدال[٣٧٥]. والدوائر الموازية لها[٣٧٦] يقال لها[٣٧٧] **المُقَنْطرات**[٣٧٨].

[٧] **دائرة نصف النهار**[٣٧٩] [٣٨٠]دائرة عظيمة تمرّ بقطبي العالم وسمتي[٣٨١] الرأس والقدم. وقطباها نقطتا[٣٨٢] المشرق والمغرب. وتنصّف[٣٨٣] دائرة الأفق بنقطتين تُدعى إحداهما[٣٨٤] **نقطة الجنوب** والأخرى **نقطة الشمال** ويقال للخطّ الواصل بينهما **خطّ نصف النهار**. وهذا الخطّ وخطّ المشرق والمغرب يُستخرجان في سطوح الرخامات.

[٨] **دائرة الارتفاع**[٣٨٥] وتسمّى[٣٨٦] أيضاً **الدائرة السمتيّة**[٣٨٧] هي[٣٨٨] دائرة عظيمة تمرّ بسمتي الرأس والقدم وبطرف الخطّ[٣٨٩] الخارج من مركز العالم إلى سطح الفلك الأعلى[٣٩٠]مارّاً[٣٩١] بمركز[٣٩٢] الكوكب[٣٩٣] أو الشمس. وتقطع دائرة الأفق على زوايا قائمة بنقطتين غير ثابتتين، بل منتقلتين[٣٩٤] على دائرة الأفق حسب[٣٩٥] انتقال [٣٩٦]الكوكب[٣٩٧] أو الشمس. تسمّى[٣٩٨] كلّ واحدة[٣٩٩] منهما **نقطة السمت**. والقوس من[٤٠٠] دائرة الأفق بينهما وبين إحدى نقطتي المشرق والمغرب تسمّى **قوس السمت**، وما بينهما وبين إحدى نقطتي[٤٠١] الجنوب [٤٠٢]والشمال يسمّى[٤٠٣] **تمام السمت**. وهذه الدائرة تنطبق على دائرة نصف النهار في اليوم بليلته[٤٠٤] مرّتين.

[٩] **دائرة أوّل السُمُوت**[٤٠٥] دائرة عظيمة تمرّ بسمتي الرأس والقدم وبنقطتي[٤٠٦] المشرق والمغرب [٤٠٧]وقطباها نقطتا الجنوب والشمال وتقاطع دائرة[٤٠٨] نصف النهار على نقطتي سمت[٤٠٩] الرأس والقدم، وإنّما سمّيت بذلك لأنّ دائرة الارتفاع إذا انطبقت عليها كانت ليس لها قوس سمت. والمدار الذي يماسّها يسمّى مدار ذلك البلد الذي هذا سمت رأس أهله.

[١٠] **دائرة المَيل**[٤١٠] [٤١١]دائرة عظيمة مارّة بقطبي معدّل النهار[٤١٢]. ويُعرف بها بُعد الكوكب عن معدّل النهار وميل فلك البروج عن معدّل النهار[٤١٣] أعني **الميل الأوّل** وستعرفه.

[11] **The latitude circle** is a great circle that passes through the two poles of the zodiac, and through the endpoint of a line extending from the World center and passing through the center of a star until the surface of the Great Orb. With it one determines the latitude of the star and the **second declination** of the zodiacal orb from the equinoctial.

[12] The imaginary circles that are traced by the rotation of points in the planetary orbs are either traced on the surfaces of spheres or else are not traced on the surfaces. The traced [circles] on the surfaces are then: that traced from the motion of the center of the Sun on the circumference of its eccentric orb; and those traced from the motions of the centers of the epicycles on the circumferences of the deferent orbs and from the motions of the centers of the planets on the circumferences of the epicycle orbs. Each of these circles bears the name of the orb on whose circumference it is traced; thus the traced [circle] from the motion of the Sun's center is called the eccentric orb, the traced [circles] from the motions of the epicycle centers [are called] deferent orbs, and the traced [circles] from the planet centers [are called] epicycle orbs. If these deferent orbs and the equator of the inclined orb are assumed to intersect the World, there occurs on the surfaces of the parecliptic orbs, the zodiacal orb, and the Great Orb circles that are called **declination orbs** due to their inclination from the zodiacal orb. Because the motions of the orbs in which [these circles] have been traced occur about poles that are neither the two zodiacal poles nor the two World poles, these being the inclined orbs, they intersect the parecliptics at two points, one of them being the crossing point of the planet on the zodiacal circle toward the north, called the **head**, and the other the **tail.**

[13] Those not traced on surfaces are traced by the deferent center of Mercury and of the Moon through the dirigent's moving of Mercury's deferent and through the inclined's moving of the Moon's deferent. These traced [circles] are called the deferent orb of the deferent center, since the deferent center revolves on its circumference.

[١١] **دائرة العرض**[٤١٤] دائرة عظيمة تمرّ بقطبي البروج وبطرف الخطّ الخارج من مركز العالم المارّ بمركز الكوكب[٤١٥] إلى سطح[٤١٦] [٤١٧]الفلك الأعظم[٤١٨]. ويُعرف بها[٤١٩] عرض الكوكب[٤٢٠] **والميل الثاني** لفلك البروج عن معدّل النهار.

[١٢] **الدوائر**[٤٢١] **المتوهّمة المرتسمة بدور النقط في أفلاك السيّارة**[٤٢٢] وهي[٤٢٣] إمّا مرتسمة على بسائط[٤٢٤] الأكر[٤٢٥] وإمّا مرتسمة[٤٢٦] لا على البسائط[٤٢٧]. فالمرتسمة[٤٢٨] على البسائط[٤٢٩] هي المرتسمة من حركة مركز الشمس على محيط فلكها الخارج المركز، والمرتسمة من حركات[٤٣٠] مراكز التداوير على محيطات الأفلاك الحاملة[٤٣١] ومن حركات مراكز الكواكب على محيطات أفلاك التداوير[٤٣٢]. وكلّ دائرة منها تسمّى باسم الفلك[٤٣٣] الذي ترتسم على محيطه، فالمرتسمة من حركة مركز الشمس تسمّى بالفلك [٤٣٤]الخارج المركز، والمرتسمة من حركات[٤٣٥] مراكز التداوير بالأفلاك الحاملة[٤٣٦]، والمرتسمة من[٤٣٧] مراكز الكواكب بأفلاك[٤٣٨] التداوير. وهذه[٤٣٩] الأفلاك [٤٤٠]الحاملة[٤٤١] ومنطقة [٤٤٢]الفلك المائل إذا فُرضت قاطعة للعالم حدثت [٤٤٣]في سطوح الأفلاك الممثّلة وفلك البروج والفلك الأعظم [٤٤٤]دوائر تسمّى **الأفلاك المائلة** لميلها عن فلك البروج. ولكون[٤٤٥] حركات الأفلاك التي ارتسمت[٤٤٦] فيها على اقطاب غير قطبي البروج وقطبي[٤٤٧] العالم، وهذه الأفلاك المائلة، تقاطع[٤٤٨] الممثّلات على نقطتين، إحداهما[٤٤٩] وهي مجاز الكوكب على[٤٥٠] دائرة البروج إلى الشمال تسمّى[٤٥١] **بالرأس**، والأُخرى **بالذنب**.

[١٣] والمرتسمة لا على البسائط[٤٥٢] هي المرتسمة من مركز الحامل[٤٥٣] لعطارد والقمر بتحريك المدير حامل[٤٥٤] عطارد وبتحريك المائل[٤٥٥] حامل[٤٥٦] القمر. وتسمّى هذه المرتسمة الفلك[٤٥٧] الحامل[٤٥٨] لمركز الحامل[٤٥٩] إذ مركز الحامل[٤٦٠] يدور على محيطها[٤٦١].

Chapter 4 of Part I

On the Arcs

[1] **The arc** is a segment of the circle's circumference. If that segment is then subtracted from 90 parts, [these] parts being those by which the circumference is 360 parts in total, the excess of 90 over it is called the complement of that arc. An example is what has preceded regarding the arc of the azimuth and its complement.

[2] **The longitude of a locality** is the arc along the equinoctial between the meridian circle at the end of the inhabited region, in other words the beginning of the inhabited region's longitude in the west, which you will learn about later, and the meridian circle for that locality.

[3] **The co-ascension** of each arc along the zodiacal orb is that which rises with it along the equinoctial. The co-ascension will obviously be bounded on the equator between two declination circles since its horizon passes through the two poles of the World and so it is also one of the declination circles. In other words, what is between two declination circles along the equinoctial is the co-ascension for what is between them along the zodiacal orb.

[4] **The co-ascension of a [discrete] part** of the zodiacal orb is an arc along the equinoctial between the head of Aries and the [discrete] part of the [equinoctial] that rises with that part.

[5] **The equation of daylight** for a [discrete] part on the zodiacal orb is the difference between its co-ascension at the equator and its co-ascension at a locality.[3] Let us take an example for this: When the head of Gemini is found toward the east for a horizon other than the equator, and we assume one of the declination circles passes through it and intersects the equinoctial, there occurs a triangle. One of its sides is the declination of the head of Gemini, and you will learn about declination later. The other two sides are two arcs between the declination circle and the vernal equinox point: one of them is along the zodiacal orb and is called the **equal degrees**; the other is along the equinoctial and is the **co-ascension of the zodiacal arc** for the horizon of the equator. The horizon of the locality divides this triangle into two triangles: one of them is above the Earth and is bounded by the **ortive amplitude** (which you will learn about later), by the aforementioned zodiacal arc, and by the arc along the equinoctial between the vernal equinox point and the horizon. The other triangle is

---

[3] For further clarification of the equation of daylight, see Commentary, I.4[5], Fig. C1.

الباب الرابع من المقالة الأولى
في القسيّ

[١] **القوس** قطعة [٤٦٢]من محيط الدائرة. فإن نقصت تلك القطعة[٤٦٣] عن تسعين [٤٦٤]جزءاً بالأجزاء[٤٦٥] التي يتمّ بها[٤٦٦] المحيط ٣٦٠ [٤٦٧] جزءاً ففضل التسعين عليها يسمّى تمام تلك[٤٦٨] القوس. ومثاله [٤٦٩]ما سلف من قوس السمت وتمامها[٤٧٠].

[٢] **طول البلد**[٤٧١] قوس من معدّل النهار فيما بين دائرة نصف النهار بآخر[٤٧٢] العمارة أعني مبدأ[٤٧٣] طول العمارة من المغرب وستعرفه وبين دائرة نصف النهار في ذلك البلد.

[٣] **مطالع** كلّ قوس من فلك البروج هي[٤٧٤] ما يطلع[٤٧٥] معها من معدّل النهار[٤٧٦]. ويكون المطالع في خطّ الاستواء لا محالة محصورة[٤٧٧] بين دائرتين من دوائر الميل[٤٧٨] لأنّ أفقه مارّ بقطبي العالم فهو[٤٧٩] أيضاً دائرة من دوائر الميل، أعني يكون ما بين دائرتي الميل من معدّل النهار مطالع لما بينهما من فلك البروج.

[٤] **مطالع**[٤٨٠] **الجزء** من فلك البروج[٤٨١] قوس[٤٨٢] من معدّل النهار[٤٧٣] بين رأس الحمل والجزء الذي يطلع منه[٤٨٤] مع[٤٨٥] ذلك [٤٨٦]الجزء.

[٥] **تعديل النهار**[٤٨٧] لجزء من فلك البروج هو[٤٨٨] الفضل بين مطالعه بخطّ الاستواء وبين مطالعه بالبلد. ولِنُمثّل لذلك مثالاً: إذا كان رأس الجوزاء ممّا يلي[٤٨٩] المشرق في [٤٩٠]أفق غير[٤٩١] خطّ الاستواء، وفرضنا دائرة من دوائر الميل تمرّ به [٤٩٢]وتقاطع معدّل النهار، [٤٩٣]حدث مُثلّث. أحد أضلاعه ميل[٤٩٤] رأس الجوزاء وستعرف الميل[٤٩٥]. والضلعان الآخران [٤٩٦]قوسان بين دائرة الميل وبين نقطة الاعتدال الربيعي، إحداهما من فلك البروج ويسمّى **بدرج**[٤٩٧] **السواء** والأخرى من معدّل النهار وهي **مطالع قوس البروج** بأفق خطّ الاستواء. وأفق البلد يقسّم[٤٩٨] هذا المثلّث إلى مثلّثين[٤٩٩] إحداهما فوق الأرض وتحيط[٥٠٠] به **سعة المشرق**، وستعرفها، وقوس البروج المذكورة وقوس من[٥٠١] معدّل النهار بين[٥٠٢] نقطة الاعتدال الربيعي وبين الأفق. والمثلّث الآخر

below the Earth and is bounded by the ortive amplitude, by the declination of the head of Gemini, and by an arc along the equinoctial that is between the horizon and the intersection point between the declination circle and the equinoctial. This arc, which is on the equinoctial, is the equation of daylight for the head of Gemini at that locality. Since the sections for horizons will differ from the example of this triangle with the varying latitudes of localities, it necessarily follows that the co-ascension will vary with different latitudes.

[6] **The solar mean** is an arc along the zodiacal orb that is between the first of Aries and the tip of a line extending from the center of [the Sun's] eccentric orb that passes through the center of the Sun and terminates at the zodiacal circle. If that line is assumed to extend from the World center, then the arc along the zodiacal orb that is between its endpoint and the first of Aries is the **true position** [*taqwīm*] of the Sun. What is between the endpoints of the two aforementioned lines is its **equation**. The angle of the two lines intersecting at the Sun's center, I mean the angle subtended by the arc of the equation, is the **angle of the equation**.

[7] **The planet's mean** is an arc along the zodiacal orb that is between the first of Aries and the endpoint of the line extending from the World center that passes through the epicycle center and terminates at the zodiacal orb, this being when the epicycle center is aligned with one of the two nodal points. Then when it moves beyond [the nodal point] and obtains latitude, the position of the line will fall outside the zodiacal orb, either to the north or to the south. One then imagines a circle passing through its position and the two poles of the zodiac that intersects the zodiacal orb; then the arc along the zodiacal orb that is between the first of Aries and the intersection point of that circle and the zodiacal circle is **the planet's mean**. If we then assume a line extending from the World center terminating at the zodiacal orb and passing through the center of the planet, then the arc that is between the first of Aries and the endpoint for a planet lacking latitude, or between the first of Aries and the intersection point of the zodiacal orb and the circle passing through the two poles of the zodiac and the endpoint, is the **planet's true position.** What is between the mean and the true position along the zodiacal orb is the **equation**.

[8] On the basis of this meaning [for equation]: when the Sun is at the apogee or perigee whereupon the two extended lines coincide—one of them from the World center and the second from the center of its eccentric orb—

تحت الأرض وتحيط به سعة المشرق وميل رأس الجوزاء وقوس[٥٠٣] من معدّل النهار ما بين الأفق وبين[٥٠٤] نقطة التقاطع بين دائرة الميل وبين معدّل النهار[٥٠٥]. وهذه القوس التي هي من[٥٠٦] معدّل النهار تعديل النهار[٥٠٧] لرأس الجوزاء[٥٠٨] في ذلك البلد. ولمّا كانت الآفاق تختلف قطعها لمثل هذا المثلّث باختلاف عروض البلدان، وجب أن تكون[٥٠٩] المطالع تختلف[٥١٠] باختلاف العروض.

[٦] **وسط الشمس**[٥١١] وهو[٥١٢] قوس من فلك البروج ما بين أوّل الحمل [٥١٣]وبين رأس خطّ[٥١٤] يخرج من مركز فلكها الخارج المركز ويمرّ[٥١٥] بمركز[٥١٦] الشمس وينتهي إلى دائرة البروج. فإذا فرض ذلك الخطّ خارجاً من مركز العالم فالقوس التي [٥١٧]بين طرفه وبين أوّل الحمل من فلك البروج هي **تقويم الشمس**. وما بين طرفي الخطّين المذكورين هو [٥١٨]**تعديلها**. وزاوية الخطّين إذا تقاطعا[٥١٩] عند مركز الشمس، أعني الزاوية التي تُوترها قوس التعديل هي **زاوية التعديل**[٥٢٠].

[٧] **وسط الكوكب** قوس[٥٢١] من فلك البروج ما بين أوّل الحمل وطرف[٥٢٢] الخطّ الخارج من مركز العالم المارّ بمركز التدوير المنتهي إلى فلك البروج، وذلك يكون عند مسامتة مركز التدوير إحدى نقطتي الجوزهريْن. فإذا جاوزها وحصل له عرض كان موقع الخطّ خارجاً عن[٥٢٣] فلك البروج إمّا إلى الشمال وإمّا إلى الجنوب. فيتوهّم[٥٢٤] دائرة [٥٢٥]مارّة على موقعه وقطبي البروج مقاطعة[٥٢٦] لفلك البروج؛ فالقوس[٥٢٧] التي هي من فلك البروج ما بين أوّل الحمل وبين نقطة التقاطع [٥٢٨]بين [٥٢٩]تلك الدائرة ودائرة البروج هي **وسط** [٥٣٠]**الكوكب**. فإن فرضنا الخطّ الخارج من مركز العالم المنتهي إلى فلك البروج مارّاً بمركز الكوكب، فالقوس التي بين أوّل الحمل وبين طرفه[٥٣١] مع عدم عرض الكوكب أو بين أوّل الحمل وبين نقطة[٥٣٢] التقاطع من[٥٣٣] فلك البروج والدائرة المارّة بقطبي البروج وبطرفه هي **تقويم الكوكب**. وما بين [٥٣٤]الوسط والتقويم من فلك البروج هو **التعديل**.

[٨] ولهذا المعنى: إذا كانت الشمس في الأوج أو الحضيض حيث ينطبق الخطّان الخارجان — أحدهما[٥٣٥] من مركز العالم والثاني من مركز فلكها الخارج المركز —

both passing through [the Sun's] center; or [when] the planets are at the apices of their epicycles or at their lowest [points], whereupon the two lines extending from the World center coincide—one of them passing through the epicycle center and the second through the center of the planet—there is thereupon no equation.

[9] They divided each one of the eccentric orbs and the epicycles into four disparate parts, two of them being lower and equal, and two upper and equal, that they called **sectors**. They differed concerning the initial [points] of these divisions. Some of them took distance into account, so the eccentric was divided by two lines, one of them extending from the World center to the apogee and perigee, and the other passing through the two mean distances, these being two facing points on the circumference of the eccentric orb where two produced lines are equal, one being from the World center and the other from the eccentric center, that terminate at either [point]. This latter line passes through the midpoint between the two centers. The epicycle is divided by two lines, one of them extending from the deferent center and passing through the epicycle's perigee and center to its apex, and the other passing through the two points of intersection between the epicycle and the deferent.

Illustration of Sectors with Respect to Distance

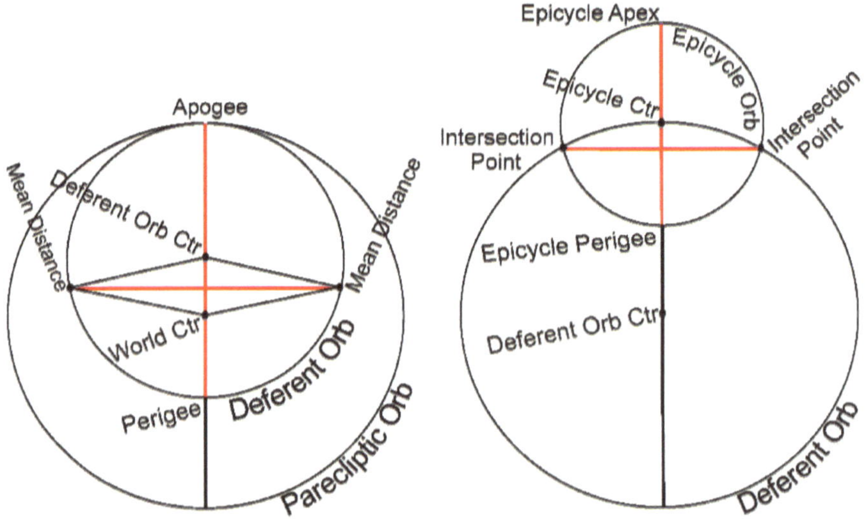

[Figure 6]

المارّان[٥٣٦] بمركزها؛ أو كانت[٥٣٧] الكواكب[٥٣٨] في ذرى تداويرها أو في أسافلها حيث ينطبق[٥٣٩] الخطّان الخارجان من مركز العالم — المارّ أحدهما بمركز التدوير والثاني بمركز الكوكب — [٥٤٠]لم يكن هناك تعديل.

[٩] وقد قسموا الأفلاك الخارجة المراكز والتداوير كلّ واحد منها أربعة أقسام مختلفة، اثنان منها سفليّان متساويان [٥٤١]واثنان[٥٤٢] علويّان متساويان[٥٤٣]، سمّوها **نطاقات**[٥٤٤]. واختلفوا في مبادئ هذه الأقسام: فمنهم من اعتبر الأبعاد فقسم الخارج المركز بخطّين، يخرج أحدهما من مركز العالم إلى الأوج والحضيض، والآخر يمرّ[٥٤٥] بالبعدين الأوسطين[٥٤٦]، وهما نقطتان متقابلتان على محيط الفلك الخارج المركز[٥٤٧] حيث يستوي الخطّان الخارجان[٥٤٨]، أحدهما[٥٤٩] من مركز العالم والآخر من مركز الخارج المركز المنتهيان إلى أيّتهما كانت. ويمرّ هذا الخطّ عند منتصف ما بين المركزين. وقسم التدوير بخطّين، يخرج [٥٥٠]أحدهما من مركز الحامل مارّاً بحضيض [٥٥١]التدوير ومركزه[٥٥٢] إلى ذروته، والآخر يمرّ[٥٥٣] بنقطتي التقاطع بين التدوير والحامل[٥٥٤].

صورة النطاقات باعتبار الأبعاد

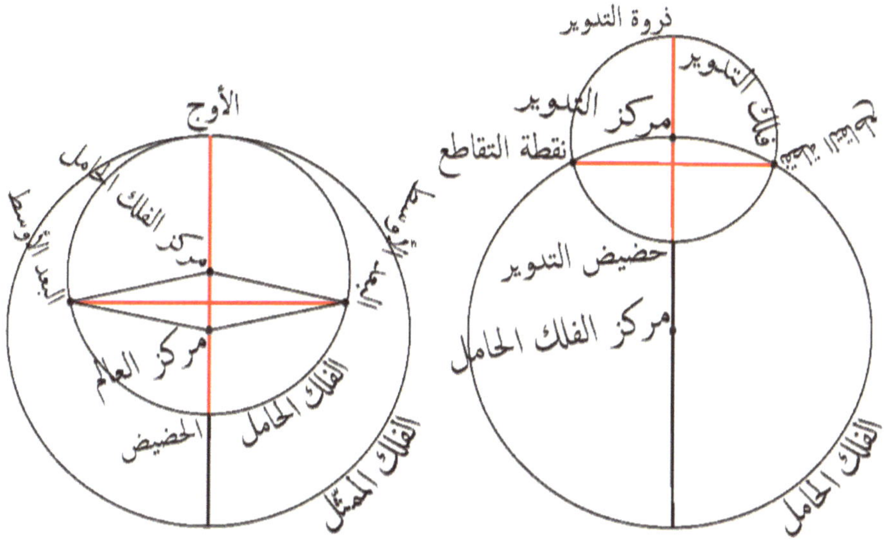

[شكل ٦]

[10] Some of them took variable movement into account [in determining the sectors]. Thus the eccentric was divided by two lines, one of them extending from the World center to the apogee and perigee, and the other passing through where the angle of equation is greatest, this being on the apogee side at a distance of 90 parts from it along the parts of the zodiacal orb. The epicycle was divided by two lines, one of them extending from the deferent center and passing through the epicycle's apex and perigee, and the other being perpendicular to it, its two endpoints terminating at the two tangent points between the epicycle circumference and two lines extending to it from the deferent center. Here too is the maximum equation with respect to the epicycle. So the first sector is what the planet reaches after it crosses the apogee or the epicycle's apex, and the second, third, and fourth [sectors] are in the sequence of its motion. As long as the planet moves from the highest part to the lowest part, i.e., when it is in the first and second sectors of the eccentric or the epicycle, it is descending; and as long as it moves from the perigee to the apogee, i.e., when it is in the other two sectors, it is ascending.

Illustration of Sectors with Respect to Variable Movement

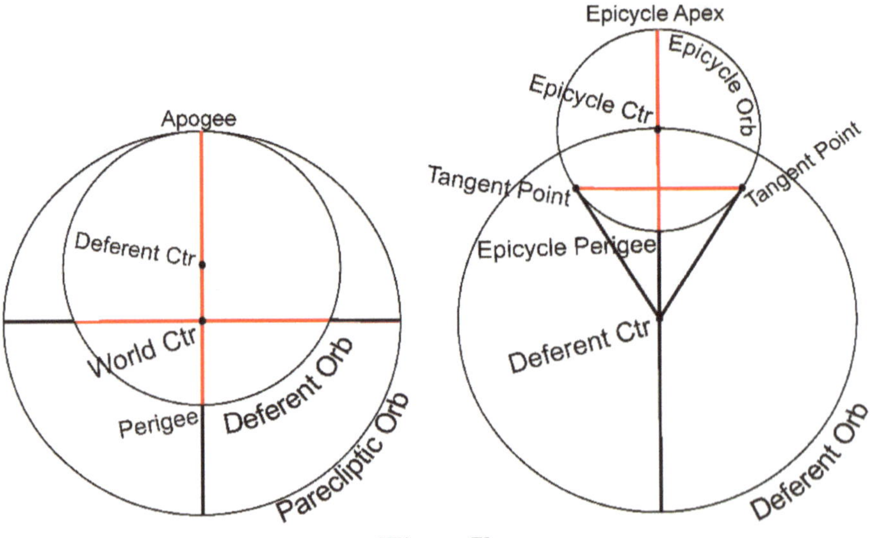

[Figure 7]

[١٠] [٥٥٥]ومنهم من اعتبر [٥٥٦]اختلاف المسير. فقسم الخارج المركز بخطّين، يخرج أحدهما من مركز العالم إلى الأوج والحضيض، والآخر[٥٥٧] يمرّ[٥٥٨] بحيث[٥٥٩] تكون زاوية التعديل أعظم، وذلك من جانب[٥٦٠] الأوج على بعد تسعين جزءاً[٥٦١] عنه من أجزاء فلك البروج. وقسم التدوير بخطّين، يخرج [٥٦٢]أحدهما من مركز الحامل[٥٦٣] ويمرّ[٥٦٤] بالذروة والحضيض من التدوير، والآخر يقوم عليه وينتهي طرفاه إلى نقطتي[٥٦٥] التماسّ بين محيط التدوير وبين خطّين يخرجان إليه من مركز الحامل[٥٦٦]. وهناك أيضاً غاية التعديل [٥٦٧]من جهة التدوير. فالنطاق الأوّل هو ما يصل إليه الكوكب بعد مجاوزته[٥٦٨] الأوج أو ذروة التدوير والثاني والثالث والرابع على توالي حركته[٥٦٩]. فما دام الكوكب يتحرّك من الأعلى إلى الأسفل أي كان في النطاق الأوّل والثاني من الخارج المركز [٥٧٠]أو التدوير[٥٧١] [٥٧٢]فهو هابط؛ وما دام يتحرّك من الحضيض إلى الأوج أي كان في النطاقين[٥٧٣] الآخرين فهو صاعد[٥٧٤].

صورة النطاقات باعتبار اختلاف المسير

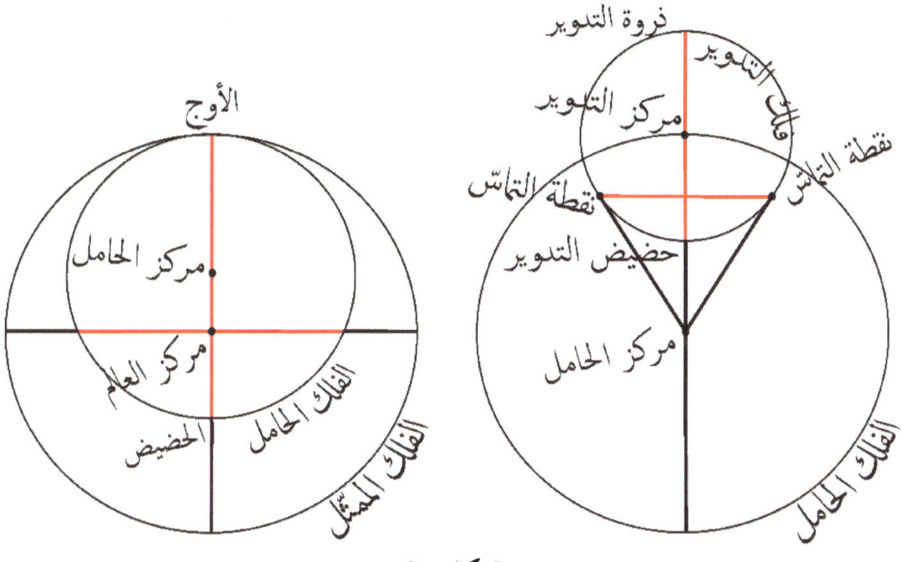

[شكل ٧]

[11] **The latitude of a locality** is an arc along the meridian circle between the equinoctial and the zenith, and it is equivalent to what is between the horizon and the pole that is on the meridian circle, which is the altitude of the pole, i.e., the nearest of the two World poles to that locality.

[12] **The declination** is an arc along a declination circle between the equinoctial and the zodiacal circle, this being the **first declination. Declination**, when used by itself, means the first declination. The **second declination** is an arc along a latitude circle between the two of them, I mean between the equinoctial and the zodiacal circle. **The maximum declination**, called the **total obliquity** [lit., the complete declination] and the **greatest declination**, is an arc between them along the solstitial colure circle; it falls under the definition of [either] the first or second declination. It is the limit of the inclination of the zodiacal circle from the equinoctial, and its amount is **23;35**.

[13] **The planet's latitude** is an arc along the latitude circle between the zodiacal circle and the tip of a line extending from the World center, passing through the planet's center, and terminating at the zodiacal orb. If the arc is along a declination circle between the equinoctial and the aforementioned tip of the line, then it is the planet's distance from the equinoctial.

[14] **The planet's altitude** is an arc along the altitude circle between the tip of the previously mentioned line and the horizon. If the altitude circle coincides with the meridian circle, then this arc is the maximum altitude of the planet.

[15] **Parallax** [lit., divergence of sight] is the arc along the altitude circle between the positions of two lines passing through a planet's center and terminating at the zodiacal orb, one [line] extending from the World center and the other from the perspective of sight, I mean [from] the Earth's surface. This can be found below the Sun's orb, it being small for the Sun's orb, and it is not found beyond it, since the Earth does not have a perceptible ratio with respect to what is beyond it.

[١١] **عرض البلد**[٥٧٥] هو[٥٧٦] قوس من دائرة نصف النهار ما بين معدّل النهار وسمت الرأس وهي مساوية لما بين الأفق [٥٧٧]والقطب من دائرة نصف النهار وهو[٥٧٨] ارتفاع القطب[٥٧٩]، أعني أقرب قطبي العالم إلى ذلك البلد.

[١٢] **الميل**[٥٨٠] قوس من دائرة الميل بين معدّل النهار ودائرة البروج، وهو **الميل الأوّل**. والميل[٥٨١] إذا أطلق يراد به الميل الأوّل. **والميل**[٥٨٢] **الثاني** قوس من دائرة العرض[٥٨٣] بينهما، أعني بين معدّل النهار ودائرة البروج[٥٨٤]. **غاية الميل**[٥٨٥]، ويقال لها **الميل الكلّي**[٥٨٦] **والميل الأعظم**، قوس[٥٨٧] بينهما من دائرة المارّة[٥٨٨] بالأقطاب الأربعة وهي تدخل تحت حدّ[٥٨٩] الميل الأوّل والثاني. [٥٩٠]وهي نهاية ميل دائرة البروج عن معدّل [٥٩١]النهار ومقدارها كجـ له[٥٩٢].

[١٣] **عرض الكوكب**[٥٩٣] قوس من دائرة العرض ما بين دائرة البروج وبين رأس الخطّ [٥٩٤]الخارج من مركز العالم المارّ بمركز الكوكب[٥٩٥] المنتهي إلى فلك البروج. فإن كانت[٥٩٦] القوس من[٥٩٧] دائرة الميل[٥٩٨] بين معدّل النهار وبين [٥٩٩]رأس الخطّ المذكور فهو بعد الكوكب عن معدّل النهار.

[١٤] **ارتفاع الكوكب**[٦٠٠] قوس من دائرة الارتفاع ما بين رأس الخطّ المذكور آنفاً وبين الأفق. فإن انطبقت دائرة الارتفاع على[٦٠١] دائرة نصف النهار، فتلك القوس[٦٠٢] هي غاية ارتفاع الكوكب.

[١٥] **اختلاف المنظر**[٦٠٣] قوس من دائرة الارتفاع ما بين موقعي [٦٠٤]الخطّين المارّين بمركز الكوكب المنتهيين إلى فلك البروج، الخارج أحدهما من مركز العالم والآخر[٦٠٥] [٦٠٦]من منظر الأبصار، أعني[٦٠٧] سطح الأرض[٦٠٨] ويوجد هذا فيما[٦٠٩] تحت فلك الشمس وهو قليل في فلك الشمس، ولا يوجد فيما وراءه إذ ليس للأرض إلى ما وراءه نسبة محسوسة[٦١٠].

[16] **The ortive amplitude** is an arc along the horizon circle between the planet's circuit and the rising place of the equinox. Since circuits are parallel to the equinoctial, the ortive amplitude for each planet is the same as its occasive amplitude. The ortive and occasive amplitudes increase with the increase of local latitude.

[17] The **azimuth** and its complement have come before.[4]

[18] The **ascendant azimuth** is an arc along the horizon between the zodiacal orb and an altitude circle.

[19] The ***qibla* azimuth** for a locality is an arc along the horizon between the meridian circle of the locality and a circle passing through the zenith for its residents and through the zenith for the residents of Mecca.

[20] The **arc of daylight** is an arc along the Sun's circuit above the Earth between its setting and rising points. The arc that is between them below the Earth along this circle is the **arc of night**. The **planet's arc of daylight** is an arc along its circuit above the Earth between its rising and setting points. The arc along it that is between them below the Earth is **the [planet's] arc of night**.

[21] **The turning of the orb** is an arc along the circle of the Sun's circuit between its [zodiacal] part and the eastern horizon during daylight; and [an arc] between its facing part and the eastern horizon along the circle of the circuit of its facing part during the night.

[22] The measure of each one of these six arcs[5] is similar to its [corresponding arc] along the equinoctial.

---

[4] This was discussed in I.3: On the Circles (see zenith and azimuth).

[5] The six arcs are: (1) the arc of daylight; (2) the arc of night; (3) the planet's arc of daylight; (4) the planet's arc of night; (5) the turning of the orb [daylight]; and (6) the turning of the orb [night].

[١٦] **سعة المشرق**[٦١١] قوس من دائرة[٦١٢] الأفق ما بين مدار الكوكب ومطلع الاعتدال. ولمّا كانت المدارات موازية لمعدّل النهار كانت سعة مشرق [٦١٣] كلّ كوكب كسعة مغربه. وسعة المشرق[٦١٤] والمغرب تزيد بزيادة عرض البلد.

[١٧] **السمت** وتمامه قد[٦١٥] سلفا[٦١٦].

[١٨] **السمت**[٦١٧] **من الطالع**[٦١٨] قوس من[٦١٩] الأفق ما بين فلك البروج ودائرة الارتفاع.

[١٩] **سمت القبلة** للبلد[٦٢٠] قوس من الأفق ما بين دائرة نصف[٦٢١] نهار البلد[٦٢٢] والدائرة المارّة بسمت رؤوس[٦٢٣] أهله ورؤوس[٦٢٤] أهل مكّة[٦٢٥].

[٢٠] **قوس النهار**[٦٢٦] قوس من دائرة مدار الشمس فوق الأرض ما بين نقطتي مغربها [٦٢٧]ومشرقها[٦٢٨]. والقوس التي بينهما تحت الأرض من هذه الدائرة هي **قوس الليل**. **قوس نهار الكوكب**[٦٢٩] قوس من دائرة مداره بين نقطتي مشرقه[٦٣٠] ومغربه فوق الأرض[٦٣١]. والقوس التي[٦٣٢] بينهما منها[٦٣٣] تحت الأرض[٦٣٤] **قوس ليله**.

[٢١] **الدائر من الفلك**[٦٣٥] [٦٣٦]قوس من دائرة مدار الشمس ما بين جزئها وأفق المشرق بالنهار[٦٣٧] وما بين نظير جزئها[٦٣٨] وأفق المشرق بالليل من دائرة مدار نظير[٦٣٩] جزئها.

[٢٢] ومقدار كلّ واحدة[٦٤٠] من هذه القسّي الستّ شبيهتها من معدّل النهار.

Chapter 5 of Part I

On What Occurs to the Planets in Their Motions

[1] **Among what occurs to the planets is longitudinal anomaly:**

[2] **The Sun** has a single anomaly: since it revolves on the circumference of a circle whose center is eccentric from the World center, more than half of it being in one of the halves of the zodiacal orb, namely the half that contains its apogee, and less than half of it in the other half of the zodiacal orb, namely the half of the perigee, and [since the Sun] will only traverse each half of the zodiacal orb by traversing what is on its own circle, it follows that the period in which it traverses one of the halves of the zodiac will differ from the period in which it traverses the other half. Thus its motion is seen in one-half of the zodiac, this being the apogee half, to be slower than its [motion] in the perigee half, since the period in which it traverses the former is longer than the period in which it traverses the perigee half. Its motion on its eccentric orb, this being its mean, does not vary, so therefore one needs to add the equation to or subtract it from its mean in order to ascertain its position on the zodiacal orb.[6]

[3] As for **the remaining planets**, they have numerous anomalies in longitude. One of them, called the **first anomaly**, is what occurs to them on account of their motion along the epicycle circumference. When they [i.e., the planets] are at the epicycle apex or its perigee, the two lines extending from the World center—one of them passing through the center of the epicycle and the other through the center of the planet—will coincide one on the other, and so there will be no difference between the planet's mean and its true position, according to what has come before. As for when [the planets] depart from the apex or perigee, the location of the two aforementioned lines on the zodiacal orb will be different, so there will result a difference between the mean and the true position. The maximum of this anomaly [lit., difference] is where the maximum equation is in the epicycles, and you have learned about this in the chapter on the sectors.[7] This anomaly will obviously be in the amount of the radius of the epicycle. The radii of the epicycles at their mean distances are:

[6] The equation is defined and discussed in I.4: On the Arcs. See also Commentary, I.5[2], Fig. C2.

[7] See I.4.

[641]الباب الخامس من المقالة الأولى
فيما يعرض للكواكب في حركاتها

[١] ممّا يعرض للكواكب الاختلاف في الطول:

[٢] **للشمس** اختلاف واحد، وهو [642]أنّها لمّا كانت تدور على محيط دائرة مركزها خارج عن[643] مركز[644] العالم، كان[645] [646]في أحد نصفي فلك البروج أكثر[647] من نصفها، وهو النصف الذي فيه أوجها، وفي النصف[648] الآخر من فلك البروج أقلّ من نصفها، وهو نصف الحضيض[649]، وكانت[650] لا تقطع كل نصف من فلك البروج إلّا بقطعها ما فيه من دائرتها، لزم أن يخالف[651] زمان قطعها أحد نصفي البروج زمان قطعها النصف الثاني[652]. فترى[653] حركتها[654] في أحد نصفي البروج، وذلك نصف الأوج[655]، أبطأ منها في نصف الحضيض لكون زمان قطعها[656] إيّاه أطول من زمان قطعها نصف الحضيض. وحركتها[657] في فلكها الخارج المركز، وهي وسطها، لا تختلف فلذلك يُحتاج[658] إلى زيادة التعديل أو[659] [660]نقصانه على [661]وسطها[662] أو منه[663] لِيتحقّق[664] موضعها من فلك البروج.

[٣] **وأمّا سائر الكواكب**، فلها عدّة من الاختلافات[665] في الطول[666]. أحدها[667] ويُسمّى **الاختلاف[668] الأوّل** [669]ما يقع لها[670] من جهة حركتها على محيط التدوير. وهو أنّها إذا كانت على ذروة التدوير أو حضيضه، كان الخطّان الخارجان من مركز العالم المارّ أحدهما بمركز التدوير والآخر بمركز الكوكب[671] انطبق أحدهما على الآخر فلم يكن اختلاف بين وسط الكوكب[672] وتقويمه[673] كما سلف. وأمّا إذا[674] زايلت الذروة[675] أو الحضيض اختلف موقع الخطّين المذكورين من فلك البروج فحصل اختلاف[676] بين[677] الوسط والتقويم. وغاية هذا الاختلاف[678] حيث تكون[679] غاية التعديل في التداوير[680]، وقد عرفته في فصل النطاقات. ويكون هذا الاختلاف لا محالة بقدر نصف[681] قطر التدوير. وأنصاف[682] أقطار التداوير[683] في أبعادها الوسطى:

for Saturn:	**6;30**
for Jupiter:	**11;30**
for Mars:	**39;30**
for Venus:	**45;0**
for Mercury:	**25;0**
for the Moon:	**6;20**.

[4] **The second anomaly** for the planets is what occurs to them on account of the nearness of the epicycle center to the Earth, and its farness from it on account of the deferent being an eccentric. So the epicycle radius is seen to be greater when it is closer, and its anomaly is greater; and when it is farther away, it is the opposite.

[5] The **third anomaly** is when the centers of the epicycles are at the apogee or perigee, their diameters thereupon coinciding with the line passing through the centers of the World, the deferent, and the epicycle; [but the diameters] do not remain coincident with [this line] when [the centers] depart from the apogee or perigee. Neither do they remain directed toward the deferent center nor the World center, but rather are directed toward another point on that line, which is called the **alignment point** for the Moon and the center of the dirigent line or center of the equant orb for the vacillating [planets]. You will come to know the meaning of this [later] in this chapter.

[6] As for the upper planets and Venus, [the epicycle diameters] are directed toward a point on the side of the apogee whose distance from the deferent center is equal to the distance of the deferent center from the World center, in other words the deferent center is between [that point] and the World center.

[7] As for Mercury, [the epicycle diameter] is directed toward a point midway between the World center and the dirigent center. I will explain this further to you [later] in this chapter.

لزحل	و ل[٦٨٤]
للمشتري	يا ل[٦٨٥]
[٦٨٦]للمرّيخ	لط ل[٦٨٧]
للزهرة	مه ٠[٦٨٨]
[٦٨٩]لعطارد	كه ٠[٦٩٠]
للقمر	و ك[٦٩١].

[٤] **اختلاف**[٦٩٢] [٦٩٣]**ثانٍ**[٦٩٤] للكواكب[٦٩٥]، وهو ما يقع لها بسبب قرب[٦٩٦] مركز التدوير من [٦٩٧]الأرض، وبعده عنها بسبب كون الحامل[٦٩٨] خارج المركز. فيرى[٦٩٩] نصف قطر التدوير حال قربه أعظم، واختلافه أعظم[٧٠٠]؛ [٧٠١]وحال بعده بالخلاف.

[٥] **اختلاف ثالث**[٧٠٢] وهو أنّ مراكز التداوير إذا كانت على الأوج أو الحضيض، فأقطارها المنطبقة حينئذ[٧٠٣] على الخطّ[٧٠٤] المارّ بمراكز[٧٠٥] العالم والحامل[٧٠٦] والتدوير لا تبقى منطبقة عليه إذا[٧٠٧] زايلت الأوج أو الحضيض[٧٠٨] ولا تبقى[٧٠٩] على صوب مركز الحامل ولا[٧١٠] مركز العالم، بل[٧١١] على صوب[٧١٢] نقطة أخرى من ذلك الخطّ تسمّى في القمر **نقطة المحاذاة** وفي المتحيّرة[٧١٣] مركز الخطّ المدير أو[٧١٤] مركز الفلك المعدّل للمسير، وستعرف معنى هذا في هذا الفصل[٧١٥].

[٦] أمّا في العلوية والزهرة، فعلى صوب نقطة ممّا يلي الأوج بُعدها عن مركز الحامل[٧١٦] كبُعد مركز الحامل[٧١٧] عن مركز [٧١٨]العالم، [٧١٩]أعني أنّ مركز الحامل فيما بينها[٧٢٠] وبين مركز العالم.

[٧] وأمّا في عطارد، فعلى صوب نقطة[٧٢١] في منتصف ما بين مركز العالم وبين[٧٢٢] مركز المدير. وأُزيدُكَ لهذا[٧٢٣] بياناً في هذا الفصل[٧٢٤].

[8] As for the Moon, [the epicycle diameter] is directed toward a point on the side of the nearest distance whose distance from the World center toward the perigee is equal to the distance of the deferent center from it, in other words from the World center toward the apogee. So when the deferent and its center rotate about the World center by the rotation of the inclined [orb], this point and the deferent center will revolve in alignment upon the circumference of a single circle, i.e., they are at the endpoints of one of its diameters.

[9] So the aforementioned diameters of the epicycles are directed towards these aforementioned points, always being in alignment with them, however they rotate; in other words, if lines are extended from these points to the epicycle centers, every line from them will coincide with the aforementioned epicycle diameter, not separating from it, however it rotates. For the vacillating [planets], this line is called the **dirigent line**; the imagined circle that is traced through the rotation of this line by the epicycle center is called the **equant orb** [lit., the orb that equalizes the movement], since the movement of the vacillating [planets] is equalized in relation to it, i.e., [the line] describes equal arcs on its circumference in equal times. The place this line falls at the upper part of the epicycle is the **mean apex**, and the place the line extending from the World center passing through the epicycle center falls is the **apparent apex**.

[10] **The distances of these points and their centers, one from the other:**

[11] The distance of the eccentric center from the World center is: for the Sun **2;29,30**; for the Moon **10;19**, and it is equal to the distance of the alignment point from [the World center] in the other direction.

[12] For the vacillating [planets], with the exception of Mercury, [the distance of the eccentric center from the World center] is equal to half the distance of the equant center to it, and that, namely the distance of the equant center from the World center, is:

for Saturn:	**6;50**
for Jupiter:	**5;30**

[٨] وأمّا في القمر، فعلى صوب نقطة ممّا يلي البعد الأقرب بعدها عن مركز العالم ممّا يلي الحضيض كبعد مركز الحامل عنه، أعني عن[٧٢٥] مركز العالم ممّا يلي[٧٢٦] الأوج. فإذا دار[٧٢٧] [٧٢٨]الحامل[٧٢٩] ومركزه حول مركز العالم[٧٣٠] بدوران المائل دارت هذه النقطة ومركز الحامل على محيط دائرة واحدة متقاطرين، أي يكونان على طرفي قطر من أقطارها.

[٩] فهذه النقط[٧٣١] المذكورة[٧٣٢] تكون[٧٣٣] الأقطار المذكورة[٧٣٤] للتداوير على صوبها مسامتةً لها دائمة كيف ما دارت، أعني لو أخرج من هذه النقط[٧٣٥] خطوط إلى مراكز التداوير يكون كلّ خطّ منها منطبقاً[٧٣٦] [٧٣٧]على القطر المذكور للتدوير، لا ينفكّ عنه[٧٣٨] كيف ما[٧٣٩] دار. وهذا الخطّ في المتحيّرة يسمّى **الخطّ المدير**، والدائرة المتوهّمة التي ترتسم بدوران[٧٤٠] هذا الخطّ مع مركز [٧٤١]التدوير تسمّى[٧٤٢] **الفلك المعدّل للمسير** إذ يعتدل[٧٤٣] مسير المتحيّرة بالنسبة إليها، أي[٧٤٤] يقطع من محيطها قسيّاً متساوية في أزمنة متساوية. وموقع هذا الخطّ من أعلى التدوير هو[٧٤٥] **الذروة الوسطى**[٧٤٦]، وموقع الخطّ الخارج من [٧٤٧]مركز العالم المارّ بمركز التدوير هو[٧٤٨] **الذروة**[٧٤٩] **المرئيّة**[٧٥٠].

[١٠] وأبعاد[٧٥١] هذه النقط[٧٥٢] والمراكز بعضها عن بعض:

[١١] **أمّا بعد مركز الخارج المركز**[٧٥٣] **عن مركز العالم:** للشمس ب كط ل[٧٥٤] للقمر[٧٥٥] ي يط[٧٥٦]، وهو[٧٥٧] مثل بعد نقطة المحاذاة عنه عن[٧٥٨] الجهة الأخرى.

[١٢] للمتحيّرة[٧٥٩] ما خلا[٧٦٠] [٧٦١]عطارد مثل نصف بعد مركز المعدّل للمسير عنه وذلك[٧٦٢]، أعني بعد مركز المعدّل للمسير[٧٦٣] عن مركز العالم:

[٧٦٤]لزحل	و ن [٧٦٥]
للمشتري[٧٦٦]	ه ل [٧٦٧]

for Mars:	**12;0**
for Venus:	**2;5.**

[13] For Mercury, the center of its equant orb is at the midpoint between the center of its dirigent and the World center. The distance of its deferent center from its dirigent center is equal to half the distance of its dirigent center from the World center. As a result, when the dirigent line on the side of the nearest distance coincides with line passing through the centers, the deferent center point falls on the equant center. When it coincides with it on the side of the farthest distance, the centers will be arranged along the line passing through them: first the World center, then the equant center, then the dirigent center, then the deferent center. The distances between them are equal, each of their distances being **3;10**, so what is between the centers of the World and the deferent is **9;30.**

[14] **Among what occurs to the planets is latitudinal anomaly:**

[15] **The Sun** has no latitude, since in its motion it adheres to the plane of the zodiacal orb.

[16] **The remaining planets** incline from the zodiacal orb to the north and south due to inclination of the inclined orb from it, and it is called the **eccentric latitude**. Its maximum is:

for Saturn:	**2;30**
for Jupiter:	**1;30**
for Mars:	**1;0**
for Venus:	**0;10**
for Mercury:	**0;45**
for the Moon:	**5;0.**

للمرّيخ[٧٦٨]	يب [٧٦٩]
للزهرة[٧٧٠]	ب ه [٧٧١].

[١٣] **وأمّا عطارد**، فمركز فلكه[٧٧٢] المعدّل للمسير على منتصف ما بين مركز مديره وبين مركز العالم. وبعد مركز حامله[٧٧٣] عن مركز مديره[٧٧٤] مثل نصف بعد مركز مديره عن مركز العالم. حتّى[٧٧٥] إذا[٧٧٦] انطبق الخطّ المدير ممّا[٧٧٧] يلي البعد الأقرب [٧٧٨]على الخطّ المارّ بالمراكز، وقعت نقطة مركز الحامل على مركز المعدّل[٧٧٩] للمسير. وإذا انطبق عليه ممّا يلي البعد الأبعد انتظمت المراكز على الخطّ المارّ بها: أوّلها مركز العالم ثمّ مركز المعدّل للمسير ثمّ مركز المدير ثمّ مركز الحامل. والأبعاد ما بينها متساوية كلّ[٧٨٠] بعد منها جـ ي[٧٨١]، فيكون ما بين مركزي العالم والحامل ط ل[٧٨٢].

[١٤] وممّا يعرض للكواكب الاختلاف[٧٨٣] في العرض:

[١٥] **الشمس**[٧٨٤] لا عرض لها لأنّها لازمة في حركتها لسطح فلك[٧٨٥] البروج.

[١٦] **وسائر الكواكب** تميل عن فلك البروج إلى الشمال والجنوب لميل الفلك[٧٨٦] المائل عنه، ويسمّى[٧٨٧] **عرض الخارج**[٧٨٨] [٧٨٩]**المركز** وغايته:

لزحل	ب ل [٧٩٠]
للمشتري	ا ل [٧٩١]
للمرّيخ	ا ٠ [٧٩٢]
للزهرة	٠ ي [٧٩٣]
[٧٩٤]لعطارد	٠ مه [٧٩٥]
للقمر	ه ٠ [٧٩٦].

The Moon has only this latitude because its inclined, deferent, and epicycle orbs are in a single plane; by these orbs we mean circles, which you have already learned about.

[17] The vacillating [planets] have another anomaly, namely the inclination of its epicycle apex and its perigee from the inclined orb. It is called the **epicycle latitude** and its maximum is:

for Saturn:	**0;32**
for Jupiter:	**0;38**
for Mars:	**6;16**
for Venus:	**1;2**
for Mercury:	**1;45**.

[18] The two lower planets have another proper anomaly, and it is the inclination from the inclined orb of the diameter that passes through the two mean distances of the epicycle orb. It is called the **latitude of the slope** [*wirāb*], **the slant** [*inḥirāf*], and the **twist** [*iltiwāʾ*]. Its maximum for both [i.e., Mercury and Venus] is **2;30**.

[19] As for the inclination of the inclined orb from the zodiacal orb, it is fixed for the upper planets and the Moon and does not change. It is not fixed for Venus and Mercury; rather, whenever the epicycle center reaches one of the two nodal points, the inclined [orb] will coincide with the zodiacal orb. Then when it crosses it, half the inclined [orb], i.e., the half upon which is the epicycle center, will begin to incline for Venus to the north and for Mercury to the south; its other half is the opposite. The inclination then continues to increase until the [epicycle] center reaches midway between the two [nodal] points, and then the inclination begins to decrease until the inclined [orb] coincides again with the zodiacal orb when the center reaches the other [nodal] point. Then when it crosses it, it returns to its original situation. So it follows that the epicycle center is always north of the zodiacal orb for Venus, and south of it for Mercury.

وليس للقمر [٧٩٧]عرض غير هذا لأنّ أفلاكه[٧٩٨] المائل والحامل والتدوير في سطح واحد ونعني[٧٩٩] بهذه الأفلاك الدوائر وقد[٨٠٠] عرفتها.

[١٧] وللمتحيّرة اختلاف آخر وهو ميل ذروة التدوير وحضيضه عن الفلك المائل ويسمّى[٨٠١] **عرض**[٨٠٢] [٨٠٣]**التدوير**[٨٠٤] وغايته:

لزحل	‾٠ ‾لب [٨٠٥]
للمشتري	‾٠ ‾لح [٨٠٦]
للمرّيخ	‾و ‾يو [٨٠٧]
للزهرة	‾ا ‾ب [٨٠٨]
لعطارد	‾ا ‾مه [٨٠٩].

[١٨] وللسفليين خاصّة اختلاف آخر وهو ميل القطر المارّ بالبعدين الأوسطين لفلك التدوير عن الفلك المائل ويسمّى[٨١٠] **عرض الوراب** و**الانحراف** و**الالتواء** وغايته في كلّ واحد منهما ‾ب ‾ل[٨١١].

[١٩] أمّا ميل الفلك المائل عن فلك البروج، فثابت في الكواكب العلوية والقمر لا يتغيّر. [٨١٢]وغير ثابت في الزهرة وعطارد؛ بل كلّما بلغ مركز التدوير إحدى نقطتي الجوزهريْن انطبق المائل على فلك البروج. فإذا جاوزها[٨١٣] [٨١٤]ابتدأ[٨١٥] نصف المائل، أعني نصفه الذي عليه[٨١٦] مركز التدوير، في الميل للزهرة إلى الشمال ولعطارد إلى الجنوب، ونصفه الآخر[٨١٧] بالخلاف. ثمّ لا يزال يزداد الميل حتّى ينتهي المركز إلى منتصف ما بين النقطتين ثمّ يأخذ الميل في النقصان حتّى ينطبق المائل أيضاً على فلك[٨١٨] البروج عند بلوغ المركز [٨١٩]النقطة الأخرى. فإذا[٨٢٠] جاوزها عادت[٨٢١] إلى[٨٢٢] الحالة[٨٢٣] الأولى ويلزم[٨٢٤] أن يكون [٨٢٥]مركز التدوير أبداً للزهرة شماليّاً عن فلك البروج ولعطارد جنوبيّاً عنه[٨٢٦].

[20] As for the inclination of the epicycle diameter, I mean the diameter passing through its apex and its perigee, it also is not fixed; rather, for the upper planets it will coincide with the zodiacal orb when the center, i.e., the epicycle center, is at either the point of the head or the tail. Then when the [epicycle] center crosses the head, the apex begins inclining to the south, and it will continue to increase until it attains its maximum when the center reaches midway between the two [nodal] points. Then it begins to decrease until it coincides a second time with the zodiacal orb when the [epicycle] center reaches the tail. Then when it crosses it, the apex begins inclining to the north, its increase, maximum, and decrease as described. It follows that the inclination of the apex will always be toward the zodiacal orb, and the inclination of the perigee away from it.

[21] For the lower planets, [the epicycle diameter] coincides with the inclined orb when the epicycle center reaches midway between the two [nodal] points, i.e., the two points of the head and the tail; and this [occurs] when the inclination of the inclined orb from the zodiacal orb is at maximum, either at the apogee or at the perigee. Then at the apogee, the epicycle apex begins inclining: for Venus toward the north and for Mercury toward the south; at perigee it is the opposite for each of them. It attains its maximum at the two [nodal] points; and its increase, decrease, and coincidence [occurs] according to the aforementioned description.

[22] As for the slant, it starts when the epicycle center reaches either the point of the head or the tail, and its maximum is when it is midway between the two of them. If the midway [point] is the apogee, the eastern endpoint of the diameter passing through the two mean distances will be at its maximum inclination for Venus to the north and for Mercury to the south, and the western [endpoint] for Venus to the south and for Mercury to the north. And if the midway [point] is the perigee, it is the opposite for both of them.

[23] It has become evident from all of this that the period of rotation for the deferent orb and for the two aforementioned epicycle diameters[8] are equal to one another, and the four quarters of their rotation are equal to one another.

[8] I.e., one passing through the apex and perigee, and one through the two mean distances.

[٢٠] وأمّا ميل قطر التدوير، أعني القطر المارّ بذروته وحضيضه، فغير ثابت أيضاً؛ بل يصير منطبقاً على فلك البروج[٨٢٧] في العلويّة عند كون المركز، أعني مركز التدوير، في إحدى نقطتي الرأس والذنب. ثمّ إذا جاوز المركز الرأس، أخذت الذروة في الميل إلى الجنوب[٨٢٨]، ولا يزال يزداد حتّى يبلغ[٨٢٩] [٨٣٠]غايته عند بلوغ المركز منتصف ما بين النقطتين. ثمّ يأخذ في الانتقاص[٨٣١] إلى أن ينطبق ثانياً[٨٣٢] على فلك البروج[٨٣٣] عند بلوغ المركز[٨٣٤] الذنب. فإذا جاوزه[٨٣٥] أخذت الذروة في الميل [٨٣٦]إلى الشمال[٨٣٧] وازدياده ومُنتهاه وانتقاصه[٨٣٨] على الرسم ويلزم أن يكون ميل الذروة أبداً إلى فلك البروج وميل الحضيض عنه.

[٢١] وفي السفليين[٨٣٩] ينطبق على فلك[٨٤٠] المائل[٨٤١] عند بلوغ مركز التدوير منتصف ما بين النقطتين، أعني نقطتي الرأس والذنب وذلك عند غاية ميل الفلك المائل عن فلك البروج إمّا عند الأوج وإمّا عند الحضيض. [٨٤٢]فعند الأوج تبتدئ[٨٤٣] ذروة التدوير في الميل للزهرة إلى الشمال ولعطارد[٨٤٤] إلى[٨٤٥] الجنوب، [٨٤٦]وعند الحضيض بالخلاف فيهما. ويبلغ[٨٤٧] غايته[٨٤٨] عند النقطتين وازدياده[٨٤٩] وانتقاصه[٨٥٠] والانطباق[٨٥١] على الرسم المذكور.

[٢٢] وأمّا الانحراف[٨٥٢]، فابتداؤه[٨٥٣] عند بلوغ مركز التدوير إحدى نقطتي الرأس والذنب، وغايته عند منتصف [٨٥٤]ما بينهما. فإن كان المنتصف هو الأوج، كان الطرف [٨٥٥]الشرقي من القطر المارّ بالبعدين الأوسطين في غاية ميله[٨٥٦] في الزهرة إلى الشمال وفي[٨٥٧] عطارد إلى الجنوب والغربي في الزهرة[٨٥٨] إلى الجنوب[٨٥٩] وفي عطارد إلى الشمال. وإن[٨٦٠] كان المنتصف هو الحضيض فعلى الخلاف فيهما.

[٢٣] وقد ظهر من هذا كلّه أنّ[٨٦١] مدّة[٨٦٢] الدور للفلك الحامل ولقطري[٨٦٣] التدوير المذكورين متساوية، وأزمان أرباع دوراتها[٨٦٤] متساوية.

[24] **Let us mention here the apogees and the nodes:**

[25] As for the apogees and the nodes that move with the motion of the orb of the fixed stars: Saturn's apogee is **50 degrees** beyond the midpoint between its two nodal points, i.e., from the maximum inclination of the inclined orb from the zodiacal orb, in the sequence of the signs; and Jupiter's apogee is **20 degrees** in advance of the midpoint in the counter-sequence [of the signs]. "In advance" means the planet reaches it in advance of reaching the midpoint; as opposed to this is the meaning of "beyond." The apogee of the remaining planets is at the midpoint [between the nodes].

[26] As for the position of the apogees, they are for the beginning of the year **1517 of Dhū al-Qarnayn** [the two-horned, i.e., the era of Alexander the Great]:[9]

for the Sun:	Gemini **27;10,33**
for Saturn:	Sagittarius **9;23,33**
for Jupiter:	Virgo **19;23,33**
for Mars:	Leo **11;53,46**
for Venus:	Gemini **27;10,33**
for Mercury:	Libra **26;23,33**.

[27] As for the positions of the nodes for that date, the head node is:

for Saturn:	Cancer **19;23,33**
for Jupiter:	Cancer **9;23,33**
for Mars:	Taurus **11;53,46**
for Venus:	Pisces **27;10,33**
for Mercury:	Capricorn **26;23,33**.

[9] The date 1517 and Jaghmīnī's parameters for the apogee and nodes are important in establishing that he was alive in 1205 CE (=602 H). Some copyists and commentators misdate the year as 1317 due to reading ش (300) for ث (500). Rudloff and Hochheim omitted the year altogether in their German translation ("Die Astronomie des Maḥmûd ibn Muḥammed ibn 'Omar al-Ǵagmînî," 253). See Commentary to I.5[26] for more detail.

[٢٤] ولنذكر ههنا[٨٦٥] الأوجات والجوزهرات:

[٢٥] أمّا[٨٦٦] الأوجات والجوزهرات المتحرّكة بحركة فلك الثوابت: فأوج زحل متأخّر عن منتصف ما بين [٨٦٧]نقطتي جوزهريه، أعني عن غاية ميل المائل عن فلك البروج على التوالي بخمسين جزءاً، وأوج المشتري متقدّم على المنتصف لا على التوالي بعشرين جزءاً[٨٦٨]، [٨٦٩]ومعنى التقدّم أنّ بلوغ الكوكب إليه يتقدّم على[٨٧٠] بلوغه إلى المنتصف، وعلى هذا معنى التأخّر. وأوج الكواكب[٨٧١] الباقية في المنتصف.

[٢٦] [٨٧٢]أمّا[٨٧٣] موضع[٨٧٤] الأوجات فهي لأوّل[٨٧٥] سنة غثيز[٨٧٦] لذي القرنين:

للشمس في الجوزاء	كز ي لج [٨٧٧]
لزحل[٨٧٨] [٨٧٩]في القوس	ط كج لج [٨٨٠]
للمشتري في السنبلة	يط كج لج [٨٨١]
للمرّيخ في الأسد	يا نج مو [٨٨٢]
للزهرة في الجوزاء	كز ي لج [٨٨٣]
لعطارد في الميزان	كو كج لج [٨٨٤].

[٢٧] وأمّا مواضع الجوزهرات لذلك التاريخ فرأس الجوزهري[٨٨٥]:

لزحل في السرطان	يط كج لج [٨٨٦]
للمشتري[٨٨٧] في السرطان	ط كج لج [٨٨٨]
للمرّيخ في الثور	يا نج مو [٨٨٩]
للزهرة في الحُوت[٨٩٠]	كز ي لج [٨٩١]
لعطارد في الجَدْي	كو كج لج [٨٩٢].

[28] Then for every year, one adds to their positions what the orb of the fixed stars moves in the year, and this you have already learned.

[29] **What occurs to the vacillating planets regarding retrogradation, direct motion, and stations:**

[30] When the planet is in the upper part of its epicycle, the motion of its center corresponds to the motion of the epicycle center in the sequence of the signs, so it is seen in direct motion, moving swiftly. Then when [the planet] approaches the lower part of the epicycle, it starts to incline counter-sequentially, according to what you have learned regarding the motion of the epicycle about its center.[10] However, as long as the motion of [the planet's] center is counter-[sequential] by a lesser [amount] than the motion of the epicycle center [moving] sequentially, it is seen in direct motion, but slow in speed. Then when the two [opposite motions] are equal, it is seen to be stationary. Then when the [counter-sequential] motion of the [planet's] center is greater than the motion of the epicycle center, it is seen retrograding. Then after retrogradation, [the planet] is stationary a second time and [then] moves in direct motion for the same reason [as before]. Despite this, [the planet] completes its rotation on its [epicycle] orb without variation occurring to it with respect to its orb. Its stationary position before retrogradation is called the **first station**, and its stationary position after retrogradation is called the **second station**.

[31] The motion of the Moon's center on the circumference of the epicycle orb is less than the motion of the epicycle center on the deferent circumference; on account of this, the Moon is not seen retrograding at all, rather it may be seen to be slow[er] in speed.

[32] **What occurs to [the vacillating planets] in relation to the Sun:**

[33] As for the upper [planets], the distance of their centers from the apices of their epicycles is always equal to the distance of their epicycle centers from the Sun; thus they will always be in conjunction with the Sun when they are at the apices of the epicycles. So as the Sun moves away from the epicycle center, the planet's center moves away from the epicycle apex in the amount of the [Sun's] distance [from the epicycle center], so when the Sun is in opposition to the epicycle center, the planet will have descended to the epicycle perigee. Thus their combusts will always be when they are at the epicycle apex, and their oppositions to the Sun will be when they are at the perigee. It has been said that when Mars is in conjunction with the Sun, the distance between it and the Sun is greater than the distance between it and the Sun when [Mars] is at opposition because the diameter of its epicycle is greater than the diameter of the Sun's parecliptic [orb].

[10] See I.2: On the Motions of the Orbs.

[٢٨] ثمّ يُزاد[٨٩٣] على مواضعها لكلّ سنة ما يتحرّك فلك الثوابت في السنة، وقد عرفت ذلك.

[٢٩] وممّا يعرض للمتحيّرة الرجوع[٨٩٤] والاستقامة والإقامة:

[٣٠] وذلك أنّ الكوكب[٨٩٥] إذا كان في أعلى تدويره، [٨٩٦]كانت[٨٩٧] حركة مركزه موافقة لحركة مركز التدوير على [٨٩٨]توالي البروج، فيُرى[٨٩٩] مُستقيماً سريع الحركة. فإذا قرب من أسفل التدوير، جعل يميل[٩٠٠] إلى خلاف التوالي، لما[٩٠١] تعرف[٩٠٢] من حركة التدوير على مركزه. لكنّه ما دام حركة مركزه إلى الخلاف أقلّ من حركة مركز التدوير إلى التوالي، يُرى مستقيماً لكن بطء السير. فإذا تساويا[٩٠٣]، يُرى مقيماً. فإذا زادت حركة مركزه على حركة مركز التدوير، يُرى راجعاً. [٩٠٤]ثمّ [٩٠٥]يقيم[٩٠٦] بعد الرجعة ثانياً ويستقيم لهذا[٩٠٧] المعنى بعينه. مع أنّه يُتمّ دورته[٩٠٨] في فلكه من غير اختلاف يقع له بالنسبة إلى فلكه. وإقامته قبل الرجعة تسمّى[٩٠٩] **المقام الأوّل**، وإقامته بعد الرجعة تسمّى[٩١٠] **المقام الثاني**.

[٣١] وحركة مركز القمر على محيط فلك التدوير أقلّ من حركة مركز التدوير على محيط الحامل، فلهذا لا يُرى القمر البتّة راجعاً بل قد يُرى بطء السير.

[٣٢] وممّا يعرض لها بالقياس إلى الشمس:

[٣٣] أمّا في العلوية، فإنّ بعد[٩١١] مراكزها عن ذُرى تداويرها [٩١٢]أبداً[٩١٣] كبعد مراكز تداويرها عن [٩١٤]الشمس، فتقارن الشمس[٩١٥] أبداً وهي في ذُرى التداوير. فكما تبعد[٩١٦] الشمس عن مركز التدوير يبعد[٩١٧] بمقدار[٩١٨] بعدها مركز الكوكب[٩١٩] عن ذروة[٩٢٠] التدوير[٩٢١] حتّى إذا قابلت الشمس مركز التدوير[٩٢٢]، كان الكوكب قد نزل إلى حضيض التدوير[٩٢٣]. فتكون احتراقاتها[٩٢٤] أبداً وهي[٩٢٥] في[٩٢٦] ذروة التدوير [٩٢٧]ومقابلاتها للشمس وهي[٩٢٨] في الحضيض. ويقال إنّ المرّيخ إذا قارن الشمس، كان البعد بينه وبين الشمس أعظم من البعد بينه وبين الشمس إذا قابلها لأنّ قطر[٩٢٩] تدويره أعظم من قطر ممثل[٩٣٠] الشمس.

[34] As for the two lower [planets], their two epicycle centers are always aligned with the Sun's center, so both distance themselves from [the Sun] only by the amount of the epicycle radius, i.e., by the amount of the first anomaly, as you have already learned. It follows that both are in conjunction [with the Sun] halfway through direct motion, that being at the epicycle apex, and halfway through retrogradation, that being at perigee.[11] Therefore, their mean is the same as the Sun's mean.

[35] What occurs to the Moon in relation to the Sun: The new Moon [*muḥāq*], waxing, full Moon, waning, its eclipsing of the Sun, and lunar eclipses

[36] The reason for [all] this is that the Moon's body in and of itself is opaque and dark, only becoming illuminated by the light of the Sun, like a mirror. Thus its half facing toward the Sun will always be illuminated, and the other half dark. Then at conjunction the Moon will be between us and the Sun, its dark half is facing us so we will not see any of its light, which is **the new Moon**. Then when it moves away from the Sun an amount of nearly **12 degrees**, more or less according to different locations in the inhabited zone, its luminous half will incline toward us so that we see an edge of it, which is **the crescent**. Then as its distance from the Sun increases, the inclination of the luminous [part] toward us increases. So its light increases until when it is in opposition, we come to be between the two and that which faces the Sun faces us, which is **the full Moon**. Then when it departs from being in opposition, some of its dark half inclines toward us; the darkness then begins to increase and the luminous [part] decrease until it is effaced. For this reason, when the Moon is in conjunction on the Sun's path, this being at the head or tail or close to them, it is interposed between the Sun and us, thus concealing its light from us, which is **a solar eclipse**. This blackness that appears in the Sun is the color of the Moon's body; for this reason, the blackness of the Sun begins from the western side because the Moon catches up with it from the west. Then when [the Moon] proceeds to transit [the Sun], the reappearance will also begin from the western side due to the explanation we have mentioned. Similarly, when the Moon is on the path of the Sun in opposition, the Earth will interpose between them and its shadow falls on the Moon. So the Sun's light will not reach it,

[11] Istanbul, Süleymaniye Library, Laleli MS 2141, f. 74a adds in the margin: "I.e., since their two epicycle centers are always aligned with the Sun's center."

[٣٤] وأمّا السفليّان، فمركزا[٩٣١] تدويرهما أبداً يتسامتان[٩٣٢] بمركز[٩٣٣] الشمس، [٩٣٤]فلا يبعدان[٩٣٥] عنها إلاّ بمقدار نصف قطر التدوير، أعني بمقدار الاختلاف الأوّل، كما عرفت. ويلزم أن يقارناها[٩٣٦] في نصف الاستقامة، وذلك عند ذروة التدوير وفي نصف الرجوع، وذلك عند الحضيض[٩٣٧]. ولذلك[٩٣٨] يكون وسطهما مثل وسط [٩٣٩]الشمس.

[٣٥] وممّا يعرض للقمر بالقياس إلى الشمس: المُحاق[٩٤٠]، والزيادة، والكمال، والنقصان، وكسفه الشمس، [٩٤١]والخسوف[٩٤٢]

[٣٦] وذلك أنّ جرم القمر في نفسه كَدِر مُظلم، إنّما[٩٤٣] يستضيء بضياء الشمس كالمرآة[٩٤٤]. فيكون نصفه المواجه[٩٤٥] للشمس أبداً مستضيئاً[٩٤٦]، والنصف[٩٤٧] الآخر مظلماً. فعند الاجتماع يكون القمر بيننا وبين الشمس، فيكون نصفه المظلم[٩٤٨] مواجهاً لنا فلا نرى[٩٤٩] شيئاً من ضوئه وهو **المحاق**. فإذا[٩٥٠] بَعُد عن الشمس مقداراً[٩٥١] قريباً من اثني عشر جزءاً أو [٩٥٢]أقلّ أو أكثر على اختلاف أوضاع المساكن، مال نصفه المُضيء إلينا فنرى[٩٥٣] طرفاً[٩٥٤] منه وهو **الهلال**. ثمّ كلّما ازداد بعده عن الشمس، ازداد[٩٥٥] ميل المضيء إلينا. فازداد ضياؤه حتّى إذا قابلها[٩٥٦]، صرنا بينهما، وصار ما يواجه الشمس يواجهنا وهو **الكمال**. فإذا انحرف عن المقابلة مال إلينا شيء من نصفه المظلم؛ ثمّ يأخذ الظلام في الزيادة [٩٥٧]والضياء في النقصان حتّى يمحق. ولذلك [٩٥٨]إذا كان القمر عند الاجتماع على طريقة الشمس، وذلك عند الرأس أو الذنب[٩٥٩] أو بقربهما، حال بين الشمس وبيننا فيستر ضوءها [٩٦٠]عنّا، وهو **كسوف**[٩٦١] **الشمس**. وهذا السواد الذي يظهر في الشمس هو لون[٩٦٢] جرم القمر؛ ولهذا يبتدئ سواد الشمس من جهة المغرب لأنّ القمر يلحقها من المغرب. ثمّ إذا أخذ يمرّ بها[٩٦٣] يبتدئ الانجلاء أيضاً[٩٦٤] [٩٦٥]من جهة المغرب لما ذكرنا من[٩٦٦] المعنى. وإذا كان القمر كذلك على طريقة الشمس عند الاستقبال، حال بينهما الأرض ووقع ظلّها على القمر[٩٦٧]. فلم يصل إليه[٩٦٨] ضوء

and it then remains in its original darkness, which is **a lunar eclipse**. The beginning of a lunar eclipse and its reappearance will be from the eastern side because the Earth's shadow catches up with it from the western side; so [the Moon's] eastern edge will arrive first into the shadow, and then proceed to blacken first. Similarly, the transit of the [Moon's] eastern edge through the shadow will be first; then it begins to reappear from it.[12]

[37] Among what occurs to the Moon is that the Sun is always in the middle between the [Moon's] apogee and the center of its epicycle. The reason for this is that when the [Moon's] epicycle center at apogee is in conjunction with the Sun's center at a point on the zodiacal orb, say, for example, the head of Aries, and then the apogee moves away from it over a day and its night due to the motion of the inclined [orb] **11;9,7,43** and to the motion of the *jawzahar* **0;3,10,37**, then its [combined] motion in the counter-sequence of the signs becomes **11;12,18,20**. The Sun moves from it approximately a degree, and the epicycle center moves due to the motion of the deferent **24;22,53,22**. Both the motions of the Sun and the [epicycle] center are in the sequence [of the signs]; however, the inclined [orb] turns back the deferent counter-sequentially by the amount of its motion, namely, **11;12,18,20**, so there remains for the [epicycle] center sequentially approximately **13;10,35,2**, which is the mean motion of the Moon in a day and its night. Then when the solar mean is subtracted from it, and [when the solar mean is] added to the inclined [plus *jawzahar*] motion, the result after the subtraction is the distance of the [epicycle] center from the [mean] Sun, and after the addition, the distance of the apogee of the Moon from [the mean Sun], both being approximately **12;11,26,41**.[13] So the Sun is midway between the two; for this reason, the motion of the center is called the **double elongation**, because when the distance between the [epicycle] center and the [mean] Sun is doubled, it equals the distance between the center and the apogee. It follows that the [epicycle] center at its quadrature to the Sun will be at the perigee, and at opposition and conjunction at the apogee; so the center will have reached the apogee and perigee twice for every rotation.

[12] See Commentary for illustrations of the illumination of the Moon in relation to the Sun (Fig. C3), a solar eclipse (Fig. C4), and a lunar eclipse (Fig. C5).

[13] 13;10,35,2 [lunar mean] minus 0;59,8,20 [solar mean] equals 12;11,26,42, and 11;12,18,20 [inclined motion and *jawzahar* motion] plus 0;59,8,20 [solar mean] equals 12;11,26,40.

الشمس، فيبقى[٩٦٩] على ظلامه الأصلي، وهو **خسوف القمر.** [٩٧٠] وابتداء خسوف القمر[٩٧١] وانجلائه من جهة المشرق لأنّه يلحقه[٩٧٢] ظلّ الأرض من جهة المغرب؛ فيصل طرفه الشرقي اوّلاً إلى الظلّ فيأخذ في السواد اوّلاً. وكذلك[٩٧٣] يكون مُرور طرفه الشرقي بالظلّ أوّلاً؛ فيبتدئ منه[٩٧٤] الانجلاء.

[٣٧] [٩٧٥]وممّا يعرض[٩٧٦] للقمر توسّط الشمس بين أوجه ومركز تدويره أبداً. وذلك أنّ[٩٧٧] مركز تدويره[٩٧٨] إذا قارن في أوجه مركز الشمس عند نقطة من فلك البروج، ولتكن[٩٧٩] مثلاً رأس الحمل[٩٨٠]، ثمّ تحرّك عنه[٩٨١] الأوج يوماً وليله بحركة[٩٨٢] المائل [٩٨٣]يا ط ز مجـ[٩٨٤] وبحركة[٩٨٥] الجوزهر[٩٨٦] ٠ جـ ي لز[٩٨٧]، فتصير[٩٨٨] حركته إلى خلاف التوالي يا يب يح كـ[٩٨٩]. وتحرّك[٩٩٠] عنه الشمس قريباً من الدرجة، [٩٩١]وتحرك مركز التدوير بحركة الحامل كد كب نجـ كب[٩٩٢]. وكلتا حركتي الشمس والمركز إلى التوالي؛ لكنّ المائل[٩٩٣] يردّ الحامل إلى خلاف التوالي مقدار حركته، وهو يا يب يح كـ[٩٩٤]، فيبقى[٩٩٥] للمركز[٩٩٦] إلى التوالي يجـ ي له ب[٩٩٧] بالتقريب[٩٩٨]، وهو وسط القمر في اليوم بليلته. فإذا نُقص وسط الشمس [٩٩٩]منه وزيد على حركة المائل، كان الحاصل بعد النقصان بُعد المركز عن[١٠٠٠] الشمس، وبعد الزيادة بُعد أوج القمر عنها، وكلاهما [١٠٠١]يب يا كو ما[١٠٠٢] بالتقريب. فتكون[١٠٠٣] الشمس متوسطة بينها، ولذلك يقال لحركة المركز **البعد المضاعف** لأنّه إذا ضوعف البعد بين المركز والشمس، كان مثل البعد بين المركز والأوج. ويلزم أن يكون المركز عند تربيعه[١٠٠٤] للشمس[١٠٠٥] في الحضيض[١٠٠٦] وعند الاستقبال والاجتماع[١٠٠٧] في الأوج، فيكون[١٠٠٨] المركز يبلغ[١٠٠٩] الأوج والحضيض في كلّ دورة[١٠١٠] دفعتين.

[38] Similar to this is what occurs to Mercury's epicycle center because the motion of its epicycle center, due to the deferent motion, is twice the motion of its apogee that is due to the dirigent motion. However, the dirigent, in the amount of its motion, turns back the deferent; so what remains from the excess motion of the [epicycle] center sequentially is equal to the dirigent motion counter-sequentially. Thus if the two are in conjunction—I mean the [epicycle] center and the apogee that is in the dirigent—in Libra with the other apogee, the parecliptic one, [and] they then both move away from [the parecliptic apogee], then whatever distance the [dirigent] apogee reaches counter-sequentially will be reached by the center sequentially. Thus [it follows that] they will both be in conjunction twice per rotation, once in Libra and once in Aries; and they will be in opposition twice when one of them reaches Capricorn, and the other Cancer.

[٣٨] ومثل هذا يعرض [١٠١١]لمركز تدوير عطارد لأنّ حركة مركز تدويره بحركة الحامل [١٠١٢]ضعف حركة أوجه بحركة المدير. لكنّ[١٠١٣] المدير بمثل حركته يردّ الحامل، فيبقى فضل حركة المركز إلى التوالي مثل حركة المدير إلى غير التوالي. فإذا تقارنا — أعني المركز والأوج الذي في المدير — في الميزان عند الأوج الآخر[١٠١٤] المثلّي، ثمّ تحرّكا[١٠١٥] عنه، [١٠١٦]فأيّ بعد يحصل عنه للأوج[١٠١٧] إلى غير التوالي يحصل للمركز إلى التوالي. حتّى أنّهما يقترنان في الدورة[١٠١٨] مرتين مرّة في الميزان ومرّة في الحمل؛ ويتقاطران[١٠١٩] مرّتين [١٠٢٠]عند بلوغ أحدهما الجَدْي والآخر[١٠٢١] السرطان[١٠٢٢].

The Second Part

On an Explanation of the Earth and What Pertains to It in Three Chapters

Chapter 1 of Part II

On the Inhabited Part of the Earth and Its Latitude, Its Longitude, and Its Division into the Climes

[1] The Earth is circular in shape as has been [mentioned] before, and we assume three circles upon it: one of them is in the plane of the equinoctial, and it is the equator as you know; the second [circle] is in the plane of the equator's horizon; and the third is in the plane of the meridian circle that is in the middle of the habitable land through the equator. Then the first cuts the Earth into two halves, a southern and a northern. The second bisects its two halves, so it becomes quarters. The inhabited part of it is one of the two northern quarters as one observes in it mountains, deserts, pastures, seas, and so on, including wastelands. The remaining quarters are uninhabitable. The third circle cuts the inhabited part into two halves, a western and an eastern. The intersection point between the first and third [circles] is called the **cupola of the Earth**.

[2] The latitude of the inhabited part is **66 degrees**, and its beginning is from the equator; however, **Ptolemy**, after writing the ***Almagest***, claimed that he found habitation below the equator to a distance of **16;25**.[14] So according to this claim of his, the latitude of habitable land is **82;25**. The longitude of the inhabited part is **180;0**, and its beginning is from the west; however, some of them take it to be from the coast of the enclosing ocean and some of them from islands well into this ocean, their distance from its coast being **10;0**.[15]

---

[14] Jaghmīnī is referring to Ptolemy's *Geography*, which Ptolemy wrote after his *Almagest*. For more details, see the Commentary to II.1[2].

[15] These are the Eternal Islands (*al-khālidāt*), sometimes called the Fortunate Islands (*su'adā'*). They are also referred to as the Isles of the Blest, and usually thought to be the Canary Islands.

المقالة الثانية

في بيان الأرض[١٠٢٣] وما يتعلّق بها

وهي[١٠٢٤] ثلاثة أبواب

الباب الأوّل

في المعمور من الأرض وعرضه وطوله وقسمته إلى الأقاليم[١٠٢٥]

[١] الأرض كريّة الشكل كما سلف، ونفرض[١٠٢٦] عليها ثلاث دوائر: إحداها[١٠٢٧] في سطح معدّل النهار وهي خطّ الاستواء[١٠٢٨] كما تعرف[١٠٢٩]؛ والثانية في سطح أفق[١٠٣٠] الاستواء؛ [١٠٣١]والثالثة[١٠٣٢] في سطح دائرة نصف النهار في منتصف العمارة بخطّ الاستواء. فالأولى تقطع[١٠٣٣] الأرض بنصفين، جنوبي وشمالي. والثانية تنصّف[١٠٣٤] نصفيها[١٠٣٥] فيصير[١٠٣٦] أرباعاً. والمعمور منها أحد الرُبعين الشماليين[١٠٣٧] على ما يرى[١٠٣٨] فيه من الجبال، والصحارى، والمروج، والبحار، ونحوها من المواضع الخربة. وسائر الأرباع خراب. والدائرة [١٠٣٩]الثالثة تقطع[١٠٤٠] المعمور[١٠٤١] بنصفين، غربي وشرقي. ونقطة [١٠٤٢]التقاطع بين[١٠٤٣] الأولى والثالثة تسمّى **قُبّة الأرض**.

[٢] وعرض المعمور سو [١٠٤٤] درجة وابتداؤه من خطّ الاستواء، إلّا أنّ **بطلميوس** بعد ما صنّف **المجسطي** زعم أنّه وجد وراء خطّ الاستواء عمارة إلى[١٠٤٥] بعد[١٠٤٦] يو[١٠٤٧] كه [١٠٤٨]. فيكون عرض العمارة على زعمه هذا[١٠٤٩] فب كه [١٠٥٠]. وطول[١٠٥١] المعمور[١٠٥٢] قف ٠ [١٠٥٣]، وابتداؤه من المغرب؛ إلّا أنّ بعضهم يأخذه[١٠٥٤] من ساحل البحر المحيط[١٠٥٥] وبعضهم[١٠٥٦] من جزائر واغلة[١٠٥٧] في هذا البحر، بُعدها عن[١٠٥٨] ساحله[١٠٥٩] ي ٠ [١٠٦٠].

[3] This inhabited part was then divided into seven longitudinal sections parallel to the equator. The **first clime** begins from it, and daylight there is always **12** hours, as you will learn; for some of them, it is from where daylight, I mean the longest daylight of the year, is **12;45** [hours] and the latitude is **12;30** [degrees]. By consensus, [the clime's] midpoint is where [maximum] daylight is **13;0** [hours] and the latitude is **16;27** [degrees].

[4] The beginning of the **second clime**, and obviously it is the end of the first clime, is where [maximum] daylight is **13;15** [hours] and the latitude is **20;14** [degrees]; and its midpoint is where [maximum] daylight is **13;30** [hours] and the latitude is **23;51** [degrees].

[5] The beginning of the **third** is where [maximum] daylight is **13;45** [hours] and the latitude is **27;12** [degrees]; and its midpoint is where [maximum] daylight is **14;0** [hours] and the latitude is **30;22** [degrees].

[6] The beginning of the **fourth** is where [maximum] daylight is **14;15** [hours] and the latitude is **33;18** [degrees]; and its midpoint is where [maximum] daylight is **14;30** [hours] and the latitude is **36;0** [degrees].

[7] The beginning of the **fifth** is where [maximum] daylight is **14;45** [hours] and the latitude is **38;35** [degrees]; and its midpoint is where [maximum] daylight is **15;0** [hours] and the latitude is **40;56** [degrees].

[8] The beginning of the **sixth** is where [maximum] daylight is **15;15** [hours] and the latitude is **43;51** [degrees] and its midpoint is where [maximum] daylight is **15;30** [hours] and the latitude is **45;1** [degrees].

[9] The beginning of the **seventh** is where [maximum] daylight is **15;45** [hours] and the latitude is **46;51** [degrees]; and its midpoint is where [maximum] daylight is **16;0** [hours] and the latitude is **48;32** [degrees].

[٣] ثم [١٠٦١] قسم[١٠٦٢] هذا[١٠٦٣] المعمور[١٠٦٤] سبع قطاع مستطيلة [١٠٦٥] على موازاة خطّ الاستواء[١٠٦٦]. وابتدأ[١٠٦٧] **الإقليم الأول** منه[١٠٦٨]، والنهار هناك أبداً يب [١٠٦٩] ساعة، كما ستعرف؛ وعند بعضهم من حيث النهار، أعني النهار[١٠٧٠] الأطول من السنة يب مه [١٠٧١] والعرض يب ل [١٠٧٢]. ووسطه بالاتفاق حيث النهار يج ٠ [١٠٧٣]، والعرض [١٠٧٤] يو كز [١٠٧٥].

[٤] وابتدأ **الإقليم**[١٠٧٦] **الثاني**، وهو لا محالة آخر الإقليم الأوّل، حيث النهار يج يه [١٠٧٧] والعرض ك يد [١٠٧٨]؛ ووسطه حيث النهار يج ل [١٠٧٩] والعرض كج نا[١٠٨٠].

[٥] وابتدأ **الثالث** حيث النهار يج مه[١٠٨١] والعرض كز يب [١٠٨٢]؛ ووسطه حيث النهار يد ٠ [١٠٨٣] والعرض ل كب [١٠٨٤].

[٦] وابتدأ[١٠٨٥] **الرابع** حيث النهار يد يه [١٠٨٦] والعرض لج يح [١٠٨٧]؛ ووسطه حيث النهار يد ل [١٠٨٨] والعرض لو ٠ [١٠٨٩].

[٧] وابتدأ[١٠٩٠] **الخامس** حيث النهار يد مه [١٠٩١] والعرض لح له [١٠٩٢]؛ ووسطه حيث النهار يه ٠ [١٠٩٣] والعرض [١٠٩٤] م نو [١٠٩٥].

[٨] وابتدأ[١٠٩٦] **السادس** حيث النهار يه يه [١٠٩٧] والعرض مج نا [١٠٩٨]؛ ووسطه حيث النهار يه ل [١٠٩٩] والعرض مه ا [١١٠٠].

[٩] وابتدأ[١١٠١] **السابع** حيث النهار يه مه [١١٠٢] والعرض مو نا [١١٠٣]؛ ووسطه حيث النهار يو ٠ [١١٠٤] [١١٠٥] والعرض مح لب [١١٠٦].

[10] According to some of them, its end is at the end of the habitable land; according to others, it is up to where the latitude is **50;25** [degrees]. The latitude, though, from the beginning of the first clime to its midpoint and what is between the middle of the seventh to its end [in both cases] turns out to be greater due to the difference in habitable land in them. For this reason, they do not count that part of the habitable land below the equator as part of the climes; for this [reason] as well, some of them do not count what is between the equator and latitude **12;30** nor what is between latitude **50:25** to the end of the habitable land. But beyond this latitude are habitations, according to what they have claimed, namely: in latitude **63** is an inhabited island whose residents live in bath-houses due to the severity of the cold[16]; in latitude **64** is a habitation whose residents are an unknown Slavic people[17]; and [from there] up to latitude **66** are habitations whose residents resemble wild animals.

Chapter 2

On the Characteristics of the Equator and Locations Having Latitude

[1] Among the characteristics of the equator are: that the equinoctial is directly overhead for its inhabitants, similarly for the Sun when it reaches the two equinox points; and that its horizon, called the **horizon of the right orb** and the **horizon of the erect sphere**, bisects the equinoctial and all the day-circuits at right angles. The turning of the orb there is wheel-like, I mean similar to the buckets of waterwheels emerging from the surface of the water at right angles. There is no star or point on the orb that does not rise or set, with the exception of the two poles of the World since they are both on the horizon. The arcs of the visible day-circuits are always the same as those below Earth. For this reason, daytime and nighttime are always equal, each of them **12** hours, and the daytime of each star is the same as its nighttime. The greatest inclination of the Sun from the zenith is the same amount northward and southward, this being in the amount of the maximum obliquity of the zodiacal orb from the equinoctial.

---

[16] Jaghmīnī is probably referring to the Thule Island, usually thought to be the Shetland Islands.

[17] The unknown Slavic people (the *Ṣaqāliba*) could be a reference to Ptolemy's "unknown Scythian peoples" at 64;30 degrees (*Almagest*, II.6[30], Toomer, *Ptolemy's Almagest*, 89). See Commentary to II.1[10] for more details.

[١٠] وآخره آخر العمارة عند بعضهم، [١١٠٧]وعند بعضهم إلى حيث العرض نَ كَه [١١٠٨]. وإنّما صار عرض[١١٠٩] ابتداء[١١١٠] الإقليم الأوّل إلى وسطه[١١١١] وما بين وسط السابع إلى آخره أكثر لتفرّق[١١١٢] العمارة فيهما. ولهذا المعنى لا يعدّون من الأقاليم[١١١٣] ما وراء[١١١٤] خطّ[١١١٥] الاستواء من العمارة، ولهذا أيضاً لا يعدّ بعضهم ما بين خطّ الاستواء إلى عرض[١١١٦] يَب لَ [١١١٧] ولا[١١١٨] ما بين عرض نَ كَه [١١١٩] إلى آخر العمارة. فإنّ[١١٢٠] وراء هذا العرض عمارات على ما زعموا أنّ في عرض سَجَ [١١٢١] جزيرة[١١٢٢] معمورة أهلها يسكنون[١١٢٣] الحمّامات لشدّة البرد، وفي عرض سَدَ [١١٢٤] عمارة أهلها [١١٢٥]قوم من الصقالبة[١١٢٦] لا يُعرفون، وإلى عرض سَوَ [١١٢٧] عمارات سكّانها[١١٢٨] شبيهة[١١٢٩] الوحوش[١١٣٠].

الباب الثاني
في خواصّ خطّ الاستواء [١١٣١]والمواضع التي[١١٣٢] لها عرض[١١٣٣]

[١] أمّا خطّ الاستواء، فمن خواصّه أنّ معدّل النهار يُسامت رؤوس أهله، وكذا[١١٣٤] الشمس عند بلوغها نقطتي الاعتدالين؛ وأنّ أفقه، ويسمّى [١١٣٥]**أفق الفلك المستقيم وأفق الكرة المنتصبة**، ينصّف معدّل[١١٣٦] النهار وجميع المدارات[١١٣٧] على [١١٣٨]زوايا قائمة. ويكون[١١٣٩] هناك دور الفلك دولابيّاً، أعني كما يخرج العصامير من سطح الماء على زوايا قائمة. ولا يكون كوكب ولا[١١٤٠] نقطة في [١١٤١]الفلك إلّا وهو يطلع ويغرب[١١٤٢]، إلّا قطبي العالم فإنّهما يكونان على الأفق. وتكون[١١٤٣] القسيّ الظاهرة للمدارات[١١٤٤] كالتي[١١٤٥] تحت الأرض. فلذلك[١١٤٦] يكون النهار والليل أبداً متساويين، كلّ منهما يَب ساعة، ويكون نهار كلّ كوكب كليله[١١٤٧]. ويكون أكثر ميل الشمس عن سمت الرأس في الشمال والجنوب بقدر واحد وذلك بقدر غاية ميل فلك[١١٤٨] البروج عن معدّل النهار.

[2] **As for the oblique locations to the north of the equator whose latitude does not reach 90 degrees:** among their characteristics is that their horizons, called **oblique horizons**, bisect the equinoctial alone into two [equal] halves, but not at right angles, so the turning of the orb for them is slanted. The [horizons] cut all the day-circuits into two unequal sections; thus the visible arcs for the northern day-circuits are greater than those below the Earth, and for the southern [day-circuits] it is the opposite. For this reason nighttime and daytime are not equal for them, except when the Sun reaches the two equinox points, this being the days of Nayrūz and Mihrjān. Daytime is longer than nighttime when the Sun is in the northern signs, and shorter when it is in the southern signs. The greater the local latitude, the greater will be the difference in amount between nighttime and daytime; obviously, this is due to the zenith being inclined in these locations to the equinoctial. By the amount of its inclination, the northern pole and the day-circuits that are in its direction will be elevated, and the southern pole and the day-circuits that are adjacent to it will be depressed. As the latitude increases, the inclination of the zenith from the equinoctial will increase, so the altitude of the northern pole and the day-circuits adjacent to it increase; the excess of their visible arcs will then increase over those below the Earth. The depression of the southern pole and the day-circuits near it will then increase as [also] the excess of their arcs that are below the Earth over the visible ones. For each day-circuit whose distance from the northern pole is equal to the pole's altitude, then all that is in it and all the stars up to the northern pole that its circle contains will be permanently visible; its corresponding [day-circuit] on the southern side, with all that is in it, is permanently invisible.

[3] **Those locations whose latitude does not reach 90 degrees have divisions**, each division having characteristics: Among them are **the locations whose latitude is less than the maximum obliquity** of the zodiacal orb from the equinoctial. The Sun is directly overhead for its inhabitants twice per year, this being when it reaches two points, each on [one of] the two sides of the summer solstice point, whose declination from the equinoctial is equal to the local latitude.

[٢] **وأمّا المواضع المائلة إلى الشمال عن خطّ الاستواء التي لم يبلغ**[١١٤٩] **عرضها تسعين جزءاً:** فمن خواصّها أنّ آفاقها، وتسمّى **الآفاق**[١١٥٠] **المائلة**، تنصّف معدّل النهار وحده بنصفين[١١٥١]، لكن لا على زوايا [١١٥٢]قائمة، فيكون دور الفلك فيها حمائلياً. وتقطع[١١٥٣] المدارات[١١٥٤] كلّها بقطعتين[١١٥٥] مختلفتين[١١٥٦]؛ فالقسيّ[١١٥٧] الظاهرة للمدارات[١١٥٨] الشمالية أعظم من التي تحت الأرض، وللجنوبية[١١٥٩] بالخلاف. ولذلك[١١٦٠] [١١٦١]لا يستوي الليل والنهار فيها[١١٦٢]، إلّا عند بلوغ الشمس نقطتي الاعتدالين، وذلك في يوم النيروز والمِهْرجان. ويكون النهار أطول من الليل عند كون الشمس في البروج الشمالية، وعند كونها في البروج الجنوبية أقصر. وكلّما كان عرض[١١٦٣] البلد أكثر، كان مقدار [١١٦٤]التفاوت بين الليل والنهار أكثر؛ وذلك لأنّ سمت الرأس مائل في هذه المواضع لا محالة عن معدّل النهار. وبقدر[١١٦٥] ميله، يرتفع القطب الشمالي والمدارات[١١٦٦] التي في ناحيته، وينحطّ القطب الجنوبي والمدارات التي تليه[١١٦٧]. فكلّما[١١٦٨] ازداد العرض، [١١٦٩]ازداد ميل سمت الرأس عن معدّل [١١٧٠]النهار، فازداد ارتفاع القطب الشمالي والمدارات التي تليه؛ فازداد[١١٧١] فضل قسيّها [١١٧٢]الظاهرة على التي تحت الأرض. فازداد[١١٧٣] انحطاط القطب الجنوبي والمدارات[١١٧٤] التي عنده، وفضل قسيّها التي تحت الأرض على[١١٧٥] الظاهرة. وكلّ مدار[١١٧٦] بُعده عن القطب الشمالي مثل ارتفاع القطب[١١٧٧]، [١١٧٨]فهو بجميع ما فيه وجميع ما تحويه[١١٧٩] دائرتُه إلى القطب الشمالي[١١٨٠] من الكواكب[١١٨١] أبديّ الظهور، ونظيره من ناحية الجنوب، بجميع[١١٨٢] ما فيه، أبديّ[١١٨٣] الخفاء.

[٣] **وهذه المواضع التي لم**[١١٨٤] **يبلغ**[١١٨٥] **عرضها تسعين جزءاً**[١١٨٦] **أقسام**، يخصّ[١١٨٧] كلّ قسم منها خواصّ[١١٨٨]: منها[١١٨٩] **المواضع التي عرضها أقلّ من الميل الأعظم** الذي[١١٩٠] لفلك البروج عن معدّل النهار. فالشمس تسامت[١١٩١] رؤوس أهلها في السنة مرتين، وذلك عند بلوغها نقطتين عن جنبتي نقطة الانقلاب الصيفي ميلهما عن معدّل النهار مثل عرض البلد.

[4] Among them are **the locations whose latitude equals the maximum obliquity**. The Sun is thus directly overhead once per year, this being when it reaches the summer solstice point. Those locations from the equator to this latitude have in them two shadows, i.e., planar shadows [*umbra recta*], which you will learn about, sometimes being at noon toward the south and at other times to the north. [Locations] from this latitude to latitude 90 have one shadow, i.e., the shadow is only toward the north.

[5] Among them are **[the locations] whose latitude is greater than the maximum obliquity**. The Sun is thus not directly overhead for its inhabitants.

[6] Among them are **[the locations] whose latitude equals the complement of the obliquity, this being 66;25.**[18] When the zodiacal pole reaches the meridian circle by the motion of the Universe, it will be at the zenith, whereupon the zodiacal circle coincides with the horizon; so Aries is at the east point, Capricorn is at the south point, Libra at the west point, and Cancer is at the north point. Then when [the pole] departs from the zenith, six zodiacal signs rise in one stroke, and they are those in the eastern half along the horizon, namely from Capricorn to Cancer; and the other six set in one stroke. The circuit of Cancer[19] here does not set because of what was [said] before, so when the Sun reaches it, it does not set until it has passed it. So the longest day is **24** hours and similarly the longest night, since [the former] is in the amount that occurs for the northern day-circuits of permanent visibility; and the magnitude of the visible arcs ensues in their counterparts of permanent invisibility and in the magnitude of the arcs that are below the Earth.

[7] Among them are **[the locations] whose latitude exceeds the complement of the obliquity**, i.e., over **66;25**. So the zodiacal pole[20] is inclined away from the zenith toward the south in the amount of the excess of the latitude over **66;25**; and it follows that those [northern] parts of the zodiacal orb whose declination from the equinoctial is greater than the colatitude of the locality do not set.

---

[18] 66;25 is the complement of Jaghmīnī's value for the total obliquity, i.e., 23;35 (see his discussion on the declination in I.4).

[19] "Cancer" here should be understood as Cancer 0°.

[20] As noted by ʿAbd al-Wājid, what is meant here is the zodiacal pole on the meridian at its highest altitude (Istanbul, Süleymaniye Library, Laleli MS 2127, f. 112b).

[٤] ومنها[١١٩٢] **المواضع التي** [١١٩٣]**عرضها**[١١٩٤] **مثل الميل الأعظم**. فالشمس تسامت[١١٩٥] رؤوسهم مرّة [١١٩٦]في السنة[١١٩٧]، وذلك عند[١١٩٨] بلوغها نقطة الانقلاب الصيفي. والمواضع التي هي من[١١٩٩] خطّ الاستواء إلى هذا العرض ذوات[١٢٠٠] ظلّين، أعني أنّ الظل[١٢٠١] المستوي فيها، وستعرفه، يكون في نصف النهار تارة[١٢٠٢] إلى الجنوب وأخرى إلى الشمال. والتي من هذا [١٢٠٣]العرض إلى عرض تسعين ذوات ظلّ واحد، أعني يكون الظلّ إلى الشمال فقط.

[٥] ومنها[١٢٠٤] **التي عرضها أكثر من الميل الأعظم**. فإنّ الشمس لا تسامت[١٢٠٥] رؤوس أهلها.

[٦] ومنها[١٢٠٦] **التي عرضها مثل تمام الميل، وذلك** سو كه [١٢٠٧]. فإنّ قطب البروج إذا بلغ دائرة نصف النهار بحركة الكلّ، وقع على سمت الرأس وحينئذ[١٢٠٨] تنطبق[١٢٠٩] دائرة البروج على الأفق؛ فيكون الحمل على نقطة المشرق، والجدي على نقطة الجنوب، والميزان على نقطة المغرب، والسرطان على نقطة الشمال. فإذا زال[١٢١٠] عن سمت الرأس، طلعت [١٢١١]ستّة من البروج دفعةً، وهي التي في النصف [١٢١٢]الشرقي على[١٢١٣] الأفق، وهي من الجدي إلى السرطان؛ وغربت الستّة الأخرى دفعةً[١٢١٤]. [١٢١٥]ومدار السرطان هناك لا يغرب لما سلف، فإذا بلغته الشمس، لم تغرب حتّى تجاوزه. فيكون النهار الأطول [١٢١٦]كد [١٢١٧] ساعة وكذلك الليل الأطول، إذ بقدر[١٢١٨] ما[١٢١٩] يعرض للمدارات[١٢٢٠] الشمالية من الظهور الأبدي[١٢٢١]؛ [١٢٢٢]وعظم القسيّ الظاهرة يعرض لنظائرها[١٢٢٣] الخفاء الأبدي وعظم القسيّ التي تحت الأرض.

[٧] ومنها[١٢٢٤] **التي عرضها زائد على تمام**[١٢٢٥] **الميل**، أعني على سو كه [١٢٢٦]. فيميل قطب البروج عن سمت الرأس إلى الجنوب بقدر زيادة العرض على سو كه [١٢٢٧]؛ ويلزم أن لا تغرب[١٢٢٨] من فلك البروج الأجزاء التي ميلها عن معدّل النهار أكثر من تمام عرض البلد.

[8] A way to facilitate conceiving this is for us to assume the zodiacal pole is on the meridian circle, which is then inclined southward from the zenith on the part [of the meridian] that is toward the south; and in the amount of its inclination, the head of Capricorn will be depressed below the horizon in the south, and the head of Cancer will be elevated in the north. [That part of] the equinoctial that is toward the south [on the meridian] is above the horizon, and its altitude is in the amount of the difference of the latitude from 90 degrees, which is the colatitude, I mean [the latitude's] "completion," and it is known as the complement of the arc.[21] So the [southern] parts on the zodiacal orb whose declination from the equinoctial is less than the colatitude will therefore be, along with the equinoctial, obviously above the horizon [when] toward the south; those whose declination is equal to the colatitude will touch the horizon, not being depressed from it [at that time]; and those whose declination is greater than the colatitude will obviously be depressed [below the horizon] and will thus be permanently invisible. The permanently invisible is obviously an arc along the zodiacal orb whose midpoint is the winter solstice point. The period of time the Sun traverses this arc with its proper movement is the length of the longest nighttime for that locality; and the counterpart of this arc along the northern signs is permanently visible, as you have learned, and the period of time the Sun traverses this counterpart is the length of the longest daytime for that locality. Among these localities are those where the length of their daytime amounts to approximately six months and likewise the length of nighttime.

[9] It happens that for part of the zodiacal orb that rises there [i.e., locations whose latitude is between 66;25 and 90], it will rise in reverse order and set in regular order. This is in the half of the zodiacal orb that is from Capricorn to Cancer; so Gemini rises before Taurus, Taurus before Aries, and [continuing] according to this pattern. For part of it, it will rise in regular order and set in reverse order, this being in the other half of the zodiacal orb; so Sagittarius sets before Scorpio, Scorpio before Libra, and [continuing] according to this pattern.[22]

[10] A way to facilitate conceiving this is that if we take the zodiacal pole to be on the meridian circle toward the south from the zenith, then half of the [zodiacal] orb from Aries to Libra is visible in sequence toward the north, and the other half is invisible toward the south. The head of Aries is on the east point, and the head of Libra is on the west point. Hence, Aries will have risen before Pisces, and Libra will have set before Virgo. Then when the zodiacal pole inclines away from the meridian circle toward the west while Aries is ascending, that which is contiguous

---

[21] See I.4.

[22] Note that the vernal equinox (Aries 0°) is the midpoint in the first half of the zodiac (i.e., rising in reverse order and setting in regular order), and the autumnal equinox (Libra 0°) is the midpoint in the second half (i.e., rising in regular order and setting in reverse order) (cf. ʿAbd al-Wājid, Laleli MS 2127, f. 115a).

[٨] وممّا يُسهّل تصوّر ذلك أن نفرض[١٢٢٩] قطب البروج على دائرة نصف النهار فيكون مائلاً إلى الجنوب عن سمت الرأس ممّا يلي الجنوب[١٢٣٠]؛ وبقدر ميله ينحطّ رأس الجدي عن الأفق في الجنوب، ويرتفع[١٢٣١] [١٢٣٢]رأس[١٢٣٣] السرطان في الشمال. ويكون معدّل النهار ممّا يلي الجنوب فوق الأفق، وارتفاعه بقدر ما ينقص العرض عن[١٢٣٤] تسعين[١٢٣٥] جزءاً، وهو تمام العرض، أعني كلّه ويعرف بتمام القوس. فالأجزاء من فلك البروج التي[١٢٣٦] ميلها عن معدّل النهار أقلّ من تمام العرض فإنها تكون[١٢٣٧] لا محالة[١٢٣٨] مع معدّل النهار [١٢٣٩]فوق الأفق ممّا يلي الجنوب؛ والتي ميلها تساوي[١٢٤٠] تمام العرض فإنّها تماسّ الأفق ولا تنحطّ عنه؛ والتي ميلها[١٢٤١] أكثر من تمام العرض فإنّها تنحطّ[١٢٤٢] [١٢٤٣]لا محالة فتكون[١٢٤٤] أبديّة الخفاء. والأبديّة الخفاء تكون[١٢٤٥] لا محالة[١٢٤٦] قوساً من فلك البروج منتصفها نقطة الانقلاب الشتويّ. ومدّة قطع الشمس لتلك القوس بمسيرها الخاصّ[١٢٤٧] طول الليل الأطول لذلك البلد؛ ونظيرة تلك القوس من البروج الشمالية أبديّة الظهور، كما عرفت، [١٢٤٨]ومدّة قطع الشمس لتلك النظيرة[١٢٤٩] طول النهار الأطول لذلك البلد[١٢٥٠]. فمن هذه البلاد ما يبلغ طول نهاره[١٢٥١] قريباً من ستّة [١٢٥٢]أشهر وكذلك طول الليل[١٢٥٣].

[٩] ويعرض لبعض ما يطلع من فلك البروج هناك أن يطلع منكوساً ويغرب[١٢٥٤] مستوياً. وذلك في نصف فلك البروج الذي من[١٢٥٥] الجدي إلى السرطان؛ فيطلع الجوزاء قبل[١٢٥٦] الثور، والثور قبل الحمل، وعلى [١٢٥٧]هذا القياس[١٢٥٨]. ولبعضه أن يطلع مستوياً ويغرب منكوساً، [١٢٥٩]وذلك في النصف الآخر[١٢٦٠] من فلك البروج؛ فيغرب القوس قبل العقرب، والعقرب قبل الميزان[١٢٦١]، وعلى هذا القياس.

[١٠] وممّا يسهّل تصوّر ذلك أنّا إذا فرضنا قطب البروج على دائرة نصف النهار[١٢٦٢] ممّا يلي الجنوب عن سمت الرأس، فيكون نصف الفلك من الحمل إلى الميزان على التوالي ظاهراً ممّا يلي[١٢٦٣] الشمال، والنصف الآخر غائباً ممّا يلي الجنوب. ورأس الحمل [١٢٦٤]على نقطة المشرق، ورأس الميزان على نقطة المغرب. فيكون إذن[١٢٦٥] قد طلع الحمل قبل الحوت[١٢٦٦]، وغرب الميزان قبل السنبلة. فإذا مال[١٢٦٧] قطب البروج عن دائرة[١٢٦٨] نصف النهار إلى المغرب والحمل طالع، أخذ [١٢٦٩]في الطلوع[١٢٧٠] ما كان مُتّصلاً

with Aries toward the south, namely the end of Pisces, begins to rise counter-sequentially until the rising of Pisces is complete. Then Aquarius begins rising similarly. Setting is likewise: I mean that Libra having set, its head is at the west setting point. So when [Libra] has set and become depressed, that which is contiguous with it, namely the end of Virgo, begins to set counter-sequentially with it; and [continuing] according to this pattern.

[11] When we take the head of Cancer to be on the meridian circle toward the south, from Libra to Aries is toward the north below the horizon, and the other half is visible. Then Virgo will have risen before Libra in regular order. Then when the head of Cancer inclines away from the meridian circle, Libra begins to rise in regular order as we have stated. Since that which sets faces that which rises, then that facing what rises in reverse order will set in reverse order, and vice versa.[23] And since the rising in one of the two halves of the [zodiacal] orb in terms of order is contrary to the rising in the second [half] but matches the setting,[24] it follows that the rising of each half will be contrary to its setting, so what rises in reverse order will set in regular order, and vice versa.[25]

[12] **As for the locations whose latitude is 90 degrees**, the World pole corresponds to the zenith there. The equinoctial is coincident with the horizon circle, and the rotation of the [celestial] orb is spinning parallel with the horizon. A year there is a day and a night, being six months of daytime—this when the Sun is in the northern signs—and six months of nighttime—this when the Sun is in the southern signs. There nothing of the orb has a rising or a setting; instead, its northern half is permanently visible and its southern half is permanently below the Earth.

[13] We have only described specifically the northern locations because in them is the inhabited world. Since everything pertaining to them that we have described is due to their inclination from the equator toward the north, a comparable situation pertains to southern locations due to their inclination toward the south.

[14] Now then, instruction of the above is sufficient for understanding this [topic].

---

23 According to ʿAbd al-Wājid, "vice versa" here means "that facing what rises in regular order will set in regular order" (Laleli MS 2127, ff. 118b–119a).

24 For further clarification, see Commentary, II.2[11].

25 In other words, what rises in regular order sets in reverse order.

بالحمل ممّا يلي الجنوب، وهو آخر الحوت[١٢٧١]، على غير التوالي حتّى يتمّ[١٢٧٢] طلوع الحوت[١٢٧٣]. ثمّ يأخذ الدلو في الطلوع كذلك. والغروب كذلك: أعني أنّ الميزان كان غارباً ورأسه في نقطة[١٢٧٤] المغرب للغروب. فإذا[١٢٧٥] غرب وانحطّ[١٢٧٦]، أخذ في الغروب معه[١٢٧٧] ما هو[١٢٧٨] متصّل به[١٢٧٩] وهو آخر السنبلة على غير التوالي[١٢٨٠]، [١٢٨١]وعلى هذا القياس.

[١١] وإذا فرضنا رأس السرطان على دائرة نصف النهار ممّا يلي الجنوب، كان من الميزان إلى الحمل ممّا يلي الشمال تحت الأفق، والنصف الآخر ظاهر[١٢٨٢]. فيكون قد طلع السنبلة قبل الميزان على الاستواء. ثمّ إذا مال رأس السرطان عن[١٢٨٣] دائرة نصف النهار، أخذ الميزان في الطلوع [١٢٨٤]على الاستواء كما ذكرنا. ولمّا كان الغارب يقابل الطالع[١٢٨٥]، كان[١٢٨٦] ما يطلع منكوساً يغرب مقابله[١٢٨٧] منكوساً، وبالضدّ. ولمّا كان الطلوع في أحد[١٢٨٨] نصفي الفلك يخالف الطلوع في[١٢٨٩] الثاني في الاستواء ويوافق الغروب، لزم أن يكون طلوع كلّ [١٢٩٠]نصف يخالف[١٢٩١] غروبه، فما يطلع[١٢٩٢] منكوساً يغرب مستوياً، [١٢٩٣]وبالضدّ.

[١٢] **وأمّا المواضع التي عرضها تسعون جزءاً**، فيوافق قطب العالم سمت الرأس فيها. ومعدّل النهار منطبق[١٢٩٤] على دائرة الأفق، ودور[١٢٩٥] الفلك رحويّ مواز للأفق. وتكون[١٢٩٦] السنة هناك يوماً وليلةً، ستّة أشهر [١٢٩٧]نهاراً، وذلك إذا كانت الشمس في البروج الشمالية، وستّة أشهر [١٢٩٨]ليلة، وذلك إذا كانت الشمس في البروج[١٢٩٩] الجنوبية. وهناك لا يكون لشيء من الفلك طلوع ولا غروب؛ بل[١٣٠٠] نصفه الشمالي ظاهر[١٣٠١] أبداً ونصفه الجنوبي تحت الأرض أبداً.

[١٣] وإنّما خصّصنا المواضع الشمالية بالوصف لأنّ فيها العمارة. ولأنّ جميع ما يعرض لها ممّا وصفناه[١٣٠٢] بسبب ميلها عن خطّ الاستواء إلى الشمال، يعرض مثل ذلك للمواضع الجنوبية[١٣٠٣] بسبب ميلها إلى[١٣٠٤] الجنوب.

[١٤] فتعريف هذا يكفي في معرفة ذلك[١٣٠٥].

Chapter Three

Miscellaneous Items

[1] The **ascendant** is the part [i.e., point] of the zodiacal orb on the horizon that is toward the east. The **degree of rising of the star** is the degree of the zodiacal orb that rises with the rising star. The **degree of transit of the star** is the degree of the zodiacal orb that transits the meridian circle along with the transit of the star. Then if the star is [aligned] with one of the two solstice points or it has no latitude, its degree, i.e., [the star's projected] place on the zodiacal orb, is its degree of transit; and if it has latitude and is not [aligned] with the solstice point, then not. This is because when the star is between the first of Cancer and the end of Sagittarius, it reaches the meridian circle after its degree [of transit] if it has a northern latitude, and in advance of it if it has a southern latitude. When [the star] is in the other half of the zodiacal orb,[26] the reverse will hold since the [northern] zodiacal pole[27] [*] will be easterly when the first half is on the meridian; the circle passing through [the pole] and through the degree of the star is then inclined toward the west and will reach a star with northern latitude first and then its degree. Thus the star is farther from the meridian than its degree, so it arrives on it after it, but before it if it has southern latitude for this very same reason. What is between the star's [longitudinal] degree and its degree of transit is called **the transit difference.** You should follow this same approach for **[the star's] degree of rising**. As for the right orb, the rules for this are exactly the same. As for the inclined orbs, one needs to take into account the horizons.

[2] **The shadow** is taken either: from a gnomon erected parallel to the plane of the horizon, called the first shadow, the reversed [*umbra versa*], and the erect; or from a vertical gnomon perpendicular to the plane of the horizon, called the second shadow and the planar [*umbra recta*].[28] The gnomon is sometimes divided into twelve divisions called digits, sometimes into seven or six and a half divisions called feet, and sometimes into sixty divisions called units. When the shadow reaches its limit at the Sun's maximum altitude, it is then the start-time for the noon [*ẓuhr*] prayer. The start-time for the afternoon [*ʿaṣr*] prayer is, according to al-Shāfiʿī—may God have mercy upon him—when that limit is increased by the equivalent of the gnomon [length], and according to Abū Ḥanīfa—may God be pleased with him—it is when it has increased by twice the gnomon [length].

---

[26] I.e., between the first of Capricorn to the end of Gemini.

[27] MS F is missing a folio. The beginning and end of the omitted passage corresponds to II.3[1–4], which I have indicated (in both the Arabic edition and the English translation) using two asterisks [*]…[*].

[28] On these shadows, see Commentary, II.3[2].

الباب الثالث

في أشياء منفردة[١٣٠٦]

[١] **الطالع** جزء من فلك البروج على الأفق مما يلي المشرق. **درجة طلوع الكوكب** هي درجة من فلك البروج[١٣٠٧] تطلع[١٣٠٨] مع طلوع الكوكب[١٣٠٩]. **درجة ممرّ**[١٣١٠] **الكوكب** درجة من فلك البروج تمرّ بدائرة نصف النهار مع مرور [١٣١١]الكوكب [١٣١٢]بها. فإن كان الكوكب على إحدى نقطتي الانقلابين [١٣١٣]أو كان لا عرض له، فدرجته، أعني مكانه من فلك البروج، هي درجة ممرّه[١٣١٤]؛ وإن كان ذا عرض على غير نقطة الانقلاب، فلا. وذلك لأنّ الكوكب إذا كان فيما بين أوّل السرطان إلى آخر القوس وصل إلى دائرة نصف النهار بعد درجته إن كان شمالي العرض، وقبلها إن كان جنوبي العرض[١٣١٥]. وإن كان في النصف الآخر من فلك البروج، فعلى الخلاف لأنّ قطب البروج [١٣١٦] [*] يكون شرقياً عند كون النصف الأوّل على[١٣١٧] نصف النهار؛ فتكون[١٣١٨] الدائرة المارّة به[١٣١٩] وبدرجة الكوكب مائلةً[١٣٢٠] إلى المغرب وتنتهي[١٣٢١] إلى الكوكب الشماليّ العرض أوّلاً ثمّ إلى درجته. فيكون الكوكب أبعد من درجته عن نصف النهار، فيصل[١٣٢٢] إليه[١٣٢٣] بعدها وقبلها إن كان جنوبي العرض لهذا بعينه. وما بين درجة الكوكب ودرجة ممرّه[١٣٢٤] يسمّى **اختلاف الممرّ**. وقِس على هذا **درجة طلوعه**. أمّا في الفلك المستقيم، فالحكم هذا بعينه. وأمّا [١٣٢٥]في الأفلاك[١٣٢٦] المائلة، فيعتبر الأفق.

[٢] **الظلّ** مأخوذ إمّا[١٣٢٧] [١٣٢٨]من المِقْياس المنصوب على[١٣٢٩] موازاة سطح الأفق، ويسمّى[١٣٣٠] الظلّ الأول، والمعكوس، والمنتصب؛ [١٣٣١]وإمّا من المقياس القائم عموداً على سطح الأفق[١٣٣٢]، [١٣٣٣]ويسمّى[١٣٣٤] الظلّ الثاني[١٣٣٥] والمستوي. وقد يقسم المقياس مرّة باثني عشر قسماً وتسمّى أقسامه أصابع، ومرّة بسبعة[١٣٣٦] أقسام[١٣٣٧] أو ستّة ونصف وتسمّى أقسامه[١٣٣٨] أقداماً، ومرّة بستّين قسماً وتسمّى أقسامه أجزاء. وإذا انتهى الظلّ نهايته عند[١٣٣٩] غاية ارتفاع الشمس فهو أوّل وقت الظهر؛ وأوّل وقت العصر إذا زاد على غايته تلك بمثل[١٣٤٠] المقياس وهذا عند الشافعي رحمه الله[١٣٤١]، وعند ابي حنيفة رضي الله عنه[١٣٤٢] إذا زاد عليه(!)[١٣٤٣] بمثلي المقياس.

[3] **On determining the meridian line and the equinox line**: land is leveled in such a way that if water were poured over it, it would flow evenly in all directions. Then a circle of any size is constructed on [the land]; this circle is called the **Indian circle**. A conic gnomon is erected at its center with a height one-fourth its diameter at a right angle, which can be determined by either a plumb-line or by measuring an equal amount between the tip of the gnomon and the circumference from three points on the circumference. The tip of the shadow is observed when it arrives at the circumference on the western part before diminishing [at noon], and afterwards on the eastern part. Each of the two arrival points is marked and the arc between them is bisected; you produce a line from [the arc's] midpoint that passes through the center to whatever distance you wish. This then is **the meridian line**, and it cuts the circle into two halves. Then you produce a line from the midpoints of the [circle's] two halves that intersects the meridian line at the center at right angles; this is the **east-west line**.

Illustration of the Indian Circle

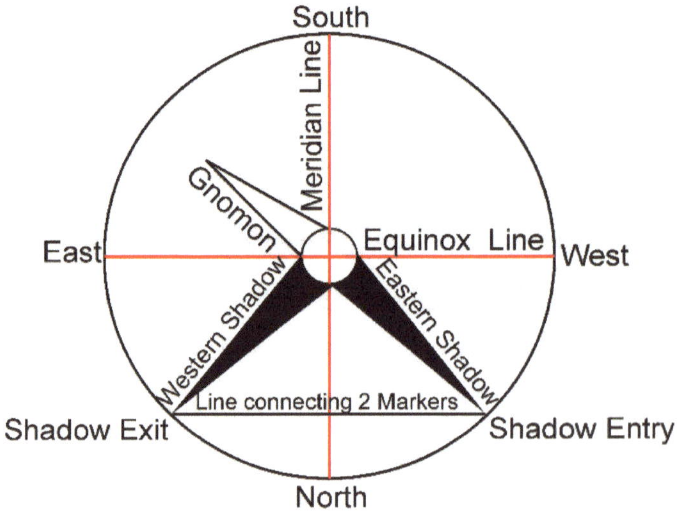

[Figure 8]

[٣] **في معرفة خطّ**[١٣٤٤] **نصف النهار وخطّ الاعتدال:** تُسوَّى أرض[١٣٤٥] بحيث لو صُبَّ فيها ماء[١٣٤٦] سال من جميع الجهات بالسويّة. ثمّ يُدار فيها دائرة بأيّ بُعد كان، وتسمّى[١٣٤٧] هذه **الدائرة**[١٣٤٨] **الهنديّة**[١٣٤٩]. ويُنصب[١٣٥٠] على مركزها **مقياس مخروطي** طوله[١٣٥١] رُبع قطرها نصباً على زاوية قائمة[١٣٥٢]، [١٣٥٣]ويُعرف ذلك إمّا[١٣٥٤] بالشاقول وإمّا بأن[١٣٥٥] يُقدّر[١٣٥٦] ما بين رأس المقياس والمحيط بمقدار واحد من ثلاث نقط[١٣٥٧] من المحيط. ويُرصد رأس الظلّ عند وصوله إلى محيطها ممّا يلي المغرب قبل الزوال وبعده مما يلي المشرق. ويُعلَّم[١٣٥٨] على كلتي نقطتي الوصول وتُنصَّف[١٣٥٩] القوس التي بينهما؛ وتُخرِج[١٣٦٠] [١٣٦١]من[١٣٦٢] منتصفها[١٣٦٣] خطّاً[١٣٦٤] يمرّ بالمركز إلى أيّ بُعد شئت[١٣٦٥]. فهو **خطّ نصف النهار** وقد قطع الدائرة[١٣٦٦] بنصفين. فتُخرِج من منتصفي النصفين خطّاً[١٣٦٧] يقطع[١٣٦٨] خطّ نصف النهار عند المركز على زوايا قائمة، وهو **خطّ المشرق والمغرب**[١٣٦٩].

صورة الدائرة الهندية[١٣٧٠]

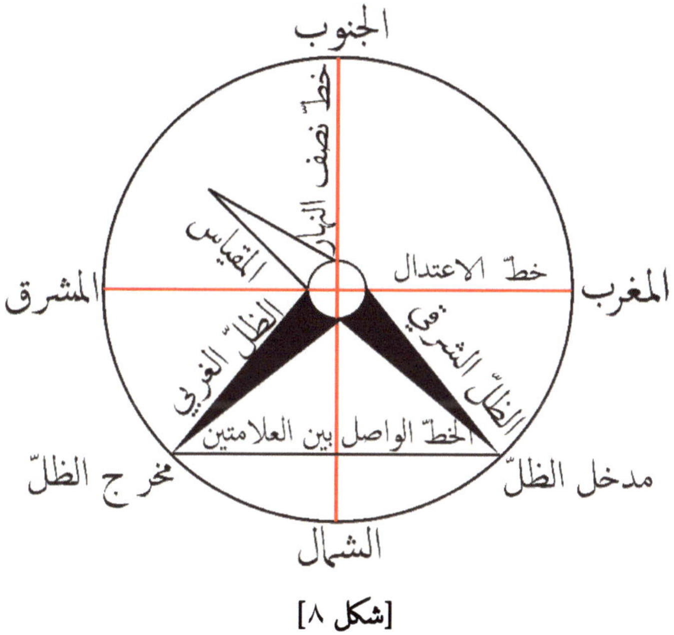

[شكل ٨]

[4] **On determining the *qibla* bearing:** we mean here by the *qibla* bearing, a point on the horizon [such] that when a person faces it he will also be facing the Kaʿba. Since the longitude and latitude of Mecca are less than the longitude and latitude of our locality, we count along the Indian circle from the south point toward the west the amount of the difference between the two longitudes, and its equivalent from the north point. We join what is between the two endpoints with a straight line. We count from the west point toward the south the amount [of the difference] between the two latitudes, and its equivalent from the east point; we join the two endpoints with a straight line. The two lines will then obviously intersect one another. We now produce a line from the center of the circle to their intersection point, and we extend it to the circumference. This line then is in the direction of the *qibla*, [*] and the arc between the [line's] endpoint and the south point is the **arc of the *qibla* bearing**, it being the amount that the worshipper should incline away from the south point. Do something comparable when the longitude or latitude, or both, of Mecca is greater.

[5] If the longitude of the locality equals the longitude of Mecca, then the *qibla* is on the meridian. If its latitude equals Mecca's latitude, then make note that the degrees of the zodiacal orb that pass overhead for the people of Mecca during a rotation are **Gemini 7;21** and **Cancer 22;39**. Place it, I mean one of the two [zodiacal degrees], on the mid-heaven line of the astrolabe that has been constructed for the latitude of the locality. Put a mark at the position of the "almuri";[29] then turn the rete [lit., spider] toward the west by the difference between the two longitudes if the locality is toward the east, and the reverse if it is toward the west. So wherever the [chosen zodiacal] degree lands on the altitude almucantars, you will observe the Sun when it reaches that altitude and erect a gnomon; then its shadow at that time is the bearing for the *qibla*.

[29] The almuri, known as the tooth (or denticle) of Capricorn, is the marker located at the head of Capricorn which juts out from the astrolabe rete, and can be used for various calculations, such as here for longitude difference. Cf. Bīrūnī, *Tafhīm*, no. 325 (194).

[٤] [١٣٧١]**في معرفة**[١٣٧٢] **سمت القبلة** ونعني[١٣٧٣] بسمت القبلة ههنا[١٣٧٤] [١٣٧٥]نقطة[١٣٧٦] في الأفق إذا واجهها الانسان كان مواجهاً للكعبة أيضاً. إذا[١٣٧٧] كان طول مكّة وعرضها أقل من طول بلدنا وعرضه عددنا من الدائرة الهنديّة من نقطة الجنوب بقدر فضل ما بين الطولين إلى المغرب ومن نقطة الشمال مثله ونصل[١٣٧٨] ما بين النهايتين بخطّ مستقيم. ونعدّ[١٣٧٩] [١٣٨٠]من[١٣٨١] نقطة المغرب إلى الجنوب بقدر ما بين العرضين ومن[١٣٨٢] نقطة المشرق مثله، ونصل[١٣٨٣] بين النهايتين بخطّ مستقيم. فيتقاطع الخطّان لامحالة. فنُخرِج من مركز الدائرة خطّاً إلى نقطة تقاطعهما[١٣٨٤] وننفذه[١٣٨٥] إلى المحيط. فذلك الخطّ هو[١٣٨٦] على صوب القبلة [*][١٣٨٧] والقوس التي بين طرفه ونقطة الجنوب هي **قوس**[١٣٨٨] **سمت**[١٣٨٩] **القبلة** وهي[١٣٩٠] مقدار ما ينبغي أن ينحرف المصلّي عن نقطة الجنوب. وقِسْ على ذلك كون طول مكّة[١٣٩١] أو عرضها أو كليهما أكثر[١٣٩٢].

[٥] وإن كان طول البلد يساوي[١٣٩٣] طول مكّة، فالقبلة على [١٣٩٤]نصف[١٣٩٥] النهار. وإن ساوى[١٣٩٦] عرضه عرض مكّة [١٣٩٧]فاعْرِف الأجزاء التي تسامت[١٣٩٨] في الدورة من فلك البروج[١٣٩٩] رؤوس أهل مكّة، وهي ز كا [١٤٠٠] من الجوزاء و كب لط [١٤٠١] من السرطان[١٤٠٢]. [١٤٠٣]وضَعْها أعني[١٤٠٤] إحداهما على خطّ وسط السماء في الأسطرلاب[١٤٠٥] المعمول لعرض البلد. وأَعْلِم على موضع المُرِيّ علامةً، ثمّ أدِر[١٤٠٦] العنكبوت بقدر ما بين الطولين إلى المغرب إن كان البلد شرقيّاً، وبالخلاف إن كان غربيّاً[١٤٠٧]. فحيث[١٤٠٨] انتهت الأجزاء[١٤٠٩] من مُقنْطرات الارتفاع، رَصَدْتَ [١٤١٠]بلوغ الشمس إلى ذلك الارتفاع، ونَصَبْتَ مقياساً؛ فظلّه في ذلك الوقت هو[١٤١١] المسامت للقبلة[١٤١٢].

Illustration of the *Qibla* bearing

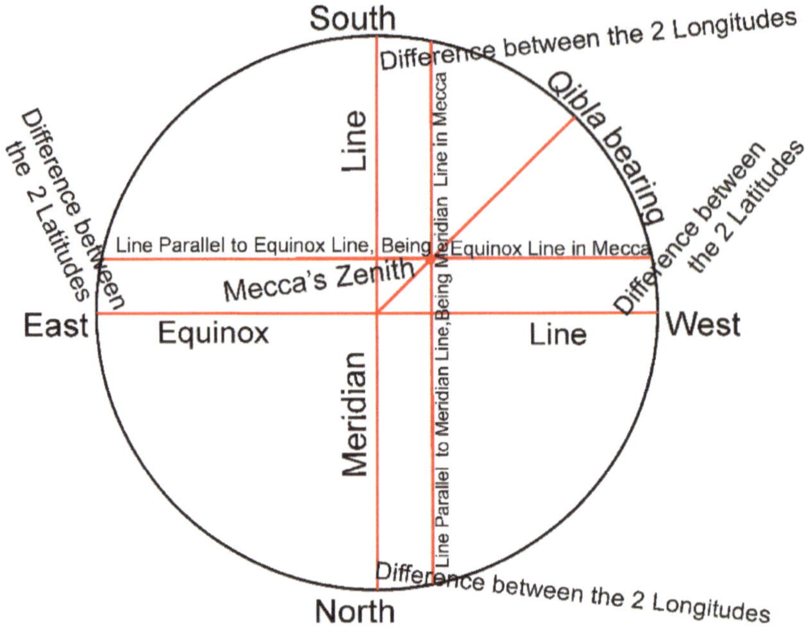

[Figure 9]

[6] **On daytime, nighttime, hours, the year, and the month:** When the Sun's light falls on the Earth, its face toward the Sun is illuminated and its shadow falls opposite the direction of the Sun. So when the Sun is above the Earth, then it is daytime since the only light distinguishing daytime is the Sun's light; when [the Sun] is below the Earth, its shadow falls above it, and it is nighttime. The [Earth's] shadow occurs in the shape of a cone, since the Sun is of a greater size than the Earth. Then when the Sun is below the Earth near the horizon, the shadow cone is inclined away from the zenith, and the air illuminated by the Sun's light is nearby so light appears on the horizon; and as the Sun comes nearer, light predominates and red appears, as is the case of dusk and dawn.

صورة المسامت للقبلة

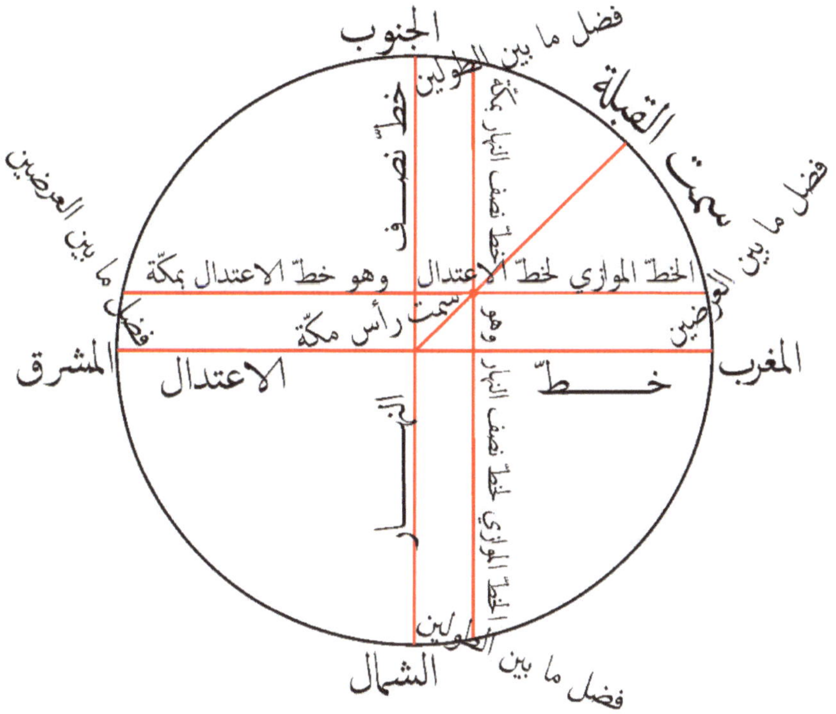

[شكل ٩]

[٦] **في النهار والليل**[١٤١٣] **والساعات والسنة والشهر**[١٤١٤]: الشمس إذا وقع ضوؤها على الأرض [١٤١٥]استضاء وجهها المواجه للشمس، ووقع[١٤١٦] ظلّها[١٤١٧] في مقابلة جهة الشمس. فإذا[١٤١٨] كانت الشمس[١٤١٩] فوق الأرض فهو النهار إذ ليس يخصّ النهار ضوء سوى ضوء الشمس؛ وإذا كانت تحت الأرض وقع ظلّها[١٤٢٠] فوقها وهو الليل. ووقوع ظلّها[١٤٢١] [١٤٢٢]يكون على شكل مخروط إذ[١٤٢٣] الشمس أعظم جرماً من الأرض. فإذا[١٤٢٤] كانت الشمس [١٤٢٥]تحت الأرض قريبةً من الأفق كان مخروط[١٤٢٦] الظلّ مائلاً عن سمت الرأس وكان[١٤٢٧] الهواء المستضئ بضياء الشمس قريباً[١٤٢٨] فيظهر في الأفق النور، وكلّما كانت [١٤٢٩]الشمس أقرب، كانت الأنوار[١٤٣٠] أغلب ويظهر[١٤٣١] الحمرة، كحال الشفق والفجر.

[7] The **nychthemeron** [lit., a day with its night] is the time between the Sun's departure from the meridian circle until it returns to it with the motion of the Universe; but according to people at large, it is from the setting of the Sun to its equivalent. Its beginning could be the departure of the Sun from any assumed point on the orb; however, the calculators and the astral practitioners have conventionally taken its beginning to be from the meridian circle because the variations in the ascensions with respect to the horizons of the inhabited regions are many, [but] it has only one variation with respect to the meridian circle because the meridian circle for all inhabited regions serves as a horizon for the equator. The time of a nychthemeron exceeds one rotation of the Universe by the co-ascension of what the Sun has traveled along the zodiacal orb; and since the Sun cuts off arcs of variable [sizes] along the zodiacal orb, then its co-ascensions will be variable. Furthermore, even if the Sun were presumed to cut off arcs of equal size, the co-ascension of the equal arcs would not be equal. So due to these factors, the nychthemerons are variable. Thus [the practitioners] classified the nychthemeron into a true and a mean. The true is the time of return of an equinoctial point to a given point plus the time to traverse the co-ascension of what the Sun has traveled during [the motion of] that given point; the mean is the time of return of an equinoctial point to a given point plus the time to traverse an equinoctial arc that is equal to the Sun's [daily] mean, namely **0;59,8,20**, during [the motion of] that point. This is recorded in the ***zījes*** [astronomical handbooks]. The difference between the true and the mean is called the **equation of the time** [lit., the equation of the days with their nights].

[8] The **duration of daytime** is from the rising of the Sun to its setting, and in law [*sharʿ*] from the rising of dawn [*al-fajr*] to the setting of the Sun; and from its setting to the Sun's rising is the **duration of nighttime**, and in law to the rising of dawn. Then [practitioners] divide the daytime and the nighttime into regularized hours and **seasonal** [lit., temporal] **hours**. Regularized hours, called **equal hours**, are in the amount by which the Universe rotates **15 degrees**. Then if one divides the arc of daylight or the arc of night or an arc of the orb's rotation by 15, the result is the number of regularized hours for that day or night, or a period within the day or night. The **seasonal hour**, called **unequal** [lit., distorted], is always one

[٧] **واليوم بليلته** هو زمان ما بين مفارقة الشمس دائرة نصف النهار إلى عودها إليها بحركة الكلّ[1432]؛ وعند العامّة من غروب الشمس[1433] إلى مثله. وابتداؤه يمكن[1434] من مفارقة كون[1435] الشمس كلّ نقطة تُفرض في[1436] الفلك؛ لكنّ الحُسّاب والمُنجّمين[1437] اصطلحوا على [1438]ابتدائه من دائرة نصف النهار لأنّ اختلافات المطالع بحسب الآفاق في المساكن[1439] كثيرة، [1440]واختلافها[1441] واحدٌ[1442] بحسب دائرة نصف النهار[1443] لأنّ دائرة نصف النهار في جميع المساكن تقوم[1444] مقام أفق خطّ الاستواء. وزمان اليوم بليلته يزيد على دور الكلّ[1445] بمطالع ما سارت الشمس من فلك[1446] البروج ولمّا كانت الشمس تقطع[1447] من فلك البروج قسيّاً مختلفةً فمطالعها مختلفة. وأيضاً لو كانت الشمس[1448] بالتقدير تقطع[1449] قسيّاً متساويةً، فليست مطالع القسيّ المتساوية متساويةً. فمن هذه الوجوه تختلف الأيّام بلياليها فقسموا اليوم بليلته إلى حقيقيّ ووسط[1450]. فالحقيقيّ هو زمان عودة نقطة من معدّل النهار إلى نقطة مفروضة [1451]مع زمان مُرور مطالع[1452] ما سارت الشمس بتلك النقطة المفروضة؛ والوسط[1453] هو زمان عودة نقطة من معدّل النهار إلى نقطة مفروضة مع زمان مرور قوس[1454] من معدّل [1455]النهار[1456] مساوية[1457] لوسط الشمس وهو ٠ نط ح ك[1458] بتلك النقطة[1459] [1460]وهو الموضوع في **الزيجات** والفضل بين الحقيقي والوسط[1461] يُسمّى **تعديل الأيّام بلياليها**.

[٨] **وزمان النهار** من طلوع الشمس إلى غروبها[1462]، وفي الشرع من طلوع الفجر إلى غروب الشمس؛ ومن غروبها[1463] إلى طلوع الشمس[1464] زمان الليل، وفي الشرع إلى طلوع الفجر. [1465]ثمّ إنّهم قسموا اليوم والليلة[1466] إلى ساعات معتدلة وزمانيّة. [1467]فالساعات[1468] المعتدلة، وتسمّى **المستوية**، هي[1469] بقدر ما يدور الكلّ **خمس عشرة درجة**[1470]. فإذا قُسمت قوس النهار أو قوس الليل أو قوس الدائر[1471] من الفلك على خمسة عشر، كان ما يخرج عدد الساعات المعتدلة لذلك اليوم أو الليلة أو ما مضى من اليوم[1472] أو الليلة[1473]. **والساعة**[1474] **الزمانية**، وتسمّى[1475] **المُعْوَجّة** فهي[1476] جزء

of **12 parts** of daytime or nighttime; so if daytime is longer than nighttime, its hours are longer than the night hours, and if [daytime is] shorter, its hours are shorter. When the arc of daylight or the arc of night is divided by 12, the result is what the orb rotates in each seasonal hour, which is in parts of the seasonal hour called **units of time.** It has thus become clear that regularized hours are those whose number varies according to the length and shortness of daytime, but their units of time do not vary; seasonal hours are those whose units of time vary, but their number does not vary.

[9] **The year** is the time from the Sun's departure from any given point on the zodiacal orb until it returns to it with its proper motion, which it has from west to east. [The practitioners] began this year from the time the Sun is situated at the head of Aries, but they differed on the duration of this year. For some of them said 365¼ days; according to **Ptolemy**, 365¼ days less 1/300 part of a day; and according to **Battānī** 365¼ days less 3 parts 24 minutes out of 360 parts of a day. What is intended here by a day is the nychthemeron. The above is a **solar year;** as for the **lunar year**, it is 12 months.

[10] **The month** is the time from the Moon's departure from any given position it has from the Sun until it returns to [that position]. The most obvious position is the crescent (*hilāl*). However, the sighting of the crescent varies according to changes in inhabited regions, so one only takes it into account for religious matters. The beginning of the month was set from the conjunction of the Sun and Moon, and [the month's] duration is between two conjunctions with the mean motion of the two luminaries: they take away the Sun's mean [motion] from the Moon's mean,[30] and they divide the remainder by the rotation of the orb, namely 360 degrees, thus resulting in **29;31,50,8** days, which is the amount of a month. They then multiplied that by 12, obtaining the days in a lunar year: 354+⅕+⅙ days. This year is less than the solar year by approximately 10 days and 20½ hours.

[30] 13;10,35,2 [Moon's mean] minus 0;59,8,20 [Sun's mean] equals 12;11,26,42 (see I.5).

من **اثني عشر جزء** من النهار أو الليل أبداً، فإذا كان النهار أطول من الليل كانت ساعاته أطول من ساعات[١٤٧٧] الليل، وإن كان[١٤٧٨] أقصر كانت ساعاته أقصر. وإذا قُسمت قوس[١٤٧٩] النهار أو قوس الليل على اثني عشر، كان ما يخرج هو ما يدور الفلك في كلّ ساعة زمانية، وهي[١٤٨٠] أجزاء الساعة[١٤٨١] الزمانية وتسمّى **الأزمان**[١٤٨٢]. فقد تبيّن أنّ الساعات المعتدلة هي التي يختلف عددها على قدر طول النهار وقصره، ولا يختلف أزمانها؛ والساعات الزمانية هي التي تختلف أزمانها[١٤٨٣] ولا يختلف[١٤٨٤] عددها.

[٩] **السنة** هي زمان[١٤٨٥] مفارقة الشمس أيّة[١٤٨٦] نقطة تُفرض من فلك البروج إلى عودها إليها بحركتها الخاصّة التي لها من المغرب إلى المشرق. وقد جعلوا ابتداء هذه السنة من حين[١٤٨٧] حُلول الشمس رأس الحمل، واختلفوا في مدّة[١٤٨٨] هذه السنة. فقال بعضهم شسه[١٤٨٩] يوماً وربع يوم؛ وعند **بطلميوس** شسه[١٤٩٠] يوماً وربع[١٤٩١] إلّا جزءاً من ثلاثمائة جزء من يوم؛ وعند **البتّاني** شسه[١٤٩٢] يوماً[١٤٩٣] وربع إلّا ثلاثة[١٤٩٤] أجزاء واربعاً وعشرين دقيقةً من ثلاثمائة وستين جزء[١٤٩٥] من يوم. والمراد باليوم هنا[١٤٩٦] هي[١٤٩٧] اليوم بليلته. وهذه هي[١٤٩٨] السنة الشمسيّة؛ وأمّا السنة[١٤٩٩] القمرية، فهي[١٥٠٠] اثنا عشر شهراً.

[١٠] **والشهر** زمان مفارقة القمر[١٥٠١] أيّ وضع[١٥٠٢] يُفرض[١٥٠٣] له من الشمس إلى عوده إليه. وأظهر الأوضاع هو الهلال لكنّ رؤية الهلال تختلف باختلاف[١٥٠٤] المساكن. فلم يلتفت[١٥٠٥] إليها إلّا في الأمور الشرعية. وجُعل ابتداء الشهر من اجتماع الشمس والقمر وزمانه[١٥٠٦] ما بين الاجتماعين[١٥٠٧] بالمسير الوسط[١٥٠٨] من النيّرين[١٥٠٩]: بأن ألقوا وسط الشمس من وسط القمر وقسموا على ما بقي[١٥١٠] دور الفلك، وهو شس[١٥١١] جزء، فخرج[١٥١٢] كط لا ن ح[١٥١٣] من الأيّام، وهو مقدار الشهر. ثمّ ضربوا ذلك في اثنى عشر؛ فحصلت أيّام السنة القمريّة: شند[١٥١٤] يوماً وخُمس يوم وسُدسه[١٥١٥]. وهذه السنة ناقصة عن السنة الشمسيّة[١٥١٦] بعشرة أيّام وعشرين ساعة ونصف ساعة بالتقريب.

[11] This is as much as allowed by [my] ignoble character, a tormented mind, thought befuddled by preoccupations beyond counting, and concerns [so overwhelming] they would make a mother neglect her child. I have gone to great lengths to elucidate and illuminate the content with concise and succinct expressions, fulfilling the obligations of obeisance and service while guarding against the tedious and the cluttered. This volume that I have presented is perhaps enough to attain what I desired, faithful to what has been indicated above. So it is best that I limit myself to that, so let this be the end of the book.[31]

[31] For the different colophons, see *§II.2: Description of the Manuscripts.*

[١١] هذا[١٥١٧] ما سمح به الطبْع[١٥١٨] الطَبِع[١٥١٩]، والخاطر المتوزّع، والفكر المشوّش[١٥٢٠] بأشغال لا يعدّ عديدها، وهُموم لا يُنادى وليدُها[١٥٢١]. وقد بذلتُ الوُسْعَ في كَشْف[١٥٢٢] المعاني وإظهارها مع إيجاز الألفاظ[١٥٢٣] واختصارها أداءً لشرائط[١٥٢٤] الامتثال والخدمة مع التحرّز[١٥٢٥] عن الإملال والزَحْمَة. ولعلّ هذا المقدار الذي[١٥٢٦] أوردتُ كافٍ[١٥٢٧] لِتحصيل ما أردتُ، وافٍ بما جرت الإشارة إليه. فالأولى[١٥٢٨] أن أقتصر عليه فليكن هذا خاتمة الكتاب.

TEXT APPARATUS

الديباجة

[١] ص٦: ب؛ ٤٨ب: س؛ ٢ب: ف؛ ١ب: ك؛ ٦١ب: ل.

[٢] الرحيم] +رب الهم بالصواب: س = +وبه نستعن: ف.

[٣] كِفاء] ف = كِفآءٍ: ب = كفأ: س.

[٤] نبيّه] س، ف = رسوله: ب.

[٥] وآله] ب، ف = وعلى اله: س(«على» مشطوبة) = +محمد: شاف.

[٦] قال الإمام ... رحمه الله] يقول الشيخُ الامام الهمام الفاضل الكامل المتبحّر شرف الدّين محمود بن محمّد بن عُمر الجَغمنيّ رحمه الله: ب = قال الامامُ البارعُ الاجلُ العلّامةُ استادُ الوري شرفُ الافاضِل عدمُ الاماثِل مَلكُ الفضلأ خاتمُ الحكماءِ محمودُ بن محمد بن عمرَ الفقيهِى الجَغْمينِيّ الخوارِزمِىّ رحمه الله: س = قال الامام البارع الاجل العلامه استاد الورى شرف الافاصل عدم الاماثل ملك الفضلا حاتم الحكمآء محمود بن محمد بن عمر ~~الفقهى~~ (؟) الجعمينى الخوارزمى رحمه له: ف.

[٧] نَقَلَ إليّ] س، ف = نُقل الى: ب.

[٨] مولانا] س = مولاى: ب، ف.

[٩] فخر الإسلام والمسلمين] س، ف = —ب.

[١٠] رحمه الله] س، ف = —ب.

[١١] يقرن] س، ف = تَقرُبُ: ب.

[١٢] المعاني] ب، ف = المعان: س.

[١٣] امتثال] س، ف = امثال: ب.

[١٤] شعراً] س = —ب، —ف.

[١٥] محلّ] س، ف = محلّى: ب.

[١٦] ٤٩آ: س.

[١٧] لي] فاس.

[١٨] المرجّي] س، ف = المروحىّ: ب.

[19] خطير] +عظم: تاس.

[20] لمثل] س، ف = بمثل: ب.

[21] امتثالاً] س، ف = امثالاً: ب.

[22] إلى] س، ف = لى: ب.

[23] مع] س، هاف(مع رمز «ح») = من: ف.

[24] مخبراً] س، ف = دالا: تاس(مع رمز «ح»).

[25] يشتمل] تشتمل: س = مشملا: ف.

[26] بسم الله ... على مقدّمة ومقالتيْن] بِسمِ ٱللهِ ٱلرَّحْمَنِ ٱلرَّحِيمْ الحمد لله كفاء(غير مقروء) افضاله والصلاة علي رسوله محمد وآلهِ قال قال(؟) الشيخ الامام الاجل البارع العلامه استاد الوري شرف الاماثل ملك الفضلا خاتمة الحكماءِ محمود بن محمد بن عمر الجعميني الخوارزمي تغمدهُ الله تعالي برحمته ان اعزة الاحباب وخالصة الاصحاب بدر الملة والدين فخر الاسلام والمسلمين عريز الملوك والسلاطين راحة الاشياخ وشفا الارواح محمد بن بهرام القلانسِي احمد الله عواقبه اشار ان اجمع في علم الهيية كتابا يقرن بين الاختصار والبيان ويجمع ايجاز اللفظ الي بسط المعاني فعددت ذلك من نعمه المتواليه وبادرت الي امتثال اشارته العاليه والفت هذا الكتاب علي حسب الامكان قاصداً للتلخيص فيه مع البيان وسميتُه الملخص في(غير مقروء) الهيئة اسمه مخبرا عن معناه وظاهره دالا علي فحواه وجعلته يشتمل علي مقدمة ومقالتين: ك = بسم الله الرّحمن الرّحيم الحمدُ للهِ كِفاَ اِفضاله والصلوة على نبيه محمد وآله نقول عبد الله الفقر الى رحمته محمودُ بن محمدِ بن عمر الجغمينى رحمه الله انى الفتُ هذا الكتابَ فى هيئه العالم تدكرةً منّى نعدِى لكل عالم مُتحرياً فه اللخيصَ مع البيان وانجازَ الالفاط الى بَسط المعانى على حسب الامكان وسميتُه الملخّص في الهيئه لكون اسمه دالا على معناه وظاهره مُخْبراً عن فحواهُ وحعلتُه يَشتمل على مقدمة ومقالتين: ل. || وألّفتُ هذا الكتاب ... على مقدّمة ومقالتيْن] فالفت هذا الكتاب فى هيئة [ص٧: ب] العالم تدكرةً منّى بَعْدى لكلّ عالمٍ مُتَحَرّيا فيه التّلحيْص مع اليان وايحازَ الالفاظ الى بسط المعانى على حَسب الامكان وسميته الملخِّص فى

الهيئة ليكون اِسمُه دالّاً على معناهُ وظاهرهُ مُحبراً عن معنا لحواه وجعلته(!) تشتمل على مقدّمة ومقالتين: ب.

٢٧ ٢آ: ك.

٢٨ المقدّمة في] اما المقدّمة ففي: ب.

٢٩ المقالة الأولى] طاك.

٣٠ آ] ف، ل = الاوّل: ب = ١: س = البَابُ الاول: ك.

٣١ ب] ف، ل = والثانى: ب = ٢: س = البَابُ الثاني: ك.

٣٢ بيان] —ل.

٣٣ حـ] ف، ل = والثالث: ب = ٣: س = البَابُ الثالث: ك.

٣٤ د] ف، ل = والرابع: ب = ٤: س = البَابُ الرابع: ك.

٣٥ هـ] ف، ل = والخامس: ب = ٥: س = البَابُ الخامس: ك.

٣٦ ٣آ: ف.

٣٧ هيئات] —ل.

٣٨ ٤٩ب: س.

٣٩ آ] ف، ل = الاوّل: ب =١: س = البَابُ الاول: ك.

٤٠ من الأرض] —ك.

٤١ ب] ف، ل = والثانى: ب = ٢: س = الباب الثاني: ك.

٤٢ حـ] ف، ل = والثالث: ب = ٣: س = الباب الثالث: ك.

٤٣ منفردة] منفرد: س.

المقدّمة في بيان أقسام الأجسام على الإجمال

٤٤ ومركّبات وهي التي تنقسم إلى أجسام مختلفة الطبائع] —ب، —ك.

٤٥ ٢ب: ك.

٤٦ ص٨: ب.

٤٧ بسيط] تاس(مع رمز «صح»).

[٤٨] وطبعه] وطبعيته: ك.

[٤٩] التشكيلات] الشكلات: ب.

[٥٠] لأسباب] –ف.

[٥١] نشاهد] نشاهدها: ب، ل.

[٥٢] من الوهاد والتلال] س، ف، ك = من الجبال والوهاد: ب، ل = +والجبال: هاس(مع رمز «خ»).

[٥٣] ٦٢آ: ل.

[٥٤] هذه] هذا: ف.

[٥٥] لا تقدح] ك، ل = لا نقدح: ب = لا يقدح: س، ف.

[٥٦] الشكل] الاسكال: ف.

[٥٧] لو] اذا: ب.

[٥٨] أُلزِقت بها حبّاتُ] أُزلِقت بها حبّاتُ: ف.

[٥٩] ٥٠آ: س.

[٦٠] شعير] شعيرة: ك.

[٦١] يقدح] +في كونها لو: شاف.

[٦٢] شكل] هال.

[٦٣] كريّ] +الشكل: ب.

[٦٤] عن سطحه] هاف(مع رمز «صح»).

[٦٥] المقعّر] المقـ المقعّر: ب.

[٦٦] أيضاً] –س، –ك.

[٦٧] الأصحّ] هاف(مع رمز «صح»).

[٦٨] يحيط] س، ك، ل = تحيط: ب = يحط: ف.

[٦٩] والأرض] ب، ك، ل = فالارض: س، ف.

[٧٠] فهو] وهو: ف.

[٧١] ٣آ: ك.

[٧٢] فهو محيط بها] –س.

٧٣ فلك عطارد] فلك العطارد: ب.

٧٤ ص٩: ب.

٧٥ الفلك الأعظم] فلك الأعظم: ف.

٧٦ لا خلاء ولا] («لا» مرتان فوق السطر في مخطوطة ل).

٧٧ ليس ورائه شيء لا خلاء ولا ملاء] ليس وراء خلائه وملاء: ب.

٧٨ وكلّ محيط يماسّ المحاط به الذي يليه في الترتيب المذكور] –ل.

٧٩ وبجملة] س، ف = وجملة: ب، ك، ل(هناك إشارة إلى الهامش حيث نجد «وعلى» مع رمز «صح» في مخطوطة ل).

٨٠ الأجرام] ب، ك، ل = الأجسام: س، ف.

٨١ عليها] ب، فاس، ك = –ف، –ل.

٨٢ وصورتها هذه] س، ف = وهذه صورة: ب = وهذه وصورتها: ك = صورة كرة العَالم: ل.

المقالة الأولى ، الباب الأوّل في هيئات الأفلاك

٨٣ ٣ب: ك؛ ٦٢ب: ل.

٨٤ وما يتعلّق بها] +وهي خمسة ابواب: ك، هال.

٨٥ المقالة] فاك.

٨٦ في هيئات] في بيان هيئات: ك.

٨٧ ص١٠: ب.

٨٨ فمركز سطحيْها] س، ف = فمركزهما: ب، ك، ل.

٨٩ هو مركزها] +الكرة: تال.

٩٠ ٦٣آ: ل.

٩١ هنا] ههنا: ب، ك = هاهنا: ل.

٩٢ الجهات] +بحيث: ب.

٩٣ حتّى] +لا: س.

٩٤ جزء] جزأً: ك.

٩٥ ١٥آ: س.

٩٦ جرم] جرمى: ب.

٩٧ مركزهما] +هما: شاف.

٩٨ تسمّى] +تلك النقطة: س.

٩٩ وتسمّى] وسمى: ف، ك = ويسمي: ك.

١٠٠ الحضيض] شاف = +الاوح ومقعر سطحه يماس لمقعر الاول على نعطة مشتركه سهما وسمى الحصض: هاف(مع رمز «صح»).

١٠١ ٤آ: ك.

١٠٢ بحيث] ب، ك، ل = حيث: س، ف.

١٠٣ تصل] يَصِلُ: ل.

١٠٤ من] —ب.

١٠٥ من مقعّره] هاف.

١٠٦ ٤آ: ف.

١٠٧ فبالضرورة] فالضروة: ل.

١٠٨ يصير] تصير: ب = تَصر: ل.

١٠٩ متوازيتيْ] موازيتي: س.

١١٠ غير متوازيتيْ السطوح بل] هال = +محا: شاك.

١١١ إحداهما] احدهما: س، ف.

١١٢ محوية] محبوبه: ك.

١١٣ فيه] له: ب.

١١٤ الحاوية] الحاوى: ب، ل.

١١٥ بالخلاف] بخلافها: ب.

١١٦ وتسمّى] ب، س = تسمى: ف = ويسمى: ك = سمّى: ل.

١١٧ واحدة] واحد: ب = —ك.

١١٨ يسمّى] ك، ل = سمّى: ب = تسمى: س = سمى: ف.

١١٩ يسمّى] ك = تسمّى: ب، ل = تسمى: س = سمى: ف.

١٢٠ ص١١: ب.

١٢١ محيطه] محيطها: ب.

١٢٢ ١٥ب: س.

١٢٣ مركوز] مركوزة: ب.

١٢٤ ويماسّ] ب، ك = وتماس: س = بماسّ: ف، ل.

١٢٥ بينها] بينهما: ب.

١٢٦ ٤ب: ك.

١٢٧ الخارجة] الخارجية: س.

١٢٨ المراكز] المركز: ك.

١٢٩ يماسّ] ك = بماس: ب، ف، ل = تماسّ: س = +سطحه: شاس(مع رمز «ح» فوق السطر).

١٣٠ حامله] حاملة: ب.

١٣١ هذه الأفلاك الصغار] س، ف = –ب، –ك، –ل.

١٣٢ فيها] –ك.

١٣٣ مصمت مركوز] ب، ك، ل = مركوز مصمت: س، ف.

١٣٤ ٦٣ب: ل.

١٣٥ التداوير] التدوير: ك.

١٣٦ منها] +وهذه صورة فلك الشمس: س، ف.

١٣٧ ٥٢آ: س؛ ٤ب: ف.

١٣٨ مشتمل] مشتملان: ك.

١٣٩ المركز] المراكز: ك.

١٤٠ أيْ] –ب.

١٤١ ص١٢: ب.

١٤٢ محدّبه] ب، ك، ل = محدّبها: س، ف = +المـ : شاك.

١٤٣ الممثّل] فاب.

١٤٤ وهي] س = وهو: ب.

١٤٥ ومقعّره] ك، ل = ومقعّرها: ب، س، ف.

١٤٦ ٥آ: ك.

١٤٧ نقطة] +كذلك: ك.

١٤٨ الخارجي المركز] ب، س، ل = الخارج المركز: ف، ك.

١٤٩ وهو المحوي] والمحوي: ك.

١٥٠ لمركز التدوير] –ك، –ل.

١٥١ جرم] هاس، –ف.

١٥٢ على الرسم] كذلك: ب، ك، ل = +اى بحيث يماس محدّب الحامل محدّب المدير على نقطةٍ مشتركةٍ بينهما وهى الاوج ومقعره مقعره على نقطه وهى الحضيض: ب.

١٥٣ والكوكب] ب، ف، ك، ل = الكواكب: س.

١٥٤ على ما ذكرنا في سائر التداوير] س، ف = على الرسم: ب، ك، ل.

١٥٥ ٥٢ب: س.

١٥٦ مركزهما] مركزها: ك.

١٥٧ يسمّى] ف، ك = سمّى: ب = وتسمى: س(الـ«و» مشطوب) = ويسمى: ل.

١٥٨ المائل] المائل وهو: هاف.

١٥٩ على نحو ما ذكرنا] س، ف = على الرسم: ب، ك، ل.

١٦٠ هذه] +الدور: شاك.

١٦١ ص١٤: ب؛ ٦٦آ: ل.

١٦٢ الثوابت] س، ف = الكواكب الثابتة: ب، ك، ل.

١٦٣ ويسمّى] ك = وسمّى: ب، ف، ل = وتسمى: س.

١٦٤ معنى] هاس(مع رمز «صح»).

١٦٥ فجرم] س، ك = وهو جرم: ب = جرم: ف، ل.

١٦٦ مركزه] هاب(مع رمز «صح»).

١٦٧ وهو] س = وهو جرم: ب = هو: ف، ك، ل.

١٦٨ سطحيْه] سطحه: ك.

١٦٩ ٥٣آ: س.

١٧٠ ومحدّبهما] ك، ل = ومحدّبها: ب، س، ف.

١٧١ والفلك الأعظم] واما الفلك الاعظم: ك.

١٧٢ مركزه] ف = +مركزة: شاف.

١٧٣ يماسّ] مماس: ب.

١٧٤ ومحدّبهما] ف، ك، ل = ومحدّبه: ب = ومحدّبها: س.

١٧٥ ٦آ: ك.

١٧٦ ملاء] +والله اعلم: ك.

المقالة الأولى ، الباب الثاني في حركات الأفلاك

١٧٧ الباب الثاني] قال الباب الثاني: ب.

١٧٨ حركات الأفلاك] —ب.

١٧٩ المشرق] +الى: شاس.

١٨٠ فأمّا] واما: ك.

١٨١ التي هي] —س = التى: هاس(مع رمز «صح»).

١٨٢ تتمّ] ف = تتمّ: ب = يتم: س، ك، ل.

١٨٣ وليلة] ولَيْلته: ك.

١٨٤ ٥ب: ف.

١٨٥ وتسمّى] وسُمّي: ب = وسمى: ف = ويسمى: س، ك، ل.

١٨٦ أوّل ما] هال(مع رمز «صح»).

١٨٧ ٥٣ب: س.

١٨٨ قطبي] قطب: ك.

١٨٩ ص١٥:ب.

١٩٠ ومنطقتها] ومِنْطَقُها: ب.

١٩١ مدير] فاس(مع رمز «صح»).

١٩٢ وتسمّى] ف، ك = وسمّى: ب = ويسمى: س.

١٩٣ لعطارد] –ف.

١٩٤ سلف] +وهي: ك.

١٩٥ ٦ب: ك.

١٩٦ وستعرفهما] ب، ل = وستعرفها: ف، ك.

١٩٧ وهي] وحركة: ب.

١٩٨ في] –ك.

١٩٩ ٠ نط ح ك] ف، ك، ل = ٥٩ دقيقةً ٨ ثوان ٢٠ ثالثةً: ب.

٢٠٠ وهو] وهي: ب.

٢٠١ وستعرفهما وهي في كل يوم بليلته ٠ نط ح ك وهو مثل وسط الشمس] هاس(مع رمز «صح»).

٢٠٢ وستعرفه] وستعرفها: ب.

٢٠٣ العالم] +وهي: ف.

٢٠٤ ٠ ج ي لز] س، ف، ك، ل = ٣ دقائق ١٠ ثوان ٣٧ ثالثه: ب.

٢٠٥ ٦٦ب: ل.

٢٠٦ ومنطقة البروج] وغير منطقة البروج: ف = البروج: ك.

٢٠٧ يا ط ز مجـ] س، ف، ل(يوجد «درجة» فوق «يا»؛ «دقايق» تحت «ط»؛ «ثوان» تحت «ز»؛ «ثالثه» فوق «مجـ») = ١١ درجة ٩ دقايق ٧ ثوان ٤٣ ثالثه: ب = يا ط ح مح: ك.

٢٠٨ التي] +هي: ب، ك، ل.

٢٠٩ فمنها] منها: ف.

٢١٠ تقطع] ف = نقطع: ب، ل = يقطع: س، ك.

٢١١ واحداً] هال.

٢١٢ وستين] +سنة: س.

[٢١٣] وستعرفها] وستعرفها: س.

[٢١٤] على] ــك.

[٢١٥] ٥٤آ: س.

[٢١٦] ومنطقة البروج] فاس(مع رمز «صح»).

[٢١٧] يسمّيان] وتُسمّيان: ب.

[٢١٨] ٧آ: ك.

[٢١٩] تقاطع] ف = تقاطع: ب، ل = يقاطع: س، ك.

[٢٢٠] ص١٦: ب.

[٢٢١] هذا] ــف = +في باب: شاك.

[٢٢٢] الدوائر] +ان شا الله تعالي: ك.

[٢٢٣] تتحرّك] ك، ل = تتحرك: ب، ف = يتحرك: س.

[٢٢٤] المركز] فال(مع رمز «صح»).

[٢٢٥] منطقة] +البروج وقطبيها: مشطوب في ل.

[٢٢٦] مسامتة] متشابهة: س.

[٢٢٧] وقطبين] +على: فاب(بين الـ«و» والـ«ق»).

[٢٢٨] فلك] ــف.

[٢٢٩] مسامته لمنطقة ... البروج] هال(مع رمز «صح»).

[٢٣٠] البروج] +وقطبيها: شاس(مع رمز «خ» فوق السطر).

[٢٣١] ٠ نط ح كَ] س، ف، ل = ٥٩ دقيقه ٨ ثوان ٢٠ ثالثه: ب = ح نط ح ك : ك.

[٢٣٢] حركات] حركة: ك.

[٢٣٣] الأفلاك] ــف = هال(مع رمز «صح»).

[٢٣٤] ٦آ: ف.

[٢٣٥] مناطق] منطق: ك.

[٢٣٦] وأقطابهما] وأقطابها: ك.

٢٣٧ ٠ ب ٠ له] س، ف، ك = دقيقتان: ب = ٠ ب ه(«٠» من تحت «ه»؟) له: ل.

٢٣٨ للمشتري] ف، ل = وللمشتري: ب، س، ك.

٢٣٩ ٠ د نط يو] ف، س، ل = ٥ دقايق: ب = مد ه(؟) نط يو(الـ«ي» مهمل): ك.

٢٤٠ للمرّيخ] ف، ل = وللمرّيخ: ب، س، ك.

٢٤١ ٠ لا كو م] س، ف، ك، ل = ٣١ دقيقةً: ب.

٢٤٢ ٧ب: ك.

٢٤٣ للزهرة] ف، س، ل = وللزهرة: ب، ك.

٢٤٤ ٠ نط ح ك] س، ف، ك، ل = ٥٩(مع رمز «٢» من تحت) دقيقه ٨ ثوان ٢٠ ثالثه: ب.

٢٤٥ لعطارد] ف، س، ل = للعطارد: ب، ك.

٢٤٦ ا نح يو م] س، ف، ك، ل = وضعف ذلك(مع رمز «٢» تحت «ذلك»، وهو إشارة إلى الـ«٢» السابق المرافق للزهرة): ب.

٢٤٧ للقمر] ف، ل = وللقمر: ب، س، ك.

٢٤٨ ٥٤ب: س.

٢٤٩ كد كب نج كب] ك، ل = ٢٤ درجة ٢٣ دقيقه: ب = كد كج نج كب: س = كد(+«كب» و « نح» في الهامش) كب: ف.

٢٥٠ وتسّمى] ك = وسمّى: ب، ل = ويسمى: س، ف.

٢٥١ الكوكب] الكواكب: ك.

٢٥٢ تسمّى] ب = ويسمّى: س = سمى: ف، ل = نسمي: ك.

٢٥٣ بعينها] ب، ك = تعينها: س = نعسها: ف = نعنها: ل.

٢٥٤ وسنزيد] ل = وسيزيد: ب = وسيرد: س = وسرىد: ف = سَينزيد: ك.

٢٥٥ تسمّى] سمّى: ب، ف = ويسمى: س، ل = يسمي: ك.

٢٥٦ أيضاً] –ب = هال(مع رمز «صح»).

٢٥٧ أمّا] س، ف = وأما: ب، ك، ل.

٢٥٨ أفلاك] هاس(مع رمز «صح»).

٢٥٩ مخالفة] +في: ب.

٢٦٠ ص١٧: ب.

٢٦١ ٦٧آ: ل.

٢٦٢ المغرب] +فحركة الاسفل: شاس.

٢٦٣ لتداوير] ب، ف، ل = كتداوير: س = التداوير: ك.

٢٦٤ حركة الأعلى] –ك.

٢٦٥ لتدوير] ب، ف، ل = كتدوير: س = التداور: ك.

٢٦٦ المذكور] المسير: ف.

٢٦٧ المعتبر] –ب.

٢٦٨ من] منه: ك.

٢٦٩ وهو] وهي: ب.

٢٧٠ ٨آ: ك.

٢٧١ ٥٥ آ: س.

٢٧٢ لزحل] +النحس: شاك.

٢٧٣ ٠ نز ز مد] س، ف، ك، ل = ٥٧ دقيقةً: ب.

٢٧٤ للمشتري] وللمشتري: ب.

٢٧٥ ٠ ند ط جـ] س، ف، ك، ل = ٥٤ دقيقةً: ب.

٢٧٦ للمرّيخ] وللمريخ: ب.

٢٧٧ ٠ كز ما م] س، ف، ك، ل = ٢٨ دقيقةً: ب.

٢٧٨ للزهرة] وللزهرة: ب.

٢٧٩ • لو نط كط] س، ف، ل(يتغير الـ«لو» إلى «لز» والـ«نط» إلى «يط» في الثلاث نسخ) = ٣٧ دقيقةً: ب = مد لو نط كط : ك.

٢٨٠ لعطارد] ولعطارد: ب.

٢٨١ جـ و كد ز] ف، ك، ل = ٣ اجزاء ٦ دقايق: ب = ج و كد كب: س(+«و» و«كد» و«ز» من تحت «و» و«كد» و«كب»).

٢٨٢ يج جـ نجـ نو] س، ف، ل = ١٣ درجة ٤ دقايق: ب = حح ح نو : ك.

٢٨٣ الاختلاف] الاخلاف: س.

٢٨٤ والحركة] حركة: ب.

٢٨٥ للكوكب] س، ف = للكواكب: ب، ك.

٢٨٦ والله اعلم] ــب، ــك.

المقالة الأولى ، الباب الثالث في الدوائر

٢٨٧ تنصّف] ينصف: ف.

٢٨٨ لا] ــب.

٢٨٩ تنصّفه] ينصّفه: ف.

٢٩٠ ولتسمَّ] س، ف = وتسمى: ب، ك = وسمى: ل.

٢٩١ معدّل النهار] اما الدواير العظام فمنها معدل النهار: ك.

٢٩٢ وتسمّى] ب، ف، س = ويسمي: ك = وسمى: ل.

٢٩٣ الفلك المستقيم] فلك المستقيم: ك.

٢٩٤ اعتدل] اعتدال: ب.

٢٩٥ ص١٨: ب.

٢٩٦ ٨ب: ك.

٢٩٧ الأرض] الا:ب(+«رض» فوق).

٢٩٨ ٦ب: ف.

٢٩٩ توهّمنا] توهينا: ك.

٣٠٠ للعالم] لعالم: ب(+«للعالم» في الهامش).

٣٠١ الموازية] المتوازيه: ك.

٣٠٢ اليوميّة] +بل المعدل ابضا يسمى مدار يوميا: هال(مع رمز «صح»).

٣٠٣ تفرض] ب، ل = يفرض: س، ك = نفرض: ف.

٣٠٤ دائرة البروج] ومنها دايرة البروج: ك.

٣٠٥ وتسمّى] +هذة: شاك.

٣٠٦ ٥٥ب: س.

٣٠٧ والدوائر] فالدّوائر: ب.

٣٠٨ تسمّى] ب، ك، ل = يسمى: س= سمى: ف.

٣٠٩ الدائرة] الدواير: ك.

٣١٠ تقدّر] يقدر: س.

٣١١ طول] —ف.

٣١٢ والشمس] —ك.

٣١٣ ٦٧ب: ل.

٣١٤ بمراكز الكواكب] بمركز الكوكب: ب.

٣١٥ وحينئذ] وحـ : ف = وح: ك.

٣١٦ للكوكب] له: فاس.

٣١٧ البروج] فلك البروج: ك.

٣١٨ وطرف] وتطرف: ك.

٣١٩ مقاطعة] قاطعة: ب.

٣٢٠ ٩آ: ك.

٣٢١ عرض حينئذ] س، ل = حسئذ عرض: ب = عرض حـ : ف، ك.

٣٢٢ ص١٩: ب.

[٣٢٣] النقطة] تلك النقطه: ب.

[٣٢٤] وهذا هو] س، ف = وهو: ب، ك، ل.

[٣٢٥] ٥٦آ: س.

[٣٢٦] والدوائر الموازية لها تسمّى] وتسمي الدواير الموازية لها : ك. تسمّى] ب، ك، ل = يسمى: س = سمى: ف.

[٣٢٧] من] هال(مع رمز «صح»).

[٣٢٨] أن] طاب = هاب.

[٣٢٩] يأخذ] ب، ك، ل = تأخذ: س، ف.

[٣٣٠] عنه] ب، س = —ف، —ك، —ل.

[٣٣١] ٩ب: ك.

[٣٣٢] نقطة] بنقطة: ب.

[٣٣٣] نقطة] بنقطة: ب.

[٣٣٤] فتُعيّن] س، ف = فيتعين: ب، ك، ل.

[٣٣٥] نقط] نقطة: ب.

[٣٣٦] ومدّة] وقد: شاف = مده: تاف(مع رمز «صح»).

[٣٣٧] ٧آ : ف.

[٣٣٨] أربعة] اربع: ب.

[٣٣٩] نتوهّم] ب، ك، ل = يتوهّم: س، ف.

[٣٤٠] كلّ] فاف.

[٣٤١] واحدة] س، ف، ك = واحدٍ: ب، ل.

[٣٤٢] عن] من: ب.

[٣٤٣] ٥٦ب: س.

[٣٤٤] ثمّ نتوهّم] ك، ل = فيتوهم: ب = ثمّ يتوهم: س، ف.

[٣٤٥] تتقاطع] س، ك، ل = يتقاطع: ب = سقاطع: ف.

[٣٤٦] البروج] +احديهما: ل.

[٣٤٧] ٦٨آ: ل.

٣٤٨ تمرّ] ثم: ك.

٣٤٩ ص ٢٠: ب.

٣٥٠ العالم وبقطبي البروج] البروج وقطبى العالم: ب.

٣٥١ وبنقطتي] ونقطى: ب.

٣٥٢ نقطتا] نقطى: ب.

٣٥٣ تمرّ] ثم: ك.

٣٥٤ بنقطتي] ىمصـ ىقطه: ب.

٣٥٥ نقطتا] نقطى: ب.

٣٥٦ والأخرى تمرّ بنقطتي الاعتدالين وقطباها نقطتا الانقلابين] —ف.

٣٥٧ بالنقط] س، ل = بالمقطه: ب، ف = لنقط: ك.

٣٥٨ الربعين] ربعين: ك.

٣٥٩ نقط] نقطهٍ: ب.

٣٦٠ أخر] أُخرى: ب.

٣٦١ هي على] وهي: ف.

٣٦٢ ١٠آ: ك.

٣٦٣ المقابلين] ف، ك، ل = المتقابلين: ب، س = +للمفـ: ب

٣٦٤ كلّ قسم منها يسمّى] ك، ل = كل قسم منها تسمّى: س، ف = تسمّى كل قسم: ب.

٣٦٥ منها] منهما: ب.

٣٦٦ تسمّى] ب، ل = يسمّى: س، ف، ك.

٣٦٧ تسمّى] ب = يسمّى: س = تسمى: ف، ك، ل.

٣٦٨ الموهومة] المتوهّمة: ب.

٣٦٩ تنقسم] س، ك = ىنقسم: ب، ف = ينقسم: ل.

٣٧٠ أيضاً] —ب.

٣٧١ تفصل] +ما: ك.

٣٧٢ ٥٧آ: س.

٣٧٣ سَمْتا] سمتى: ب.
٣٧٤ لإحداهما] لاحدهما: س.
٣٧٥ وخطّ الاعتدال] —ب.
٣٧٦ لها] فاس(مع رمز «صح»).
٣٧٧ يقال لها] تسمّى: ب.
٣٧٨ المُقَنْطرات] مقنطرات: ك.
٣٧٩ النهار] هاك.
٣٨٠ ص٢١: ب؛ ١١آ: ك.
٣٨١ وسمتي] وبستي: ك.
٣٨٢ نقطتا] نقطةَ: ب.
٣٨٣ وتنصّف] تتصف: س.
٣٨٤ إحداهما] احدهما: س، ف.
٣٨٥ دائرة الارتفاع] ومنها دارة الارتفاع: س.
٣٨٦ وتسمّى] ب، ك = ويسمى: س = وسمى: ف، ل.
٣٨٧ أيضاً الدائرة السمتيّة] الدائرة السمتيه ايضا: ك.
٣٨٨ هي] س = وهي: ب، ف = —ك، —ل.
٣٨٩ الخطّ] فاس(مع رمز «صح»).
٣٩٠ ٥٧ب: س؛ ٧ب: ف.
٣٩١ مارّاً] +بكوك: شاس.
٣٩٢ بمركز] بمراكز: ف.
٣٩٣ الكوكب] الكواكب: ك.
٣٩٤ منتقلتين] س، ف، ل = متقالتين: ب = منقلتن: ك.
٣٩٥ حسب] ك، ل = بحسب: ب = على حسب: س(الـ«على» مشطوب؟)، ف.
٣٩٦ ٦٨ب: ل.
٣٩٧ الكوكب] الكواكب: ل.
٣٩٨ تسمّى] ك = سمى: ب، ف = يسمى: س، ل.

٣٩٩ واحدة] واحد: س.

٤٠٠ من] التى من: ب.

٤٠١ نقطتي] +البروج: ك.

٤٠٢ ١١آ: ك.

٤٠٣ يسمّى] ب، ك = تسمّى: س = سمى: ف، ل.

٤٠٤ بليلته] س، ك، ل = بلمليه: ب = بلمله: ف.

٤٠٥ دائرة أوّل السُمُوت] ف، ك، ل = دائرة اول السموة وهى: ب = ومنها دارة اول السموت: س .

٤٠٦ وبنقطتي] وبنقطه: ب.

٤٠٧ ص٢٢: ب.

٤٠٨ دائرة] هاس(مع رمز «صح»).

٤٠٩ سمت] سمتي: ك.

٤١٠ دائرة المَيل] +وهي: ب.

٤١١ ٥٨آ: س.

٤١٢ معدّل النهار] العالم: ب.

٤١٣ عن معدّل النهار] عنه: ب.

٤١٤ دائرة العرض] +وهى: ب.

٤١٥ الكوكب] ب، ك = الكواكب: س، ف = —ل.

٤١٦ سطح] +الفلك: ك.

٤١٧ ١١ب: ك.

٤١٨ وبطرف الخطّ الخارج ... إلى سطح الفلك الأعظم] هاس(مع رمز «صح») = —ل.

٤١٩ ويُعرف بها] وبها يعرف: ب.

٤٢٠ الكوكب] هاك(مع رمز «صح»).

٤٢١ الدوائر] ومنها الدّواير: ب = الدائرة: ك = الدوار(مع «و» تحت السطر): ل.

٤٢٢ بدور النقط في أفلاك السيّارة] هاس(مع رمز «صح»)، ك، ل = —ب = بدور النقطه فى افلاك السياره: هاف(مع رمز «صح»).

[423] وهي] هی: ل.

[424] بسائط] +«سطح» تحت السطر: س.

[425] الدوائر المتوهّمة ... الاُكر] ومنها الدّواير المتوهّمةُ المرتسمةُ بحركه مركز الكوكب او فلك التدوير وهی اما مرتسمةٌ على سطح الأُكر: ب.

[426] على بسائط الاُكر وامّا مرتسمة] –ك.

[427] البسائط] السطح: ب.

[428] فالمرتسمة] والمرتسمه: ل.

[429] البسائط] السطح: ب.

[430] حركات] حركة: ف.

[431] الحاملة] الحاصله: ك.

[432] على محيطات الأفلاك الحاملة ومن حركات مراكز الكواكب على محيطات أفلاك التداوير] –ب.

[433] الفلك] هاس(مع رمز «صح»).

[434] ص٢٣: ب.

[435] حركات] ب، س = حركة: ف، ك، ل.

[436] الحاملة] الحاصله: ك.

[437] من] +حركات: ب.

[438] بأفلاك] بالافلاك: ف.

[439] وهذه] وهی («هی» مشطوب و«هذه» مكتوب فوق السطر بخط غير الناسخ).

[440] ١٠آ: ف.

[441] الحاملة] الحاصله: ك.

[442] ٥٨ب: س.

[443] ١٢آ: ك.

[444] ٦٩آ: ل.

[445] ولكون] +ولكون: هاف.

[446] ارتسمت] اوارسمت: ف.

٤٤٧ البروج وقطبي] هال(مع رمز «صح»).
٤٤٨ تقاطع] س، ل = تقطعُ الدّوائر المسمّاة بالافلاك: ب = تقاطع: ف = بقاطع: ك.
٤٤٩ إحداهما] –ك.
٤٥٠ على] عن: ب.
٤٥١ تسمّى] س، ك = يسمّى: ب = سمّى: ف، ل.
٤٥٢ البسائط] السطوح: ب.
٤٥٣ الحامل] الحاصل: ك.
٤٥٤ حامل] حاصل: ك.
٤٥٥ المائل] هاف(مع رمز «صح»).
٤٥٦ حامل] الحامل: ف = حاصل: ك.
٤٥٧ الفلك] س، ك، ل = بالفلك: ب، ف.
٤٥٨ الحامل] الحاصل: ك.
٤٥٩ لمركز الحامل] ب، ف، ل = هاس («الحامل» غير مقروء) = لمركز الحاصل: ك.
٤٦٠ اذ مركز الحامل] –ك.
٤٦١ محيطها] +ومنها الفلك المعدّل للمسير وهى دايرةٌ ترتسم بحركة الخطّ الخارج من نقطهٍ بكون قطر التدوير على صوبها دايماً كيف ما دارتْ وسيزيد وضوح هذا فى باب الدوائر[!]: ب.

المقالة الأولى ، الباب الرابع في القسيّ

٤٦٢ ص٢٤: ب.
٤٦٣ القطعة] القوس: ب.
٤٦٤ ١٢ب: ك.
٤٦٥ بالأجزاء] س، ف، ك = من الاجزآء: ب، ل.
٤٦٦ يتمّ بها] س = بها تمّ: ب = بم بها: ف = –ك = بها: ل.
٤٦٧ ٣٦٠] س، ف، ك، ل = ثلثمائة وستين: ب، +هال.

٤٦٨ تلك] –ك.

٤٦٩ ٥٩آ: س.

٤٧٠ وتمامها] +ومنها:هاس(مع رمز «صح»).

٤٧١ البلد] +وهي: ب، س(تحت السطر مع رمز «صح»).

٤٧٢ بآخر] آخر وباحى: ف.

٤٧٣ مبدأ] بهذا: ك.

٤٧٤ هي] وهي: س(الـ«و» مشطوب).

٤٧٥ يطلع] س = بطلع: ف = تطلع: ك، ل.

٤٧٦ مطالع كلّ قوس من فلك البروج هي ما يطلع معها من معدّل النهار] س، ف، ك، ل = منها(ومنها: هاس) المطالع اذا طلع قوسٌ من دآيرة البروج فلا بد وان يطلع معهُ(معها: هاس) قوسٌ من معدّل النهار وهذا القوس(+الذى هو: هاس) من معدّل النهار يقال له مطالع تلك القوس التى هى من فلك البروج: ب، هاس(مع رمز «خ»).

٤٧٧ محصورة] +مقصورة: تاس(مع رمز «خ»).

٤٧٨ الميل] +الميول: تاس(مع رمز «خ»).

٤٧٩ فهو] ب، س، ف = هي: ك = فهى: ل.

٤٨٠ مطالع] ومنْها مطالع: ب = ومطالع: س، ف.

٤٨١ من فلك البروج] –س، –ف، هال(مع رمز «صح»).

٤٨٢ قوس] وهى قوس: ب.

٤٨٣ النهار] +ما: ب.

٤٨٤ منه] ب = –س، –ك = هاف(مع رمز «صح»)، هال(مع رمز «صح»).

٤٨٥ مع] +من: تاس(مع رمز «خ»).

٤٨٦ ١٣آ: ك.

٤٨٧ تعديل النهار] ومنها تعديل النهار: ب.

٤٨٨ هو] وهي: ب، +تاس.

٤٨٩ يلي] يل: ك.

٤٩٠ ص٢٥: ب.

٤٩١ غير] –س، –ف.

٤٩٢ ١٠ب: ف.

٤٩٣ ٥٩ب: س.

٤٩٤ ميل] مثل: ك.

٤٩٥ الميل] +والجوزا: شاك.

٤٩٦ ٦٩ب: ل.

٤٩٧ بدرج] بفلك: ك.

٤٩٨ يقسّم] يَقْسم: ب = يقسم: ك = يُقسِّمُ: س = يُقسم: ف = يُقسِّم: ل.

٤٩٩ مثلّثين] هاس = المثلّثَ: س.

٥٠٠ وتحيط] ب = وىحيط: س، ف، ل = ويحيط: ك.

٥٠١ من] –ك.

٥٠٢ بين] هاس(مع رمز «صح»).

٥٠٣ وقوس] قوس: ك.

٥٠٤ بين] –س.

٥٠٥ ما بين الأفق وبين نقطة التقاطع بين دائرة الميل وبين معدّل النهار] ف، ك، ل = –ب = هاس(مع رمز «صح»).

٥٠٦ من] –ف.

٥٠٧ النهار] هاف(مع رمز «صح»).

٥٠٨ النهار لرأس الجوزاء] ب، س = طاف = نهار راس الجوزا: ك، ل.

٥٠٩ تكون] س = ىكون: ب، ف، ل = يكون: ك.

٥١٠ تختلف] ل = ىحتلف: ب = ىختلف: س = ىحلف: ف = ىختلفه: ك.

٥١١ وسط الشمس] ومنها وسط الشمس: ب، س(«ومنها» في الهامش مع رمز «صح»).

٥١٢ وهو] –ف، –ك، –ل.

٥١٣ ٦٠آ: س.

٥١٤ ما بين أوّل الحمل وبين رأس خطّ] وما بين اول ولراس: ك.

٥١٥ المركز ويمرّ] المركز الي: س(«الي» مشطوب و«ويمر» فوق السطر مع رمز «صح») = المركز الى: ف.
٥١٦ ويمرّ بمركز] بمركز: س(«ويمر بمركز» تحت السطر مع رمز «ح صح») = مركز: ف.
٥١٧ ص٢٦: ب.
٥١٨ ١٤آ: ك.
٥١٩ تقاطعا] هال(مع رمز «صح»).
٥٢٠ التعديل] +ومنها: ب، هاس(مع رمز «صح»).
٥٢١ قوس] هو قوس: ب.
٥٢٢ وطرف] س، ك، ل = وبين طرف: ب، ف.
٥٢٣ عن] +من: تاس(مع رمز «ح»).
٥٢٤ فيتوهّم] فنتوهّم: ل.
٥٢٥ ١٣آ: ف.
٥٢٦ مقاطعة] قاطعة: ب.
٥٢٧ فالقوس] والقوس: ف.
٥٢٨ ٦٠ب: س.
٥٢٩ ٧٠آ: ل.
٥٣٠ ١٤ب: ك.
٥٣١ طرفه] طرفي: ب.
٥٣٢ نقطة] نقطتى: ب.
٥٣٣ من] س، ف = بين: ب، ك، ل = +س: تاس(مع رمز «ح»).
٥٣٤ ص٢٧: ب.
٥٣٥ أحدهما] احديهما: ب.
٥٣٦ المارّان] ب = +المارّو: شاب.
٥٣٧ أو كانت] –ك.
٥٣٨ الكواكب] الكوكب: ف.
٥٣٩ ينطبق] مطبق: ف.

٥٤٠ ١٥آ: ك.

٥٤١ ٦١آ: س.

٥٤٢ واثنان] –ك.

٥٤٣ متساويان] متساابان: ل.

٥٤٤ نطاقات] نطاقان: ك.

٥٤٥ يمرّ] ف = تمر: ب، س = بمر: ك، ل.

٥٤٦ الأوسطين] الاولين: ف.

٥٤٧ المركز] تاس(مع رمز «صح»).

٥٤٨ الخارجان] الخارج: ف.

٥٤٩ أحدهما] اهاحدهما: س.

٥٥٠ ص٢٨: ب.

٥٥١ ١٥ب: ك.

٥٥٢ ومركزه] +التقاطع: شاك.

٥٥٣ يمرّ] ك = تمر: ب، س = بمر: ف، ل.

٥٥٤ والحامل] +وهذه صورتُهُ: ب = +بهذه الصورة: ك.

٥٥٥ ٦١ب: س؛ ٧٠ب: ل.

٥٥٦ ١٣ب: ف.

٥٥٧ والآخر] +هو: س.

٥٥٨ يمرّ] تمر: ب، س = بمر: ف، ك، ل.

٥٥٩ بحيث] به حث: ب.

٥٦٠ جانب] جانبي: ب.

٥٦١ جزءاً] +درجة: هاس(مع رمز «ح»).

٥٦٢ ١٦آ: ك.

٥٦٣ الحامل] الحاصل: ك.

٥٦٤ يمرّ] ك = تمر: ب، س = بمر: ف، ل.

٥٦٥ نقطتي] نقطه: ب.

٥٦٦ الحامل] الحاصل: ك.

٥٦٧ ص٢٩: ب.

٥٦٨ مجاوزته] مجاوزيه: ب.

٥٦٩ على توالي حركته] +بهذه الصوره: ك.

٥٧٠ ٦٢آ: س.

٥٧١ أو التدوير] س، ك = او التداوير: ب = والتدوير: ف، ل.

٥٧٢ ١٦ب: ك.

٥٧٣ النطاقين] النطاقتين: ك.

٥٧٤ فهو صاعد] +قد تُجعل صاعداً فى النطاق الاول والرابع هابط فى الاخرين وهذه صورتُهُ: ب.

٥٧٥ عرض البلد] (بياض في ب).

٥٧٦ هو] (مصحّح من «هى» في س).

٥٧٧ ص٣٠: ب.

٥٧٨ وهو] وذلك: ب = +ودلك: تاس(مع رمز «ح»).

٥٧٩ القطب] +من دايرة نصف النهار: ك.

٥٨٠ الميل] (بياض في ب).

٥٨١ والميل] –ك.

٥٨٢ والميل] الميل: ل.

٥٨٣ من دائرة العرض] –ب.

٥٨٤ البروج] +من دائرة العرض: ب.

٥٨٥ غاية الميل] –ب.

٥٨٦ الكلّي] ب، ك، ل = الكل: س، ف.

٥٨٧ قوس] هال(مع رمز «صح»).

٥٨٨ دائرة المارّة] الدايرة الماره: ك.

٥٨٩ حدّ] –ل.

٥٩٠ ١٧آ: ك.

٥٩١ ١١آ: ف.

٥٩٢ كجـ له] (بياض في ب).

٥٩٣ عرض الكوكب] (بياض في ب).

٥٩٤ ٦٢ب: س.

٥٩٥ الكوكب] الكواكب: ف.

٥٩٦ كانت] كان: ك.

٥٩٧ من] +من: ك.

٥٩٨ الميل] –ك.

٥٩٩ ٧١ آ: ل.

٦٠٠ ارتفاع الكوكب] (بياض في ب).

٦٠١ على] هي: ب = +على: هاب.

٦٠٢ وبين الأفق فإن انطبقت دائرة الارتفاع على دائرة نصف النهار فتلك القوس] هاف(مع رمز «صح»).

٦٠٣ اختلاف المنظر] (بياض في ب).

٦٠٤ ص٣١: ب.

٦٠٥ والآخر] +من: ك.

٦٠٦ ١٧ب: ك.

٦٠٧ أعني] +بحط: ف.

٦٠٨ سطح الأرض] هاف(مع رمز «صح»).

٦٠٩ هذا فيما] –ك.

٦١٠ محسوسة] مخصوصة: ب.

٦١١ سعة المشرق] (بياض في ب).

٦١٢ قوس من دائرة] طاس.

٦١٣ ٦٣آ: س.

٦١٤ المشرق] هاك(مع رمز «صح»).

٦١٥ قد] هاب.

٦١٦ سلفا] +والله اعلم: ك.

٦١٧ السمت] (بياض في ب).

٦١٨ الطالع] ف، ل =المطالع: ب، س، ك.

٦١٩ من] —س.

٦٢٠ سمت القبلة للبلد] (بياض في ب).

٦٢١ نصف] +النهار: شاس.

٦٢٢ البلد] —ب.

٦٢٣ رؤوس] راوس: ك.

٦٢٤ ورؤس] الروس وروس: ف(الروس مشطوب) = راوس: ك.

٦٢٥ مكّة] +شرفها الله: ك.

٦٢٦ قوس النهار] (بياض في ب).

٦٢٧ ١٨آ: ك.

٦٢٨ مغربها ومشرقها] مشرقها ومغربها: ب.

٦٢٩ قوس نهار الكوكب] (بياض في ب).

٦٣٠ مشرقه] مشرقى: ب.

٦٣١ فوق الأرض] —ك.

٦٣٢ التي] ب، هاس(مع رمز «صح»)، ك = —ف، —ل.

٦٣٣ منها] ف، ل = —ب، —ك = شاس.

٦٣٤ الأرض] +هي: ك.

٦٣٥ الدائر من الفلك] ف، ل = (بياض في ب) = الدايره من الفلك: س(الـ«ه» مشطوب و«الداير» تحت السطر مع رمز «ح»)، ك.

٦٣٦ ص٣٢: ب.

٦٣٧ بالنهار] +ما سن نطير حراها وافق(١١ب) وافق المشرق بالنهار: شاف.

٦٣٨ جزئها] +من معدل النهر: شال.

٦٣٩ نظير] هال(مع رمز «صح») = نطيره: س.

٦٤٠ واحدة] ل = واحد: ب، س، ف، ك.

المقالة الأولى ، الباب الخامس فيما يعرض للكواكب في حركاتها

٦٤١ ٦٣ب: س.
٦٤٢ ٧١ب: ل.
٦٤٣ مركزها خارج عن] فال(مع رمز «صح»). عن] هاس(مع رمز «صح»).
٦٤٤ مركز] مركزها: ك(الـ«ها» مشطوب).
٦٤٥ كان] كانت: ب.
٦٤٦ ١٨ب: ك.
٦٤٧ أكثر] واكثر: س(الـ«و» مشطوب).
٦٤٨ النصف] هال(مع رمز «صح»).
٦٤٩ نصف الحضيض] ف، ك، ل = النصف الحضيض: ب، س = +الذى فيه: هاس(مع رمز «صح»).
٦٥٠ وكانت] س، ف = ولمّا كانت: ب، ك، ل(«لمّا» في الهامش مع رمز «صح»).
٦٥١ يخالف] يختلف: ب.
٦٥٢ الثاني] الآخر: ب.
٦٥٣ فترى] س(نقطتان تحت الـ«تـ» مشطوبتان) = فَنُرى: ب = فيرى: ف، ك، ل.
٦٥٤ حركتها] حركاتها: ب.
٦٥٥ الأوج] الاوجى: ب.
٦٥٦ قطعها] –ب.
٦٥٧ وحركتها] وحرتها: ب.
٦٥٨ يُحتاج] نُحتاج: ب، س، ف، ل = تحتاج: ك.
٦٥٩ أو] و: ل.
٦٦٠ ص٣٣: ب.
٦٦١ ٦٤آ: س.

٦٦٢ وسطها] +الوسط: تاس(مع رمز «ح»).

٦٦٣ أو منه] ب، تاس(مع رمز «ح»)، ك، هال(مع رمز «صح») = —ف.

٦٦٤ لِيتحقّق] ك = لتتحقق: ب = لتحقق: س = لسحقق: ف = لحقو: ل.

٦٦٥ الاختلافات] الاحلاف: ف.

٦٦٦ الطول] —ك.

٦٦٧ أحدها] احدهما: ب.

٦٦٨ الاختلاف] +الاختلاف: ك.

٦٦٩ ١٩آ: ك.

٦٧٠ لها] —ك.

٦٧١ الكوكب] الكواكب: س.

٦٧٢ الكوكب] الكواكب: س.

٦٧٣ وتقويمه] والتقويم: ب.

٦٧٤ وأمّا إذا] س، ف، ل = فامّا اذا: ب = فاذا: ك.

٦٧٥ الذروة] الدورة: س.

٦٧٦ اختلاف] ب، ف، ل = الاختلاف: س(الـ«لا» فوق «خـ»)، ك.

٦٧٧ بين] +بين: ب.

٦٧٨ الاختلاف] هاف(مع رمز «صح»).

٦٧٩ تكون] ب = يكون: س، ك = نكون: ف، ل.

٦٨٠ التداوير] التدوير: ب.

٦٨١ نصف] —ف.

٦٨٢ وأنصاف] وانصفاف: س

٦٨٣ التداوير] التدوير: ب.

٦٨٤ لزحل و لَ] س، ف، ك، ل = ٦ اجزاءٍ و ١١٢: ب.

٦٨٥ للمشتري يا لَ] س، ف، ك، ل = —ب.

٦٨٦ ٦٤ب: س.

٦٨٧ للمرّيخ لط ل] س، ل = –ب = ل ط ل : ف، كـ.

٦٨٨ للزهرة مه ٠] س، ف، ل = وللزهرة ٤٣ جزءً و ١١٦: ب = م ٥ : كـ.

٦٨٩ ١٢آ: ف.

٦٩٠ لعطارد كه ٠] س، ف، ل = ولعطارد ٢٢ جزءاً و ١١٢ جزءٍ: ب = كب ل: كـ.

٦٩١ للقمر و كـ] س، ف، كـ، ل = وللقمر ٥ اجزآءٍ ١١٤: ب = +من كل ذلك بالاجزآء التى تكون بها نصف قطر حامل الكوكب او مايل القمر ستين جزءاً الى: ب (يوجد خطّ أفقي فوق السطر من «كل» إلى «جزءاً»).

٦٩٢ اختلاف] الاختلاف: ب.

٦٩٣ ١٩ب: كـ.

٦٩٤ ثانٍ] الثانى: ب.

٦٩٥ للكواكب] –ب.

٦٩٦ قرب] قريب: كـ.

٦٩٧ ص٣٤: ب.

٦٩٨ الحامل] الحاصل: كـ.

٦٩٩ فيرى] س(يوجد «تـ» فوق «يـ»)، ف، كـ، ل = ويُرى: ب.

٧٠٠ واختلافه أعظم] هال(مع رمز «صح»).

٧٠١ ٧٢آ: ل.

٧٠٢ اختلاف ثالث] الاختلاف الثالث: ب.

٧٠٣ حينئذ] ب، س، ل = ح : ف = علي: كـ.

٧٠٤ الخطّ] +المستقيم: هاس(مع رمز «صح»).

٧٠٥ بمراكز] بمركز: ب.

٧٠٦ والحامل] والحاصل: كـ.

٧٠٧ إذا] اذ: ب.

[٧٠٨] أو الحضيض] س، ف = والحضيض: ب، ك، ل.
[٧٠٩] ولا تبقى] +مطمه عليه: شاف.
[٧١٠] ولا] +على: هاس(مع رمز «صح»).
[٧١١] بل] +هو: ب.
[٧١٢] مركز الحامل ولا مركز العالم بل على صوب] –ك.
[٧١٣] المتحيّرة] +؟: ب.
[٧١٤] أو] و: ك.
[٧١٥] هذا الفصل] ب، ف = هذا لفصل: س = هذا لفضل: ك.
[٧١٦] الحامل] الحاصل: ك.
[٧١٧] الحامل] الحاصل: ك.
[٧١٨] ٢٠آ: ك.
[٧١٩] ٦٥آ: س.
[٧٢٠] بينها] ب، ك، ل = بينهما: س = طاف.
[٧٢١] نقطة] +مما يلى الاوج بعدها: شاف.
[٧٢٢] بين] س، ف = –ب، –ك، –ل.
[٧٢٣] لهذا] ب، هاس(مع رمز «صح»)، ك، ل = بهذا: ف.
[٧٢٤] الفصل] الفضل: ك.
[٧٢٥] عن] –ك.
[٧٢٦] ممّا يلي] +الحضيض كبعد: شاف.
[٧٢٧] فإذا درا] فادار: ف.
[٧٢٨] ص٣٥: ب.
[٧٢٩] الحامل] الحمل: ب.
[٧٣٠] العالم] +على دايرة واحدة: ب.
[٧٣١] النقط] س، ف، ل = النقطة: ب، ك.
[٧٣٢] المذكورة] المذكوة: س.
[٧٣٣] تكون] ل = نكون: ب، ك = يكون: س(يوجد«تـ» مشطوب فوق الـ«يـ»)، ف.

٧٣٤ المذكورة] هال(مع رمز «صح»).

٧٣٥ النقط] ف، ك(؟)، ل = النقطة: ب، س.

٧٣٦ منطبقاً] +علي: ك.

٧٣٧ ٢٠ب: ك.

٧٣٨ عنه] عنها: ك.

٧٣٩ كيف ما] +دارت اعنى لو احرج من هذه النقط خطوط الى مراكز التداوير يكون كل خطّ (١٢ب) منها منطبقا على القطر المذكور للتدور لا ىفك عنه كىف: ف.

٧٤٠ بدوران] بدروان: ب.

٧٤١ ٦٥ب: س.

٧٤٢ تسمّى] ك = سمّى: ب، ف، ل = وتسمى: س(يوجد «تـ» فوق الـ«يـ» والـ«و» مشطوب).

٧٤٣ يعتدل] ب، ك = تعدل: س(يوجد «يعتدل» تحت السطر مع رمز «ح») = نعدل: ف = نعتدل: ل.

٧٤٤ أي] +اى: شاف.

٧٤٥ هو] وهو: س(الـ«و» مشطوب).

٧٤٦ الوسطى] والوسطي: س(الـ«و» مشطوب).

٧٤٧ ٧٢ب: ل.

٧٤٨ هو] س، ك، ل = يسمّى: ب = وهو: ف.

٧٤٩ الذروة] +المرتعبة(؟): شاك.

٧٥٠ المرئيّة] +وليذكر: ب.

٧٥١ وأبعاد] س، ف، ل = ابعاد: ب، ك.

٧٥٢ النقط] لنقطه: ب.

٧٥٣ المركز] –ب.

٧٥٤ ب كط ل] س، ف، ك، ل = بحزان ٢٩ دقيقه ٣٠ ثانيةً: ب.

٧٥٥ للقمر] وللقمر: ب.

٧٥٦ ي يط] ك، ل = ١٠ اجزاءٍ ١٩ دقيقه: ب = ى ىط(مع رمز «ح»): س = ى ىط ٠: ف.
٧٥٧ وهو] وهو هو: ل.
٧٥٨ عن] س، ف = من: ب، ك، ل.
٧٥٩ للمتحيّرة] س، ف، ل = وللمتحيّرة: ب، ك.
٧٦٠ خلا] هاس(يوجد «ح» تحت الـ«خـ »).
٧٦١ ص٣٦: ب.
٧٦٢ وذلك] شاس = +وذلك: س.
٧٦٣ للمسير] +عنه وذلك اعنى بعد مركز المعدل للمسر: شاف.
٧٦٤ ٢١آ: ك.
٧٦٥ و ن] س، ف، ك، ل = ٦ اجزاءٍ ٥٠ دقيقه: ب.
٧٦٦ للمشتري] وللمشترى: ب.
٧٦٧ ه ل] س، ف، ك، ل = ٥ اجزاءٍ و ١\١٢: ب.
٧٦٨ للمرّيخ] وللمريخ: ب.
٧٦٩ يب] ف، ك، ل = ١٢ جزءاً: ب = ىت ح ٠ : س.
٧٧٠ للزهرة] وللزهرة: ب.
٧٧١ ب ه] ف، ك، ل = جزءآن و٥ دقايق: ب = ىاه: س.
٧٧٢ فلكه] فلكها: ب.
٧٧٣ حامله] خاملة: ب.
٧٧٤ وبين مركز العالم وبعد مركز حامله عن مركز مديره] –ك.
٧٧٥ حتّى] تاس(مع رمز «صح»).
٧٧٦ إذا] هاف(مع رمز «صح»).
٧٧٧ ممّا] ما: ب.
٧٧٨ ٦٦آ: س.

٧٧٩ المعدّل] لمعدل: ب.

٧٨٠ كلّ] وكل(الـ«و» مشطوب): س.

٧٨١ جـ ي] س، ف، ك، ل = (بياض في ب).

٧٨٢ ط ل] س، ف، ك، ل = (بياض في ب) = +الله اعلم: ك.

٧٨٣ الاختلاف] ف، ك = الاختلافات: ب، س(الـ«ات» زائد)، ل.

٧٨٤ الشمس] فالشمس: ب.

٧٨٥ فلك] دايرة: ب.

٧٨٦ الفلك] +الممل: شاك.

٧٨٧ ويسمى] ك، ل = وسمى: ب، ف = وتسمى: س.

٧٨٨ الخارج] الفلك الخارج: ك.

٧٨٩ ٢١ب: ك.

٧٩٠ ب ل] ف = (بياض في ب) = د(؟) ل : ك = و ل: س(يوجد «ب» فوق الـ«و» بخط ولون آخر)، ل.

٧٩١ ا ل] س، ف، ك، ل = (بياض في ب).

٧٩٢ ا ٠] س، ف، ل(الـ«٠» متغيّر إلى «ه») = (بياض في ب) = ب ي : ك.

٧٩٣ ٠ ي] ف، ل = (بياض في ب) = ٠ ي: هاس(مع رمز «صح») = د ي : ك.

٧٩٤ ص٣٧: ب.

٧٩٥ ٠ مه] س، ف، ل = (بياض في ب) = و مه : ك.

٧٩٦ ه ٠] س(مع رمز «صح» تحت الـ«ه»)، ف، ك، ل = (بياض في ب).

٧٩٧ ٣٨آ: ف.

٧٩٨ أفلاكه] —ك.

٧٩٩ ونعني] س، ك، ل = وتعنى: ب = وعنى: ف.

٨٠٠ قد] هاف، فال(مع رمز «صح»).

٨٠١ ويسمّى] ف = وسمّى: ب، ل = وتسمى: س = –ك.

٨٠٢ عرض] ب، س، ل = عروض: ف = –ك.

٨٠٣ ٦٦ب: س.

٨٠٤ وحضيضه عن الفلك المائل ويسمّى عرض التدوير] –ك.

٨٠٥ ٠ لب] ف، ل = (بياض في ب) = د ل : ك = و لب : س(يوجد«٠» على الـ«و» بخط ولون غير الناسخ).

٨٠٦ ٠ لح] س، ف، ل = (بياض في ب) = ب ل : ك.

٨٠٧ و يو] و نو: ل = و نو: س، ف = (بياض في ب) = ب ل : ك.

٨٠٨ ا ب] س، ف، ل = (بياض في ب) = د(؟) ل : ك.

٨٠٩ ا مه] س، ف، ل = (بياض في ب) = ا يه : ك.

٨١٠ ويسمّى] ك = ويُسمى: ب، ف، ل = وتسمى: س.

٨١١ ب ل] س، ف، ك، ل = (بياض في ب).

٨١٢ ٧٣آ: ل.

٨١٣ جاوزها] جا: ف(«وزها» في الهامش).

٨١٤ ٢٢آ: ك.

٨١٥ ابتدأ] ابتداء: ب، س= ابتدا: ف، ك = اتبدا: ل.

٨١٦ عليه] على: ب.

٨١٧ الآخر] الآخرى: ب.

٨١٨ فلك] +الفا: شاك.

٨١٩ ص٣٨: ب.

٨٢٠ فإذا] +فإذا: ك.

٨٢١ عادت] عادة: س.

٨٢٢ إلى] س(مع رمز «صح»)، ف = –ب، –ك، –ل.

٨٢٣ الحالة] الحالت: ك.

٨٢٤ ويلزم] ب، س، ف = فيلزم: ك، ل.

٨٢٥ ٦٧آ: س.

٨٢٦ عنه] —ب.

٨٢٧ فلك البروج] +والمايل: ب.

٨٢٨ الجنوب] +عن المايل: ب.

٨٢٩ يبلغ] ب، ك(مع «نـ» فوق «يـ») = ببلغ: س، ف = سلغ: ل.

٨٣٠ ٢٢ب: ك.

٨٣١ الانتقاص] الاسقاض: ب.

٨٣٢ ثانياً] ثاثياً: ك.

٨٣٣ فلك البروج] المفلك الممثل: ك.

٨٣٤ المركز] س، ك، ل = —ب = مركز: ف.

٨٣٥ جاوزه] حاورب: ف.

٨٣٦ ٨ب: ف.

٨٣٧ الشمال] +عن المايل: ب.

٨٣٨ وانتقاصه] وانتقاضه: ب.

٨٣٩ السفليين] س، ف، ل = السفلين: ب، ك.

٨٤٠ فلك] الفلك: ل.

٨٤١ المائل] البروج: شال(+«المائل» في الهامش مع رمز «صح»).

٨٤٢ ٦٧ب: س.

٨٤٣ تبتدئ] يبتدئ ب، ك = بيتدى: س = سدى: ف، ل.

٨٤٤ ولعطارد] لعطارد: ب.

٨٤٥ إلى] —ك.

٨٤٦ ص٣٩: ب.

٨٤٧ ويبلغ] ك = وبلغ: س = وتبلغ: ب = وسلع: ف، ل.

٨٤٨ غايته] غايتها: ب.

٨٤٩ وازدياده] س، ف، ل = وازياده: ب، ك.

٨٥٠ وانتقاصه] وانقاضه: ب.

٨٥١ والانطباق] ــك.

٨٥٢ وأمّا الانحراف] +والالتواء: ب.

٨٥٣ فابتداؤه] فابتداه: س، ف.

٨٥٤ ٢٣آ: ك.

٨٥٥ ٧٣ب: ل.

٨٥٦ ميله] الميل: ب.

٨٥٧ وفي] في: ك.

٨٥٨ إلى الشمال وفي عطارد إلى الجنوب والغربي في الزهرة] هال(مع رمز «صح»).

٨٥٩ إلى الجنوب] ــب.

٨٦٠ وإن] واذا: ب.

٨٦١ أنّ] ــك.

٨٦٢ مدّة] مداره: ف(الـ«دا» مشطوب).

٨٦٣ ولقطري] ولقطر: ك.

٨٦٤ أرباع دوراتها] بلوغ دووراتها: ك.

٨٦٥ ولنذكر ههنا] ب، س = ولنذكر منها ههنا: ك(«منها» مشطوب) = ولنذكرها هي: ف = ولنذكرها هنا: ل.

٨٦٦ أمّا] واما: ك = +واما: هاك.

٨٦٧ ٦٨آ: س.

٨٦٨ جزءاً] +واوج المشرى مىقدم عن المىصف لا على التوالى بعشرىن جزا: ف.

٨٦٩ ٢٣ب: ك.

٨٧٠ على] س، ف = ــب، ــك، ــل.

٨٧١ الكواكب] الكوكب: ب.

٨٧٢ ص٤٠: ب.

٨٧٣ أمّا] واما: ك.

٨٧٤ موضع] س، ف = مواضع: ب، ك، ل.

٨٧٥ فهي لأوّل] س، ف، ل = فهي في اوّل: ب = فهي الاول: ك.

٨٧٦ غثيز] س (+«١٣١٧»(!) تحت السطر)، ف(؟) (+«غ»، «ى»، «ر» تحت السطر) = (بياض في ب) = غشيز: ك، ل.

٨٧٧ كز ي لج] س، ف، ك، ل = (بياض في ب).

٨٧٨ لزحل] +فى: ف.

٨٧٩ آ٩: ف.

٨٨٠ ط كج لج] س، ف، ك، ل = (بياض في ب).

٨٨١ يط كج لج] س، ف، ك، ل = (بياض في ب).

٨٨٢ يا نج مو] س، ف، ك، ل = (بياض في ب).

٨٨٣ كز ي لج] س، ف، ك، ل = (بياض في ب).

٨٨٤ كو كج لج] ف(«ر» فوق«و» لـ كو) ، ك، ل = (بياض في ب) = كر كح لح : س.

٨٨٥ الجوزهريْ] س، ف، ك، ل = الجوزهرّ: ب.

٨٨٦ يط كج لج] س، ف، ل = (بياض في ب) = يط لح لح : ك.

٨٨٧ للمشتري] س، ف، ك، ل = في المشترى: ب = +فيه ايضا: هاس(مع رمز «صح»).

٨٨٨ ط كج لج] س، ف، ل = (بياض في ب) = ط لح لح : ك.

٨٨٩ يا نج مو] س، ف، ك، ل = (بياض في ب).

٨٩٠ الحُوت] الحوة: ب.

٨٩١ كز ي لج] س، ف، ك، ل = (بياض في ب).

٨٩٢ كو كج لج] س، ك، ل = (بياض في ب) = لر لح لح : ف.

٨٩٣ يُزاد] يزد: ك.

[894] الرجوع] والرجوع (الـ«و» مشطوب في س).
[895] الكوكب] الكواكب: ف.
[896] ٦٨ب: س.
[897] وذلك أنّ الكوكب إذا كان في أعلى تدويره كانت] –ك.
[898] ٢٤آ] ك.
[899] فيُرى] فترى: ب.
[900] جعل يميل] اخذ بميل: ك.
[901] لما] كما: ب.
[902] تعرف] س(يوجد«يـ» تحت « تـ» وهو مشطوب) = يعرف: ب، ك = نعرف: ف = عرفت: ل.
[903] تساويا] تساوتا: س.
[904] ص٤١: ب.
[905] ٧٤آ: ل.
[906] يقيم] تقيم: ك.
[907] لهذا] لهذ: ب.
[908] دورته] س، ف = دوره: ب، ل = ذروته: ك.
[909] تسمّى] ك = نسمّى: ب، ف = يسمى: س، ل.
[910] تسمّى] ك، س(يوجد«يـ» تحت « تـ» وهو مشطوب) = نسمّى: ب، ف، ل.
[911] فإنّ بعد] فابعد: ك.
[912] ٢٤ب: ك.
[913] أبداً] ابد: ف، ك.
[914] ٦٩آ: س.
[915] الشمس] ب، ك، هال (مع رمز «صح») = –س، –ف.
[916] تبعد] ك = ببعد: ب = يبعد: س = نعد: ف = يبعد: ل.
[917] يبعد] ل = يبعد: س = نعد: ب، ف = ببعد: ك.
[918] بمقدار] بمقدارها: ب.

٩١٩ الكوكب] الكواكب: س(الـ«ا» مشطوب؟).

٩٢٠ ذروة] ذرة: ك.

٩٢١ التدوير] التداوير: س(الـ«ا» مشطوب).

٩٢٢ حتّى إذا قابلت الشمس مركز التدوير] –ك.

٩٢٣ حضيض التدوير] الحضيض التدور: س(الـ«ال» مشطوب ويوجد «س» تحت «التدور» مع رمز «صح»)، ف.

٩٢٤ احتراقاتها] اختراقاتها: ك.

٩٢٥ وهي] –ب، –ك.

٩٢٦ في] فال(مع رمز «صح»).

٩٢٧ ٩ب: ف.

٩٢٨ وهي] –ك.

٩٢٩ قطر] +فلك: هاس(مع رمز «صح»).

٩٣٠ ممثل] مميل: ك.

٩٣١ فمركزا] ب، س، ل = فمركز: ف، ك.

٩٣٢ يتسامتان] س، ف = مسامتان: ب، ك، ل.

٩٣٣ بمركز] س، ف = لمركز: ب، ك، ل.

٩٣٤ ص٤٢: ب.

٩٣٥ يبعدان] بيعدان: س.

٩٣٦ يقارناها] يقارنها: ك.

٩٣٧ الحضيض] +اى ولكون مركزى تدويرىهما ابدا مسامتان لمركز الشمس: هال(مع رمز «صح»).

٩٣٨ عند ذروة التدوير وفي نصف الرجوع وذلك عند الحضيض ولذلك] –ف.

٩٣٩ ٢٥آ: ك.

٩٤٠ المُحاق] المخاف: ك.

٩٤١ ٦٩ب: س.

٩٤٢ وكسفه الشمس والخسوف] والكسوف والخسوف: س(يوجد «الشمس»(؟) تحت «الكسوف») .
٩٤٣ إنّما] وانما: ك.
٩٤٤ كالمرآة] —ب.
٩٤٥ المواجه] المواجهة: س.
٩٤٦ مستضيئاً] مضئا: ب.
٩٤٧ والنصف] والنص: ك.
٩٤٨ المظلم] +هو هو جها: شاك.
٩٤٩ نرى] تري: ك(+«فلا ىر» مشطوب).
٩٥٠ فإذا] واذا: س، ف.
٩٥١ مقداراً] +ما: ب.
٩٥٢ ٧٤ب: ل.
٩٥٣ فنرى] فنرا: ك.
٩٥٤ طرفاً] طرقا: ك.
٩٥٥ ازداد] ازد: ك.
٩٥٦ قابلها] قابلنا: ك.
٩٥٧ ٢٥ب: ك.
٩٥٨ ص٤٣: ب.
٩٥٩ أو الذنب] ب، ل = والذنب: س، ف، ك.
٩٦٠ ٧٠آ: س.
٩٦١ كسوف] كسف: ف.
٩٦٢ هو لون] ولون: ب.
٩٦٣ بها] —ب = ها: ك.
٩٦٤ أيضاً] هاس(مع رمز «صح»).
٩٦٥ ١٤آ: ف.
٩٦٦ لما ذكرنا من] لذلك: ب، ل.

٩٦٧ على القمر] +وابتداء خسوف القمر: شاس.

٩٦٨ إليه] اليها: ك.

٩٦٩ فيبقى] فبقى: ف، ل.

٩٧٠ (يوجد صورتان لخسوف القمر في مخطوط س: ٦٩ب، ٧٠آ.)

٩٧١ وابتداء خسوف القمر] س، ف= –ب، –ك = وسدى خسوف القمر: ل.

٩٧٢ يلحقه] س، ف = يلحق: ب، ك، ل.

٩٧٣ وكذلك] لذلك: ب.

٩٧٤ منه] –ك.

٩٧٥ ٢٦آ: ك.

٩٧٦ يعرض] يعرف: ك.

٩٧٧ أنّ] لان: ك.

٩٧٨ تدويره] التدوير: ف.

٩٧٩ ولتكن] ب، س(يوجد «يـ» تحت« تـ» وهو مشطوب)، ك = وليكن: ف = ولىكن: ل.

٩٨٠ مثلاً رأس الحمل] راس الحمل مثلا: ك.

٩٨١ عنه] فاك.

٩٨٢ بحركة] تحرك: ك.

٩٨٣ ٧٠ب: س.

٩٨٤ يا ط ز مجـ] س، ف، ك، ل = (بياض في ب).

٩٨٥ وبحركة] وتحرك: ك.

٩٨٦ الجوزهر] جوزهر: س.

٩٨٧ ٠ جـ ي لز] س، ف، ك، ل = (بياض في ب).

٩٨٨ فتصير] فيصير: ب، س(يوجد « تـ» فوق«يـ»)، ك = فصير: ف، ل.

٩٨٩ يا يب يح كـ] س، ف، ك، ل = (بياض في ب).

٩٩٠ وتحرّك] وتحركت: ف.

٩٩١ ص٤٤: ب.

٩٩٢ كد كب نجـ كب] س، ف، ل = (بياض في ب) = كد كح نح لب : ك.

٩٩٣ المائل] المسايل: ب.

٩٩٤ يا يب يح ك] س، ف، ك، ل = (بياض في ب).

٩٩٥ فيبقى] فبيقى: س.

٩٩٦ للمركز] المركز: ك.

٩٩٧ يجـ ي له ب] س، ف، ك، ل = (بياض في ب).

٩٩٨ بالتقريب] بالقرب: شاس(يوجد «بالتقريب» تحت السطر مع رمز «صح»).

٩٩٩ ٧٥آ: ل.

١٠٠٠ عن] من: ف.

١٠٠١ ٢٦ب: ك.

١٠٠٢ يب يا كو ما] ف، ك، ل = (بياض في ب) = يت ىا كو ما : س.

١٠٠٣ فتكون] فيكون: ف.

١٠٠٤ تربيعه] تربيعة: س.

١٠٠٥ للشمس] س، ك، ل = الشمس: ب، ف.

١٠٠٦ الحضيض] حضيضه: س، ف.

١٠٠٧ الاستقبال والاجتماع] ب، س، ل = الاستقبال والاحما: ف = الاجتماع والاستقبال: ك.

١٠٠٨ فيكون] +الاوج: شاك.

١٠٠٩ يبلغ] بيلغ: س.

١٠١٠ دورة] ذروة: ب.

١٠١١ ٧١آ: س.

١٠١٢ ١٤ب: ف.

١٠١٣ لكنّ] ولكنّ: س(الـ«و» مشطوب).

١٠١٤ الآخر] والاخر: س(الـ«و» مشطوب).
١٠١٥ تحرّكا] تحرك: ك.
١٠١٦ ص٤٥: ب.
١٠١٧ للأوج] س، ف، ل = الاوج: ب، ك.
١٠١٨ الدورة] الذروة: ب.
١٠١٩ ويتقاطران] وتقاطران: ب.
١٠٢٠ ٢٧آ: ك.
١٠٢١ والآخر] —ل.
١٠٢٢ السرطان] +والله اعلم: ك، ل.

المقالة الثانية ، الباب الأوّل في المعمور من الأرض وعرضه وطوله وقسمته إلى الأقاليم

١٠٢٣ في بيان الأرض] في هيئة الارض: ب.
١٠٢٤ وهي] ك، فال = وفيها: ب = وهي على: س، ف.
١٠٢٥ وقسمته إلى الأقاليم] —ب.
١٠٢٦ ونفرض] س = يفرض: ب، ك، ل = نفرص: ف.
١٠٢٧ إحداها] احدها: س = احديها: ب، ف، ك، ل.
١٠٢٨ الاستواء] فال.
١٠٢٩ تعرف] ب، ك = يعرف: س، ف، ل.
١٠٣٠ أفق] الافق: ب.
١٠٣١ ٧١ب: س.
١٠٣٢ والثالثة] والثالث: س.
١٠٣٣ تقطع] ك = يقطع: ب، س = نقطع: ف، ل.
١٠٣٤ تنصّف] ب، ل = ينصف: س = نصف: ف = +الارض بنصفين: شاك.
١٠٣٥ نصفيها] ف، ل = نصفها: ب، س = نصفيهما: ك.
١٠٣٦ فيصير] ب، ف = فنصير: س = فتصير: ك = فصير: ل.

[١٠٣٧] الشماليين] ب، ك، ل = الشمالين: س، ف.

[١٠٣٨] يرى] ترى: ب.

[١٠٣٩] ٢٧ب: ك.

[١٠٤٠] تقطع] ك = نقطع: ب، ف، ل = يقطع: س.

[١٠٤١] المعمور] المعمورة: ف.

[١٠٤٢] ٧٥ب: ل.

[١٠٤٣] بين] ب، س، ل = من: ف = —ك.

[١٠٤٤] سو] س(ويوجد«٦٦٠» زائد تحت السطر)، ف، ك، ل = —ب.

[١٠٤٥] إلى] على: ك.

[١٠٤٦] عمارة إلى بعد] هاب.

[١٠٤٧] ص٤٦: ب.

[١٠٤٨] يو كه] س، ف، ل = ١٦ درجةً و٢٥ دقيقةً: ب = —ك.

[١٠٤٩] هذا] —ب.

[١٠٥٠] فب كه] س، ف، ل = ٨٢ درجة و٨٥ دقيقةً (!): ب = يو كه : ك.

[١٠٥١] وطول] فطول: ب.

[١٠٥٢] المعمور] ب، ل، ك = العمارة: س = المعموره: ف.

[١٠٥٣] قف ٠] س، ف، ك، ل = (بياض في ب).

[١٠٥٤] يأخذه] يأخذ: ب.

[١٠٥٥] المحيط] —ك.

[١٠٥٦] وبعضهم] +ياخذ: س.

[١٠٥٧] واغلة] ب، ك، ل = داخله: س(يوجد «واغلة» تحت السطر مع رمز «صح»)، ف(يوجد «واغله» تحت السطر مع رمز «نخ»).

[١٠٥٨] عن] من: ب.

[١٠٥٩] ساحله] ساحلة: ب = ساحلها: ك.

١٠٦٠ ي ٠] س، ف، ك، ل = (بياض في ب).

١٠٦١ ١٥آ: ف.

١٠٦٢ قسم] تقسم: ك.

١٠٦٣ هذا] هاف(مع رمز «صح»).

١٠٦٤ المعمور] المعموره: ف.

١٠٦٥ ٧٢آ: س.

١٠٦٦ خطّ الاستواء] +وسمّى بها الاقاليم: ب = وتسمي وتسمي الاقاليم: ك.

١٠٦٧ وابتدأ] فابتدأ: ب، ك.

١٠٦٨ وابتداء الإقليم الأول منه] هال(مع رمز «صح»). منه] من خط الاستوآء: ب.

١٠٦٩ يب] س، ف، ك، ل = ١٢: ب.

١٠٧٠ أعني النهار] –ل.

١٠٧١ يب مه] س، ف، ك، ل = ١٢ ساعةً و ٤٥: ب = +ساعة: تاس، ك.

١٠٧٢ يب ل] س، ف، ك، ل = (بياض في ب) = +درجه: تاس.

١٠٧٣ يج ٠] س، ف، ك، ل = (بياض في ب).

١٠٧٤ ٢٨آ: ك.

١٠٧٥ يو كز] س، ف، ل = (بياض في ب) = يو لز: ك.

١٠٧٦ الإقليم] س، ف، ك = –ب، –ل.

١٠٧٧ يج يه] س، ف، ك، ل = (بياض في ب).

١٠٧٨ ك يد] ل = (بياض في ب) = كد ٠ : س (الـ«كد» متغيّر إلى «ك» و«كر» تحت السطر) = كد ٠ : ف = كد يد : ك.

١٠٧٩ يج ل] س، ف، ك، ل = (بياض في ب).

١٠٨٠ كج نا] س، ف، ك، ل = (بياض في ب) = +«كد م» : هاس.

١٠٨١ يج مه] س، ف، ك، ل = (بياض في ب).

١٠٨٢ كز يب] (بياض في ب) = كط يب : س(الـ«كط» متغيّر إلى «كز»)، ف، ك، ل =كز: هال.

١٠٨٣ يد ٠] س، ف، ك، ل = (بياض في ب).

١٠٨٤ ل كب] س، ف، ك، ل = (بياض في ب).

١٠٨٥ وابتدأ] +الاقليم: ك.

١٠٨٦ يد يه] س، ف، ك، ل.

١٠٨٧ لج يح] س، ف، ل = لح لح : ك.

١٠٨٨ يد ل] س، ف، ك، ل.

١٠٨٩ وابتدأ الرابع حيث النهار يد يه والعرض لج يح ووسطه حيث النهار يد ل والعرض لو ٠] –ب. لو ٠] س، ف، ل = لو كب : ك.

١٠٩٠ وابتدأ] +الاقليم: ك.

١٠٩١ يد مه] ف، ك، ل = (بياض في ب) = يد ٠ : س(يوجد «مه» تحت السطر).

١٠٩٢ لح له] س، ف، ك، ل = (بياض في ب).

١٠٩٣ يه ٠] س، ف، ك، ل = (بياض في ب).

١٠٩٤ ص٤٧:ب.

١٠٩٥ م نو] س(يوجد «ما» متغيّر إلى « م») = (بياض في ب) = ما نو : ف = ما نه : ك = مد نو : ل(يوجد «م» في الهامش وتحته «ما يه»).

١٠٩٦ وابتدأ] +الاقليم: ك.

١٠٩٧ يه يه] س، ف، ك، ل = (بياض في ب).

١٠٩٨ مجـ نا] س(«به» زائد تحت السطر)، ف، ك، ل(يوجد نقطتان زائدان تحت الـ«نا» و«كب» أيضاً تحت السطر بخط غير الناسخ و«يه» في الهامش بخط غير الناسخ) = (بياض في ب).

١٠٩٩ يه ل] س، ف، ل = (بياض في ب) = يه لب : ك.

١١٠٠ مه ا] س، ف، ل = (بياض في ب) = مد با(؟): ك (يوجد خطّ أفقي فوق «با»).

١١٠١ وابتدأ] +الاقليم: ك.

١١٠٢ يه مه] س، ف، ك، ل = (بياض في ب).

١١٠٣ مو نا] س، ف، ل = (بياض في ب) = مو نب : ك.

١١٠٤ يو ٠] س، ف، ل = —ب، —ك.

١١٠٥ ٧٢ب: س.

١١٠٦ ووسطه حيث النهار يو ٠ والعرض مح لب] —ك. مح لب] س، ف، ل = (بياض في ب).

١١٠٧ ٢٨ب: ك.

١١٠٨ ن كه] س، ف، ل = (بياض في ب) = ن له : ك.

١١٠٩ عرض] +ما بين: ب، ك.

١١١٠ ابتداء] هاس(مع رمز «صح»).

١١١١ وسطه] وسط: ك.

١١١٢ لتفرّق] لتعرف: ك.

١١١٣ الأقاليم] س، ف، ل = الاقليم: ب = الاقليم الاول: ك.

١١١٤ وراء] تاس(مطموس في النصّ).

١١١٥ خطّ] وسط: ك.

١١١٦ من العمارة ولهذا أيضاً لا يعدّ بعضهم ما بين خطّ الاستواء إلى عرض] هاس(مع رمز «صح»). عرض] –ف = +إلي: س.

١١١٧ يب ل] س، ف، ك، ل = (بياض في ب).

١١١٨ ولا] والي: ك.

١١١٩ ن كه] س، ف، ك، ل = (بياض في ب).

١١٢٠ فإنّ] +ما: ب.

١١٢١ سجـ] س، ف، ل = (بياض في ب) = ..ح: ك.

١١٢٢ جزيرة] +تسمّى تولى: ب.

١١٢٣ يسكنون] +في: ك.

١١٢٤ سد] ف، ك، ل = (بياض في ب) = شد: س.

١١٢٥ ٧٦آ: ل.

١١٢٦ الصقالبة] الصقاليه: ك.

١١٢٧ سو] س، ف، ك، ل = –ب.

١١٢٨ سكّانها] هال(مع رمز «صح»).

١١٢٩ شبيهة] ف، ك، ل = شبة: ب = شبية: س.

١١٣٠ الوحوش] س، ف، ل = الوخوش ومن هذه الدائرة تصوّر الاقاليم: ب = بالوحوش: ك.

المقالة الثانية ، الباب الثاني في خواصّ خطّ الاستواء والمواضع التي لها عرض

١١٣١ ١٥ب: ف.

١١٣٢ التي] +لا عرض: شاس.

١١٣٣ الباب الثاني في خواصّ خطّ الاستواء والمواضع التي لها عرض] –ب.

١١٣٤ وكذا] وكذى: ف، ك.

١١٣٥ ٢٩آ: ك.

١١٣٦ معدّل] المعدل: ف(الـ«الـ» مشطوب).

١١٣٧ المدارات] المدرات: ف.

١١٣٨ ٧٣آ: س.

١١٣٩ ويكون] فيكون: ب.

١١٤٠ ولا] +تكون: ب.

١١٤١ ص٤٩: ب.

١١٤٢ ويغرب] +هو: ف.

١١٤٣ وتكون] وكون: ب، ف = ويكون: س، ك، ل.

١١٤٤ للمدارات] للمدرات: ف = +التي فوق الارض متساويه للقسي: ك.

١١٤٥ كالتي] التي: ك.

١١٤٦ فلذلك] ولذلك: ب.

١١٤٧ كليله] كليلته: ب

١١٤٨ فلك] الفلك: ب(الـ«الـ» مشطوب).

١١٤٩ يبلغ] س = تبلغ ب، ك = سلع: ف = سلغ: ل.

١١٥٠ الآفاق] هال(مع رمز «صح») = الافلاك: ل.

١١٥١ بنصفين] –ك، –ل.

١١٥٢ ٢٩ب: ك.

١١٥٣ وتقطع] ل = نقطع: ب، ف، ك = ويقطع: س.

١١٥٤ المدارات] المدرات: ف.

١١٥٥ بقطعتين] بقطين بقطعتين: ب (الـ«بقطين» مشطوب) = بنقطتين: ك.

١١٥٦ مختلفتين] هال(مع رمز «صح»).

١١٥٧ فالقسيّ] س = والقسيّ: ب، ك، ل = القسى: ف.

١١٥٨ للمدارات] للمدرات: ف.

١١٥٩ وللجنوبية] والجنوبية: ل.

١١٦٠ ولذلك] +لا: س.

١١٦١ ٧٣ب: س.

١١٦٢ يستوي الليل والنهار فيها] يستوي فيها الليل والنهار: ك.
١١٦٣ عرض] العرض اى عرض: ب.
١١٦٤ ص٥٠: ب.
١١٦٥ وبقدر] ف، ل = وبقدر: ب = ويقدر: س(الـ«يـ» مشطوب)، ك.
١١٦٦ المدارات] المدرات: ف.
١١٦٧ تليه] ك، ل = يليه: ب = بليه: س، ف.
١١٦٨ فكلّما] وكلما: ب،
١١٦٩ ١٦آ: ف.
١١٧٠ ٧٦ب: ل.
١١٧١ فازداد] وازداد: ب.
١١٧٢ ٣٠آ: ك.
١١٧٣ فازداد] س، ف = وازداد: ب، ك، ل.
١١٧٤ والمدارات] والمدرات: ف.
١١٧٥ على] +الارض: شاس.
١١٧٦ مدار] ب، تاس(مع رمز «صح»)، ك، ل = مقدار: شاس، ف.
١١٧٧ القطب] +او اقلّ: ك ، هال(مع رمز «صح» و«عن سطح دايرة الافق» تحت السطر).
١١٧٨ ٧٤آ: س.
١١٧٩ تحويه] بحويه: ب، س = بجويه: ك = يحويه: ل.
١١٨٠ مثل ارتفاع القطب فهو بجميع ما فيه وجميع ما تحويه دائرته إلى القطب الشمالي] ـف.
١١٨١ الكواكب] الكوكب: ب.
١١٨٢ بجميع] ب، ل = بجميع: س = بجمع: ف = لجميع: ك.
١١٨٣ أبديّ] ابدا: ك.
١١٨٤ لم] ـب.

١١٨٥ يبلغ] هاب، ك = تبلغ: س = سلع: ف، ل.

١١٨٦ جزءاً] ب، ف، ل = جزء: س، ك.

١١٨٧ يخصّ] +بعضها: شاس.

١١٨٨ خواصّ] س، ف، ل = بخواصّ: ب = خواض: ك.

١١٨٩ منها] –ب.

١١٩٠ الذي] +لا: شاس.

١١٩١ تسامت] ب، ك، ل = يسامت: س، ف.

١١٩٢ ومنها] (بياض في ب).

١١٩٣ ص٥١: ب.

١١٩٤ عرضها] فاب.

١١٩٥ تسامت] ب، ك، ل = يسامت: س، ف.

١١٩٦ ٣٠ب: ك.

١١٩٧ مرّة في السنة] في السنه مرة: ك.

١١٩٨ عند] –ك.

١١٩٩ من] –ك.

١٢٠٠ ذوات] ذات: ب.

١٢٠١ الظل] ظل: ك.

١٢٠٢ تارة] مارّة: ب.

١٢٠٣ ٧٤ب: س.

١٢٠٤ ومنها] (بياض في ب).

١٢٠٥ تسامت] ب، س(قد تغيّر «يسامت» إلى ««تُسامت»)، ك، ل = يسامت: ف.

١٢٠٦ ومنها] (بياض في ب).

١٢٠٧ سو كه] (بياض في ب).

١٢٠٨ وحينئذ] ب، س، ل = وح : ف، ك.

١٢٠٩ تنطبق] نطبق: ب، ل = ينطبق: س، ف، ك.

١٢١٠ زال] زالت: ك.

١٢١١ ١٦ب: ف.

١٢١٢ ٣١آ: ك.

١٢١٣ على] س، ف، ل = من: ب، ك.

١٢١٤ دفعةً] +ومدار: ل.

١٢١٥ ٧٧آ: ل.

١٢١٦ ص٥٢: ب.

١٢١٧ كد] س، ف، ل = (بياض في ب) = كه : ك.

١٢١٨ إذ بقدر] ب، س = اد بعدر: ف = اذا تقدر: ك = اذ نقدر: ل.

١٢١٩ ما] (بياض في ب).

١٢٢٠ للمدارات] للمدرات: ف.

١٢٢١ الأبدي] الابد: ك.

١٢٢٢ ٧٥آ: س.

١٢٢٣ لنظائرها] ب، س(«الابد» زائد ومشطوب)، ف، ل = لتطايرها: ك.

١٢٢٤ ومنها] (بياض في ب).

١٢٢٥ تمام] —ف.

١٢٢٦ سو كه] س، ف، ل = (بياض في ب) = سو له : ك.

١٢٢٧ سو كه] (بياض في ب).

١٢٢٨ تغرب] س(قد تغيّر «يغرب» إلى «تغرب») = نغرب: ب، ل = يغرب: ف، ك.

١٢٢٩ نفرض] ل = نفرض: ب، س = يفرض: ف، ك.

١٢٣٠ مما يلي الجنوب] س، ف، ك = —ب = شال.

١٢٣١ ويرتفع] وبعع: ف.

١٢٣٢ ٣١ب: ك.

١٢٣٣ رأس] +الجدى: شاك.

١٢٣٤ عن] ب، ك، ل = على: س، ف.

١٢٣٥ تسعين] ب، ك، ل = التسعين: س، ف.

١٢٣٦ التي] هال(مع رمز «صح»).

١٢٣٧ تكون] ك = نكون: ب، ف، ل= يكون: س.

١٢٣٨ لا محالة] +معد: شاك.

١٢٣٩ ٧٥ب: س.

١٢٤٠ تساوي] ب، ل = يساوي: س، ف، ك.

١٢٤١ تساوي تمام العرض فإنّها تماسّ الأفق ولا تنحطّ عنه والتي ميلها] تاب.

١٢٤٢ تنحطّ] ك، ل = ىنحط: ب، ف = ينحط: س.

١٢٤٣ ص٥٣: ب.

١٢٤٤ فتكون] ل = فكون: ب = فيكون: س، ف، ك.

١٢٤٥ تكون] ل = ىكون: ب، ف = يكون: س، ك.

١٢٤٦ تكون لا محالة] لا محاله يكون: ك.

١٢٤٧ بمسيرها الخاصّ] هاب، ف، ك، ل = بميرها الحناص: ب = لمسيرها الخاص: س.

١٢٤٨ ١٧آ: ف.

١٢٤٩ النظيرة] النطره: ك.

١٢٥٠ لذلك البلد] ك = كذلك: ب، س، ف = لذلك: ل.

١٢٥١ نهاره] +الاطول: ل.

١٢٥٢ ٣٢آ: ك.

١٢٥٣ الليل] ليله: ك.

١٢٥٤ ويغرب] ك = وىغرب: ب، ل = وتغرب: ف = وتغرب: س(يوجد «يـ» تحت الـ«تـ»).

١٢٥٥ من] (بياض في ف).

١٢٥٦ قبل] +السرطان: شاك.

١٢٥٧ ٧٧ب: ل.

١٢٥٨ القياس] (بياض في ف).

١٢٥٩ ٧٦آ: س.

١٢٦٠ الآخر] الاخير: ك.

١٢٦١ والعقرب قبل الميزان] هال(مع رمز «صح»).

١٢٦٢ النهار] –ف.

١٢٦٣ يلي] يلاى: س = ىلي(؟): هاس(مع رمز «صح»).

١٢٦٤ صر٥٤: ب.

١٢٦٥ إذن] اذا: ك.

١٢٦٦ الحوت] الخوت: ب.

١٢٦٧ مال] مالت: س.

١٢٦٨ دائرة] داير: س.

١٢٦٩ ٣٢ب: ك.

١٢٧٠ في الطلوع] بالطلوع: ك.

١٢٧١ الحوت] الخوت: ب.

١٢٧٢ يتمّ] ب، س، ف = تتم: ك = ىتم: ل.

١٢٧٣ الحوت] الخوت: ب.

١٢٧٤ نقطة] +الحمل: ب.

١٢٧٥ فإذا] وإذا: ب.

١٢٧٦ وانحطّ] فانحطّ: ب = +في الغروب: شال.

١٢٧٧ أخذ في الغروب معه] اخذ معه فى الغروب: ل.

١٢٧٨ هو] كان: ك.

١٢٧٩ متصّل به] ف، ل = متصل: ب، س = متصلا به: ك.

١٢٨٠ التوالي] هاس(مع رمز«صح»).

١٢٨١ ٧٦ب: س.

١٢٨٢ ظاهر] ظاهرا: ك.

١٢٨٣ عن] على: ب.

١٢٨٤ ١٧ب: ف.

١٢٨٥ الطالع] الطلوع: ف.

١٢٨٦ كان] كا: ب.

١٢٨٧ مقابله] مقابلة: ب.

١٢٨٨ أحد] +هى: ك.

١٢٨٩ في] +النصف: هاس(مع رمز «صح») = +في: ك.

١٢٩٠ ص٥٥: ب.

١٢٩١ يخالف] خالف: ك.

١٢٩٢ يطلع] بطلوع: ف.

١٢٩٣ ٣٣آ: ك.

١٢٩٤ منطبق] نطبق: ب.

١٢٩٥ ودور] ودرور: س.

١٢٩٦ وتكون] فكون: ب = ويكون: س = وكون: ف، ك، ل.

١٢٩٧ ٧٧آ: س.

١٢٩٨ ٧٨آ: ل.

١٢٩٩ الشمالية وستّة أشهر ليلة وذلك إذا كانت الشمس في البروج] –ف.

١٣٠٠ بل] +يكون: ب.

١٣٠١ ظاهر] س، ف، ل = ظاهراً: ب، ك.

١٣٠٢ وصفناه] وصفنا: ب.

١٣٠٣ للمواضع الجنوبية] للحنوسه: ل.

١٣٠٤ إلى] +الى: ف.

١٣٠٥ ذلك] +والله المستعان: ك.

المقالة الثانية ، الباب الثالث في أشياء منفردة

١٣٠٦ الباب الثالث في أشياء منفردة] س، ف، ل = (بياض في ب) = الباب الثالث من المقاله الثانيه في اشيا منفرده: ك.

١٣٠٧ البروج] –ف.

١٣٠٨ تطلع] ب، ك، ل = يطلع: س = بطلع: ف.

١٣٠٩ الكوكب] الكواكب: ف.

١٣١٠ ممرّ] ثم: ك.

١٣١١ ٣٣ب: ك.

١٣١٢ ص٥٦: ب.

١٣١٣ ٧٧ب: س.

١٣١٤ ممرّه] ممرّة: ب.

١٣١٥ العرض] هاس(مع رمز «صح») = الشكل: س.

١٣١٦ (الظاهر أنّ ورقة ناقصة من مخطوطة ف بين ١٧ب و١٨آ أي من «[*] يكون شرقياً عند كون النصف الأوّل» إلى «فذلك الخطّ هو على صوب القبلة [*]»).

١٣١٧ على] +دايرة: ب.

١٣١٨ فتكون] س = فكون: ب، ل = فيكون: ك.

١٣١٩ به] +اى القطب الشمالى: تال.

١٣٢٠ مائلةً] س، ل = مايلا: ب = المايله: ك.

١٣٢١ وتنتهي] ل = وينتهى: ب = وىلتهى: س = وىتنهي: ك.

١٣٢٢ فيصل] س، ك = فىصل: ب = فتصل: ل.

١٣٢٣ إليه] ب، تاس(مع رمز «صح»)، ل = إليها: س، ك.

١٣٢٤ ممرّه] س، ك، ل = ممرّة: ب.

١٣٢٥ ٧٨آ: س.

١٣٢٦ الأفلاك] س، ل = الأفاق: ب، ك.

١٣٢٧ إمّا] ب، س، ل = –ك.

١٣٢٨ ٣٤آ: ك.

١٣٢٩ على] +مو: شاب.

١٣٣٠ ويسمّى] ك = وسمى: ب، ل = وتسمى: س.

١٣٣١ ص٥٧: ب.

١٣٣٢ الأفق] +وسمّى: شال.

١٣٣٣ ٧٨ب: ل.

١٣٣٤ ويسمي] ك، ل = وسمّى: ب = وتسمى: س.

١٣٣٥ الثاني] +والمنبسط: ك.

١٣٣٦ بسبعة] س، ك، ل = سبعة: ب.

١٣٣٧ أقسام] س، ك، ل = —ب.

١٣٣٨ أقسامه] +البالباما اقسامه: شاس.

١٣٣٩ عند] هاس(مع رمز «صح»).

١٣٤٠ تلك بمثل] س، ك، ل = بمثل تلك: ب.

١٣٤١ وهذا عند الشافعي رحمه الله] ك = وهذا عند الشافعى رضى الله عنه: ب = عند الشافعى(مع رمز «صح») رحمه الله تعالى(مع رمز «صح»): هاس = عند الشافعى رحمه: هال(مع رمز «صح»).

١٣٤٢ وعند ابي حنيفة رضي الله عنه] س، ك = وعند ابى حنيفه رحمه الله عليه: ب = وعند ابى حسفه رحمه الله: ل.

١٣٤٣ عليه] س، ك، ل = عليها: ب.

١٣٤٤ في معرفة خطّ] (بياض في ب).

١٣٤٥ تُسوَّى أرض] يُسوّى ارض: ب = يُسَوِّى وجه الارض: س = بسوي الارض: ك = تُسَوَّى ارضٌ: ل.

١٣٤٦ ماء] مال: ك.

١٣٤٧ وتسمّى] ك = ويسمّى: س = وسمى: ب، ل.

١٣٤٨ ٧٨ب: س.

١٣٤٩ الهنديّة] هنديه: ك.

١٣٥٠ ويُنصب] وتنصب: ب.

١٣٥١ طوله] ب، فاس(مع رمز «ح»)، ك = في طول: س، ل(يوجد «في» مشطوب والـ«طول» متغيّر إلى «طوله»).

١٣٥٢ قائمة] +ويعرف: ك.

١٣٥٣ ٣٤ب: ك.

١٣٥٤ إمّا] —ك.

١٣٥٥ بأن] ب، هاس(مع رمز «صح»)، ك = —ل.

١٣٥٦ يُقدّر] س = نقدر: ب = بقد: ك = بمقدار: ل.

١٣٥٧ نقط] نقطه: ب.

١٣٥٨ ويُعلّم] س = ونعلم: ب، ل = وتعلِمُ: ك.

١٣٥٩ وتُنصَّف] ك = ونصف: ب = ويُنصّف: س = ونُنَصَّفُ: ل.

١٣٦٠ وتُخرِج] ل = ويخراج: ب = نخرح: س = بحرج: ك.

١٣٦١ ص٥٨: ب.

١٣٦٢ من] عن: ب.

١٣٦٣ منتصفها] منتصفهما: س.

١٣٦٤ خطّاً] س(الـ«ا» مشطوب و«خطا» متغيّر إلى«خطٌ»)، ك، ل = خط: ب.

١٣٦٥ شئت] يشب: ب.

١٣٦٦ الدائرة] الدايرَ: س.

١٣٦٧ خطّاً] س، ك، ل = خط: ب.

١٣٦٨ يقطع] +على: ب.

١٣٦٩ والمغرب] +والله اعلم وهده صوره الدايره: ك.

١٣٧٠ ٧٩آ: ل.

١٣٧١ ٣٥آ: ك.

١٣٧٢ في معرفة] —ب.

١٣٧٣ ونعني] س، ل = ونعى: ب = ويعني: ك.

١٣٧٤ ههنا] ب، هاس(مع رمز «صح»)، ك = هاهنا: ل.

١٣٧٥ ٧٩آ: س.

١٣٧٦ نقطة] هى نقطة هى نقطة: س.

١٣٧٧ إذا] ب، س = وإذا: ك، ل.

١٣٧٨ ونصل] س، ل = ونصل: ب = وتصل: ك.

١٣٧٩ ونعدّ] ل = يعدّ: ب = وتعد: ك = ونَعُدّه: س.

١٣٨٠ ص٥٩: ب.

١٣٨١ من] هاب.

١٣٨٢ ومن] +بطه: شاك.

١٣٨٣ ونصل] ل = وفصل: ب = ونصل ما: س = وتصل: ك.

١٣٨٤ تقاطعها] بماطعها: ب.

١٣٨٥ وننفذه] ب، ل = وتُنَفِذُ: س= وتنفذه ك.

١٣٨٦ هو] هو طه(؟): س = —ك.

١٣٨٧ ١٨آ: ف (الظاهر أنّ ورقة ناقصة من مخطوطة ف بين ١٧ب و١٨آ أي من «[*] يكون شرقياً عند كون النصف الأوّل» إلى «فذلك الخطّ هو على صوب القبلة [*]»).

١٣٨٨ قوس] +الانحراف: ب.

١٣٨٩ سمت] +سمت: ك.

١٣٩٠ وهي] وهو: ك.

١٣٩١ مكّة] +فالقبله: شال.

١٣٩٢ أكثر] +طول مكة من جزاير الخالدات(+بياض) وعرضها(+بياض) طول خوارزم(+بياض) وعرضها(+بياض): ب.

١٣٩٣ يساوي] ب، س، ف = تساوي: ك = ساوى: ل.

١٣٩٤ ٧٩ب: س.

١٣٩٥ نصف] —ك.

١٣٩٦ ساوى] تساوي: ك.

١٣٩٧ ٣٥ب: ك.

١٣٩٨ تسامت] س، ك، ل = سامت: ب = يسامت: ف.

١٣٩٩ التي تسامت في الدورة من فلك البروج] من فلك البروج التى سامت في الدورة: ب.

١٤٠٠ زَ كَاَ] س، ف، ك، ل = (بياض في ب).

١٤٠١ وكب لط] س، ف، ل = (بياض في ب) = كب كط: ك.

١٤٠٢ السرطان] +وضَعْها: شال.

١٤٠٣ ٧٩ب: ل.

١٤٠٤ أعني] على: ك.

١٤٠٥ الأسطرلاب] ب، ل = الاصطرلاب: س، ف، ك.

١٤٠٦ أدِر] ادار: ب.

١٤٠٧ إن كان غربيّاً] ب، هاس، ك، هال(مع رمز «صح») = —ف.

١٤٠٨ فحيث] فحسب: ف.

١٤٠٩ الأجزاء] اجزاء: ب.

١٤١٠ ص٦٠: ب.

١٤١١ هو] —ب.

١٤١٢ للقبلة] +بهذه الصوره: ك.

١٤١٣ في النهار والليل] ف = (بياض)+النهار والليل: ب = في النهار والليلة: س = في معرقة الليل والنهار: ك = في معرفة(في الهامش مع رمز «صح») النهار والليل: ل.

١٤١٤ والسنة والشهر] س، ف، ل = والشهور والسنة: ب = والسنة والشهور: ك.

١٤١٥ ٨٠آ: س.

١٤١٦ ووقع] وقع: ب.

١٤١٧ ظلّها] +الارض: تاس.

١٤١٨ فإذا] وإذا: ب.

١٤١٩ الشمس] ب، هاس(مع رمز «صح»)، ك، ل = —ف.

١٤٢٠ ظلّها] الظل: ب.

١٤٢١ ظلّها] الظل: ب.

١٤٢٢ ٨٠آ: ل.

١٤٢٣ إذ] ك، ل = إذا: ب، ف.

١٤٢٤ فإذا] ف، ك، ل = فان: ب.

١٤٢٥ ص٦١: ب.

١٤٢٦ إذ الشمس أعظم جرماً من الأرض فإذا كانت الشمس تحت الأرض قريبةً من الأفق كان مخروط] هاس(مع رمز «صح»).

١٤٢٧ وكان] فكان: ك.

١٤٢٨ قريباً] +في الافق: شاك.

١٤٢٩ ١٨ب: ف.

١٤٣٠ الأنوار] ب، س، ل = +وربما فيظهر فى الافق النور وكلما كانت الشمس اقرب كانت الانوار: ف = النور: ك.

١٤٣١ ويظهر] س، ف، ل = وبظهر: ب = وتظهر: ك.

١٤٣٢ الكلّ] +وهو الفلك الاطلس: تاس.

١٤٣٣ من غروب الشمس] شاس= +من غروب الشمس: س.

١٤٣٤ يمكن] ب، ك، ل = يكون: س(«يمكن» مشطوب مع رمز «صح»)، ف.

١٤٣٥ كون] —ب، —ك.

١٤٣٦ في] من: ب، ل.

١٤٣٧ والمُنجّمين] +هم: ب.

١٤٣٨ ٣٦ب: ك.

١٤٣٩ المساكن] +الاماكن: تاس(مع رمز «صح»).

١٤٤٠ ٨٠ب: س.

١٤٤١ واختلافها] واختلافاتها: ب.

١٤٤٢ واحدٌ] —ب.

١٤٤٣ النهار] +واحده: ب.

١٤٤٤ تقوم] ك = نقوب: ب، ل = يقوم: س، ف.

١٤٤٥ الكلّ] الفلك: ب.

١٤٤٦ من فلك] في الفلك: ب.

١٤٤٧ تقطع] ك، ل = يقطع: س = نقطع: ف.

١٤٤٨ تقطع من فلك البروج قسيّاً مختلفةً فمطالعها مختلفة وأيضاً لو كانت الشمس] —ب.

١٤٤٩ تقطع] ك = نقطع: ب، ف، ل = يقطع: س.

١٤٥٠ ووسط] ووسطي: ك.

١٤٥١ ص٦٢: ب.

١٤٥٢ مطالع] —ب.

١٤٥٣ والوسط] والوسطي: ك.

١٤٥٤ قوس] +متساوي: ب.

١٤٥٥ ٣٧آ: ك.

١٤٥٦ من معدّل النهار] هال(مع رمز «صح»).

١٤٥٧ مساوية] —ب.

١٤٥٨ وهو ٠ نط ح ك] س = —ب، —ف، —ك، —ل.

١٤٥٩ النقطة] +وهو: س.

١٤٦٠ ١٨١آ: س.

١٤٦١ والوسط] والوسطي: ك.

١٤٦٢ غروبها] +من غروبها: ك.

١٤٦٣ ومن غروبها] +ومن غروبها: ف.

١٤٦٤ طلوع الشمس] س، ف، ل = طلوعها: ب، ك.

١٤٦٥ ١٩آ: ف.

١٤٦٦ والليلة] بليلته: ب.

١٤٦٧ ٨٠ب: ل.

١٤٦٨ فالساعات] والساعات: ب.

١٤٦٩ هي] ب، ك، ل = هو: س، ف.

١٤٧٠ خمس عشرة درجة] ب، ل = خمسة عشر درجة: س، ف، ك.

١٤٧١ الدائر] الدايرة: ك.

١٤٧٢ اليوم] الليل: ف.

١٤٧٣ أو الليلة] والليلة: س.

[١٤٧٤] والساعة] س، ك، ل = والساعات: ب، ف.

[١٤٧٥] وتسمّى] ك، ل = ويسمّى: ب، س = وسمّى: ف.

[١٤٧٦] فهي] وهي: ب.

[١٤٧٧] ٨١ب: س.

[١٤٧٨] وإن كان] ولو كانت: ب.

[١٤٧٩] ص٦٣: ب؛ ٣٧ب: ك.

[١٤٨٠] وهي] وهو: ك.

[١٤٨١] الساعة] الساعات: س(الـ«عات» متغيّر إلى«عة»).

[١٤٨٢] الأزمان] س(«الزمان» متغيّر إلى «الازمان»)، ف، ك، ل = للزمان: ب.

[١٤٨٣] والساعات الزمانية هي التي تختلف أزمانها] —ك.

[١٤٨٤] ازمانها والساعات الزمانية هي التي تختلف ازمانها ولا يختلف] —ب.

[١٤٨٥] زمان] الزمان: ب = زمان: هاب.

[١٤٨٦] أيّة] ل = انه: ب، ف = انه: س = من اية: ك.

[١٤٨٧] حين] —س.

[١٤٨٨] مدّة] س، ف، ل = —ب، —ك.

[١٤٨٩] شسه] س، ف، ك، ل = (بياض في ب) = +٣٦٥: تاس(مع رمز «صح»).

[١٤٩٠] شسه] س، ف، ك، ل = (بياض في ب) = +٣٦٥ وربع(الـ«وربع» مشطوب): تاس(مع رمز «صح»).

[١٤٩١] وربع] +يوم: شاك.

[١٤٩٢] شسه] س، ف، ك، ل = (بياض في ب) = +٣٦٥: تاس(مع رمز «صح»).

[١٤٩٣] ٨٢آ: س.

[١٤٩٤] إلّا ثلاثة] يوم الّا ثلثه: ب.

[١٤٩٥] جزء] حز مما: س.

[١٤٩٦] هنا] هاس(مع رمز «ح»)، ك، ل = ههنا: ب = —ف.

[١٤٩٧] هي] س، ف = —ب، —ك، —ل.

١٤٩٨ هي] –ك.

١٤٩٩ ٣٨آ: ك.

١٥٠٠ فهي] +فهي: ف.

١٥٠١ القمر] +في: ك.

١٥٠٢ أيّ وضع] اى موضع: س.

١٥٠٣ ١٩ب: ف.

١٥٠٤ ص٦٤: ب.

١٥٠٥ يلتفت] بلىف: ف.

١٥٠٦ وزمانه] ك، ل = فزمانه: ب = وزمان: س، ف.

١٥٠٧ الاجتماعين] الاجتماع: ف.

١٥٠٨ ٨١ب: ل.

١٥٠٩ النيّرين] +علي ما بقي: شاك.

١٥١٠ بقي] +من: ك.

١٥١١ شس] س، ف، ك، ل = ٠ ٦ ٣ : ب، +تاس.

١٥١٢ فخرج] +ذلك في اثني عشر وخمس يوم: ك.

١٥١٣ كط لا ن ح] ب، س، ف، ك، ل.

١٥١٤ شند] س، ف، ك، ل = +٣٥٤: تاس(مع رمز «صح») = ٤ ٠ ٦ ٣ (؟) : ب.

١٥١٥ وسُدسه] ب، ف، ل = وسد وسدسه(الـ«سد» مشطوب): س = وسدس: ك.

١٥١٦ ٨٢ب: س.

١٥١٧ هذا] +هذا: ك.

١٥١٨ به الطبع] –ب.

١٥١٩ الطبع] –ك.

١٥٢٠ ٣٨ب: ك.

١٥٢١ وليدُها] ولديها: ك.

[١٥٢٢] كَشْف] الكشف: ب.

[١٥٢٣] الألْفاظ] اللفظ: ف.

[١٥٢٤] أداءً لشرائط] ب، هاس(مع رمز «صح»)، ف، ل = اذا الشرايط: س(الـ«اذا» مشطوب)، ك.

[١٥٢٥] التحرّز] ب، ك، ل = التجوز: س، ف.

[١٥٢٦] الذي] التي: ك.

[١٥٢٧] كافٍ] —ب.

[١٥٢٨] فالأولى] والاولى: ب.

FIGURE APPARATUS

المقدّمة في بيان أقسام الأجسام على الإجمال

شكل ١: صورة الأفلاك

صورة الأفلاك] ل =كرة العالم: ب = —س، —ف، —ك. فلك الأفلاك] ب، س، ك، ل = فلك الاعظم: ف. كرة النار] ب، س، ف = النار: ك، ل. كرة الهواء] ب، س، ف = الهواء: ك، ل. كرة الماء] س، ف = —ب = الما: ك، ل. كرة الأرض] ف = كرة الارض والمآء: ب = ارض: س = الارض: ك، ل. مركز العالم] ب، ف = —س، —ك، —ل. قطب البروج (موقعان)] س = قطب فلك البروج (موقعان): ب، ل = —ف، —ك. قطب العالم (موقعان)] ب، س = —ف، —ك = قطب الجنوب؛ قطب الشمال: ل. («محور العالم» زائد في ب، ك). («محور البروج» زائد في ب = «محور فلك البروج» زائد في ك). («عالم العناصر» زائد في ب.)

المقالة الأولى ، الباب الأوّل في هيئات الأفلاك

شكل ٢: صورة فلك الشمس

صورة فلك الشمس] ف، ك، ل = —ب = فلك السمس: س. الأوج] ب، ل = اوح: س، ك = —ف. («مركز الشمس» زائد في ب). جرم الشمس] ب، س، ف، ك، ل (موقعان). («الشمس» زائد في موقع آخر في ف). متتّم محوي] ف = المتمم المحوى: ب = المتمّم من الممثل: س = ممم الممثل: ك = المتمم: ل. مقعّر الخارج] معمر خارج: ف = —ب، —س، —ك، —ل. مقعّر الممثّل] مقعر ممثل: ف = —ب، —س، —ك، —ل. مركز الخارج] ب، ل = —س، —ك = مركر حارج: ف. مركز الممثّل] مركز العالم: ب = —س، —ك = مركز ممثل: ف = مركز عالم: ل. خارج المركز] ك = الفلك الخارج المركز: ب، س = حارج مركز: ف = الحامل: ل (أربعة مواقع). محدّب الخارج] محدب خارج: ف = —ب، —س، —ك، —ل. الحضيض] ب، س، ك، ل = —ف. الممثّل] ممل: ف = الفلك الممثل: ب، ل (موقعان) = —س، —ك. محدّب الممثّل] محدب ممثل: ف =

ـب، ـس، ـك، ـل. متمّ حاوي] ف = المتمم الحاوى: ب = المتمم من الفلك الممثل: س = ممم ممثل: ك = المتمم: ل. (عدد رسوم أجرام الشمس في كل مخطوطة: ١- ب ؛ ٤ - س، ل ؛ ٢- ف ؛ ناقص في ك.)

شكل ٣: صورة أفلاك الكواكب العلوية والزهرة

(الشكل ناقص في ف). صورة أفلاك الكواكب العلويّة والزهرة] ك، ل = افلاك الكواكب العلوية والزهرة: س = صورة الافلاك الكواكب (فاب) العلوية والزهرة: ب. أوج] ك = ـب = اوج زهرة: س = الأوج: ل. («ذروة التدوير» زائد في ك، ل). فلك التدوير] ب (في موقع)، س = التدوير: ب (في موقع آخر)، ل (أربعة مواقع) = ـك. الفلك الحامل] س = ـب = الحامل: ك، ل (موقعان). المتمّم] ب، ك، ل = المتمم من الممثل: س. جرم الكوكب] ب، ل = الكوكب (ثلاثة مواقع): س = ـك. حضيض] س = ـب = حضيض الذروه: ك = الحضيض: ل. المتمّم] ب، ك، ل = المتمم من الممثل: س. («مركز التدوير» زائد في ب). («الفلك الممثل» زائد في ب، ل [موقعان]). («مركز العالم» زائد في ب). («خارج المركز» زائد في س). («مركز الخط المدير» زائد في ك). («حضيض التدوير» زائد في ل). (عدد رسوم أجرام التداوير في كل مخطوطة: ٤ - ب، س، ل ؛ ناقص في ك.)

شكل ٤: صورة فلك عطارد

صورة فلك عطارد] ب، ف، ك، ل = فلك عطارد: س. الأوجان] س = الاوج: ب = ـف = اوجا عطارد: ك، ل. («ذروة التدور» زائد في ل). الكوكب] س (أربعة مواقع) = ـب، ـف، ـك، ـل. تدوير] ف = التدور: ب، ل (أربعة مواقع) = فلك التدور: س (ثلاثة مواقع) = ـك. المتمّم من المدير (موقعان)] س، ف = «المتمم» و«المتمم المدر»: ب = متمم المدير(موقعان): ك، ل. المتمّم من الممثّل(موقعان)] س، ف (موقع واحد فقط) = المتمم(موقعان): ب = ممم الممثل: ك (موقع واحد فقط) = طال (موقعان). مركز الحامل] ب = ـس، ـك، ـل = مركز حامل: ف. مركز المدر] ب = ـس، ـك، ـل = مركز مدير: ف. مركر العالم] ب = ـس، ـك، ـل

= مركز عالم: ف. («مركز معدل المسير» زائد في ب). حامل] ف = الفلك الحامل: ب، س = الحامل: ك، ل (موقعان). مدير] ف (موقعان) = ـب، ـس، ـك، ـل. ممثّل] ف = الفلك الممثل: ب = ـس، ـك، ـل. الحضيضان] س = ـب، ـف = حضيض التدوير: ك = «حضيض التدوير» و«الحصض»: ل. (عدد رسوم أجرام التداوير في كل مخطوطة: ٤ - ب، س، ل ؛ ١- ف، ك.)

شكل ٥: صورة فلك القمر

صورة فلك القمر] ك، ل = ـب، ـف = فلك العمر: س. الأوج] ب، ل = ـس، ـف، ـك. التدوير] ف، ل (أربعة مواقع) = فلك التدوير: ب = ـس، ـك. قمر] س (موقعان) = جرم القمر: ب = ـف، ـك، ـل. المتمّم من المائل (موقعان)] س = المتمم (موقعان): ب، ل = «المتمم من المائل» و«المتمم من المدير» [كذا]: ف = «المتمم المايل» و«متمم المايل»: ك. حامل] ف = الفلك الحامل: ب، س = الحامل: ك، ل (موقعان). مركز الحامل] ب، ف = ـس، ـك، ـل. مركز العالم] ب، ف = ـس، ـك، ـل. مائل] ف = الفلك المايل: ب = ـس، ـك = المايل (موقعان): ل. الحضيض] ب، س = ـف = حضيض التدوير: ك = «حضض التدور» و«الحضض»: ل. («نقطة المحاذاة» زائد في ك). الفلك الجوزهر ويسمّى الممثّل] س = الفلك الممثل: ب = «فلك الجوزهر والممثل ايضا» و«حوزهر»: ف = جوزهر وسمى الممثل: ك = الجوزهر وسمى ممثل: ل. («وهذا فلك الحامل الذى هو فى ثحن المايل» زائد في ف). (عدد رسوم أجرام التداوير في كل مخطوطة: ٢- ب، س ؛ ٤- ل ؛ ١- ف ؛ ناقص في ك.)

المقالة الأولى ، الباب الرابع في القسيّ

شكل ٦: صورة النطاقات باعتبار الأبعاد

(الشكل ناقص في ف). (التسميات ناقصة في ل). الأوج] ـك. مركز الفلك الحامل (في الخارج المركز)] س = مركز الحامل: ب، ك. البعد الأوسط (موقعان)] ـس. الفلك الحامل (في الخارج المركز)] ـك. الفلك الممثّل] ـك. («البعد الاقرب» زائد في

الخارج المركز في ك). فلك التدوير] س= —ب، —ك. نقطة التقاطع] (يوجد مرة فقط على اليسار في ك). حضيض التدوير] حضيض: ك. مركز الفلك الحامل (في التدوير)] مركز الحامل: ب، ك. (يوجد خطّ نصف قطر زائد في التدوير في ل.)

شكل ٧: صورة النطاقات باعتبار اختلاف المسير

(بياض في ف، ل). مركز العالم] —ك. الفلك الحامل (في الخارج المركز)] —ك. («والبعد الابعد» زائد في الخارج المركز في ك). («غاية التعديل» (موقعان) زائد في الخارج المركز في ك). (يوجد خطّ من مركز الحامل زائد في الخارج المركز في ك). ذروة التدوير] ذروة: ك. فلك التدوير] —ب، —ك. مركز التدوير] —ب. نقطة التماسّ (موقعان)] ب = نقطه الشمال (يساراً)؛ غاية التعديل (يميناً): ك = —س. حضيض التدوير] —ك. مركز الحامل (في التدوير)] شاس = مركز العالم: س. الفلك الحامل (في التدوير)] —ك.

المقالة الثانية، الباب الثالث في أشياء منفردة

شكل ٨: صورة الدائرة الهندية

(الشكل ناقص في ف؛ التسميات ناقصة في ب). (الخطّ الواصل بين العلامتين ناقص في ب). (المقياس ناقص في ب، س؛ يوجد مقياسان في ك). الجنوب] ل = نقطه الجنوب: س = جنوب: ك. خطّ نصف النهار] س، ك، ل. المغرب] ك، ل = نقطه المغرب: س. خطّ الاعتدال] س، ك، ل. المقياس] ل = —س = («مقياس» مكتوب مرّتان في ك). المشرق] ك، ل = نقطه: س. الظلّ الشرقي] س، ك، ل. الظلّ الغربي] س، ك، ل. الخطّ الواصل بين العلامتين] س، ك، ل. مدخل الظلّ] ل = مخرج الطل: س = —ك. مخرج الظلّ] س، ل = —ك. الشمال] ك، ل = نقطه الشمال: س.

شكل ٩: صورة المسامت للقبلة

(التسميات ناقصة في ب). الجنوب] ك، ل = حنوب: س، ف. فضل ما بين الطولين (موقعان)] ل = (موقع واحد فقط في س، ف، ك). الخطّ الموازي لخطّ نصف النهار وهو خطّ نصف النهار بمكّة] ل = —س، —ف، —ك. الخطّ الموازي لخطّ

الاعتدال وهو خطّ الاعتدال بمكّة] ل = ـــس، ـــف، ـــك. سمت رأس مكّة] ل = ـــس، ـــف، ـــك. فضل ما بين العرضين (موقعان)] ل = (موقع واحد فقط في س، ف) = ـــك. المغرب] ك، ل = نقطة المغرب: س، ف. المشرق] ك، ل = نقطة المشرق: س، ف. خطّ الاعتدال] ك، ل = ـــس، ـــف. الشمال] ك، ل = شمال: س، ف.

Commentary to the Edition and Translation

Preface

The *Mulakhkhaṣ* has three different preface versions; however, starting with **Pref.[2], "The Introduction is about an explanation of the divisions of the bodies in general terms,"** the three versions converge enough to allow them to be collated as a single version, with variants noted in the critical apparatus. The incipits, prefaces, and colophons of the five principal manuscripts that have been used to establish the Arabic edition are provided, with English translations, in *§ II.2: Description of the Manuscripts.* All five contain Jaghmīnī's original parameters; in general, these values are either Ptolemaic or ones that Jaghmīnī refers to as "Modern," usually meaning from the ninth century (**I.2[6]**). Most of the later copies of the *Mulakhkhaṣ* and commentaries have updated at least some of these parameters, the most obvious examples occurring in the listing of the climes and their parameters (**II.1**) that can be shown to derive from Ṭūsī's *Tadhkira* (see commentary on the second clime (**II.1[4]**). This provided a convenient means of differentiating witnesses that are closer to Jaghmīnī's original version from those that have been updated.

Version 1: The preface in this version contains both a dedication by Jaghmīnī to Badr al-Dīn al-Qalānisī and a poem he composed to commemorate being entrusted with such a lofty commission. The poem, which I have metered below, has a *khafīf* rhyme:[1]

[1] For the *khafīf* meter, see W. Stoetzer, "Prosody (*ʿarūḍ*)," in *Encyclopedia of Arabic Literature*, ed. Julie Scott Meisami and Paul Starkey (London: Routledge, 1998), 2:619–22 at 621. I am grateful to Prof. Emeritus Issa Boulata for his assistance in metering this.

S.P. Ragep, *Jaghmīnī's Mulakhkhaṣ*, Sources and Studies in the History of Mathematics and Physical Sciences, DOI 10.1007/978-3-319-31993-3_4

يا لَها مِن إشا رَ ةٍ صَدَرَتْ لي	رَ فَعَتْ رُتْـــبَتي وَأعْـلَـتْ مَحَلْلي
- - u u | - u - u | - - u u	- - u u | - u - u | - - u u
صَدَرَتْ لي مِنَ الْكَريـــمِ الْمُرَجْجي	بَدْرِدين الْــهُدي الإما مُ الْأَجَلْلِ
- - u u | - u - u | - - u u	- - u u | - u - u | - - u u
قَدْ رَ آ نـي أَهْلاً لِأَمْــرٍ خَطيرٍ	لَيْــسَ مَثْلي لِمِثْــلِ ذا كَ بِأَهْلِ
- - u u | - u - u | - - u u	- - u u | - u - u | - - u u
غيرَ أنْ ني بَذَلْـتُ في ذا كَ جُهْدِي	إمْتِــثالاً لِأَمْــرِهي أيْــيَ بَذْلِ
- - u u | - u - u | - - u u	- - u u | - u - u | - - u u
قَدْ دَعاني لِذاْكَ لُطْــفاً وَبَرْراً	لا افْتِقــاراً إلى بِضا عَةِ مِثْلي
- - u u | - u - u | - - u u	- - u u | - u - u | - - u u

Oh what a proposal came my way;	it raised my rank and it advanced my standing.
It came to me from the noble one who inspires hope;	the highly esteemed Imām, the full moon [Badr] of the true religion.
He considered me worthy for a momentous task;	[but] the likes of me is not worthy of such a thing as that.
Nevertheless, I expended every effort for that;	complying with his command whatever sacrifice.
He called upon me for that in kindness and piety;	not requiring the offerings of such as myself.

Unique to Version 1 is the added nisba *al-faqīhī* to Jaghmīnī's name Maḥmūd ibn Muḥammad ibn ʿUmar al-Faqīhī al-Jaghmīnī al-Khwārizmī **(Pref.[1])**. I did not translate this since it is not clear what it refers to. Most likely it simply means that he was someone who came from a family of jurists, but it could also be an indication that he was an esteemed scholar, a reciter of the Qurʾān, or a school master.

This version also contains prayers of mercy for both Jaghmīnī and Qalānisī, implying both are deceased; however, these were presumably added later by the copyists. In this version, Jaghmīnī informs us that his motivation for composing the *Mulakhkhaṣ* is that "the dearest of friends and the sincerest of companions" conveyed to him that master Badr al-Dīn proposed that he compile a book on the subject of *ʿilm al-hayʾa*.

For Version 1, I used the following 3 manuscripts:

- ب [= B] Berlin, *Staatsbibliothek,* MS or. oct. 1511, pp. 6–64; copied 1275 hijra [1858 CE]
- ف [= F] Philadelphia, University of Pennsylvania, Rare Book & Manuscript Library, LJS MS 388, ff. 2b–19b; copied 786 hijra [1384 CE]
- س [= S] Paris, Bibliothèque nationale, MS arabe 2330, ff. 48b–82b; copied 787 hijra [1385 CE]

MS B is contaminated (not unusual given its late copy date) and contains variants that distinguish it from MS F and MS S, both of which are more closely aligned. Its preface also has a part similar to MS L (Version 3), but I included it in Version 1 because it contains both the dedication and the poem.

Version 2: The preface in this version contains only Jaghmīnī's dedication to Badr al-Dīn al-Qalānisī, and not his poem. Here Jaghmīnī states that his motivation for composing a work on *ʿilm al-hayʾa* came directly from Badr al-Dīn himself (and not from an intermediary) whom he describes as "the dearest of friends and the sincerest of companions" rather than his "master." The prayer after his name ("may God find his outcomes praiseworthy") seems to imply that he is still alive.

For Version 2, I used one manuscript:

- ك [= K] Cambridge UK, Cambridge University Library, MS Or. 593, ff. 1b–38b; copied 764 hijra [1362-63 CE]

Version 3: This version contains neither the dedication nor the poem. Unique to Version 3 is that Jaghmīnī states that he composed this book on *hayʾat al-ʿālam* (Configuration of the World), rather than *ʿilm al-hayʾa* (the science of *hayʾa*), as a memento for every scholar seeking an epitome on [*hayʾa*]. The beginning to Version 3 (with minor variations on it) is the most widespread preface for the *Mulakhkhaṣ* as well as for the commentaries.

For Version 3, I used one manuscript:

- ل [= L] Istanbul, Süleymaniye Library, Laleli MS 2141, ff. 61b–81a; copied 644 hijra [1246-47 CE], and the oldest witness to date

The First Part: On an Explanation of the Orbs and What Pertains to Them

Chapter 1 of Part I: On the Configurations of the Orbs

I.1[1]. For every sphere whose two surfaces are parallel, the center of their two surfaces is the [sphere's] center: Clearly by "*kura*" Jaghmīnī means "*falak*". Cf. the *Tadhkira*, I.1[10] and I.1[15], where Ṭūsī clearly differentiates between a complete sphere and an orb, which may be either a complete or hollowed-out sphere (Ragep, *Tadhkira*, 1:96–99).

I.1[6]. The **Illustration of the Sun's Orb [Fig. 2]** is placed after paragraph **[6]** in MS F (f. 4a) and MS S (f. 52a). MS S also juxtaposes this illustration with the illustration of the orbs of the Upper Planets and Venus [**Fig. 3**], whereas MS F omits the latter illustration entirely. MS S then places the illustrations for Mercury [**Fig. 4**] and the Moon [**Fig. 5**] together on f. 52b after **I.1[9]**, whereas MS F places these last two illustrations consecutively on the next two pages (ff. 4b–5a), but also after **I.1[9]**. MSS B, K, L place all four illustrations (i.e., the orbs of the Sun, Upper Planets and Venus, Mercury and the Moon) together after **I.1[9]** (see MS B [p. 13]; MS K [f. 5a–5b]; and MS L [ff. 64a–65b]).

I.1[8]. The Moon's orb includes two orbs, their center being the center of the world, and a deferent orb: Note that the Moon should consist of four, not three, orbs. Jaghmīnī does not count the epicycle orb as the fourth orb.

Chapter 2 of Part I: On the Motions of the Orbs

I.2[3]. It is in each nychthemeron [lit., a day with its night]: For comparative technical definitions of nychthemeron, see Ṭūsī, *Tadhkira*, III.8[1], "On the Lengths of the Nychthemerons" (Ragep, *Tadhkira,* 1:286–87); and Bīrūnī, *Tafhīm*, no. 132 (51).

I.2[6]. … according to the opinion of most Moderns, one degree in sixty-six solar years or sixty-eight lunar [years]: "Moderns" here refers to the astronomers of the ʿAbbāsid caliph al-Maʾmūn (r. 813–33) and their immediate successors, in distinction to Ptolemy who has a value of 1°/100 years. In comparison, Farghānī is committed to the Ptolemaic value of 1°/100 years and states a complete revolution occurs in 36,000 years (Ch. 13, *Jawāmiʿ*, 49–50). However, al-Battānī uses 1°/66 years in his *Zīj* (Ch. 52, 3:192–93), which Jaghmīnī follows in his calculations for the values of the apogees and nodes (see below Commentary, **I.5[26–28]**). Kharaqī also gives 1°/66 years in the *Muntahā* (Ghalandari, "A Survey of the Works of '*Hayʾa*' in the Islamic Period," 180 [56]: *Bāb* 8, *Faṣl* 2) as does Ṭūsī in his *Tadhkira,* II.4[4] (Ragep, *Tadhkira*, 1:122–25). See also F. J. Ragep, "Al-Battānī, Cosmology, and the Early History of Trepidation in Islam," 282, 290.

I.2[7]. These are the motions of the apogees and nodes [*jawzahar*s], except for one of Mercury's two apogees, namely that in the dirigent, and except for the Moon's apogee, its parecliptic, and its nodes: Ṭūsī (II.7[8]) may have had Jaghmīnī in mind when he criticizes those who say "the motion of the fixed stars is indistinguishable from the other lunar motions" for failing to point out that the reason for this is that the perceptible motion is a *composed* motion of the excess of the Moon's nodes over the fixed stars (Ragep, *Tadhkira,* 1:150–53).

I.2[9]. It is also called the motion of the center: According to Ṭūsī (II.7[10]): "It is called the motion of the center because the epicycle center is moved by it this amount" (Ragep, *Tadhkira*, 1:152–53).

On the Sources

In the *Mulakhkhaṣ* Jaghmīnī mentions two authorities: Ptolemy (**II.1[2]** and **II.3[9]**) and Battānī (**II.3[9]**). Regarding Ptolemy, he specifically mentions the *Almagest,* and he alludes to his *Geography* (**II.1[2]**). For Ptolemy's parameters, I relied on G. J. Toomer, *Ptolemy's Almagest* **[=*Alm.*]** and Olaf Pedersen, *A Survey of the Almagest* (and especially convenient is his listing of many Ptolemaic values in "Appendix B: Numerical Parameters," 423–29). Ascertaining parameters attributed to Battānī proved more challenging than simply relying on Carlo Nallino's seminal 3-volume edition, translation, and commentary of Battānī's *Kitāb al-Zīj al-ṣābiʾ (Opus astronomicum)* **[=*Zīj*]**; Nallino provides Battānī's values only to sexagesimal minutes whereas Jaghmīnī also give seconds. Consequently, I used the following additional sources: E. S. Kennedy's "A Survey of Islamic Astronomical Tables" **[=Kennedy]**; Battānī's parameters as reported by other primary sources such as Kūshyār ibn Labbān (fl. late-tenth/early-eleventh centuries) in his *Jāmiʿ Zīj* (Istanbul, Fatih 3418, f. 44a); and Battānī's parameters preserved in the fourteenth-century Persian *Zīj-i Ashrafī* by Abū ʿAbd Allāh Sanjar (see Fateme Savadi and Sajjad Nikfahm-Khubravan, "The Mean Motion of the Planets in *Zīj-i Īlkhānī* and the Criticisms Regarding It," in *Ustād-i Bashar: Essays on the Life, Times, Philosophy and Scientific Achievements of Khwājah Naṣīr al-Dīn Ṭūsī*, ed. H. M. Hamedani and M. J. Anvari (Tehran, 2012), 374 [=**S-K**]. For Kharaqī, I used witness copies of his *Tabṣira* and *Muntahā* (noting any discrepancies between the two treatises), but *Muntahā* citations in parentheses (unless specified otherwise) refer to the page and bracketed paragraph of Hanif Ghalandari's Arabic edition of the *Muntahā* ("A Survey of the Works of '*Hayʾa*' in the Islamic Period").

Below is the first of several charts I have compiled that lists parameters contained in the *Mulakhkhaṣ* for various motions, and compares them with other sources. Other charts are in **I.5[3]**, **I.5[11–13]**, **I.5[16–17]**, **I.5[26–27]**, **II.1[3–10]**.

[S]: sequence of the signs (west to east)
[CS]: counter sequence of the signs (east to west)

	Jaghmīnī, *Mulakhkhaṣ* [=*Mul.*]	Ptolemy, *Almagest* [=*Alm.*]	Farghānī, *Jawāmiʿ* [=J]	Battānī, *Zīj* [=*Zīj*]	Ṭūsī, *Tadhkira* [=T]
VARIOUS MOTIONS OF MERCURY AND THE MOON (I.2[3–5])					
Mercury [dirigent] **[CS]**	**0;59,8,20** *Mul.*,**I.2[3]**	0;59,8,17, 13,12,31 *Alm.*, IX.4 (441)	Same as mean Sun J: Ch. 14 (58)	**0;59,8,20**,46 S-K (374)	Same as mean Sun T: II.8[9] (1:166–7)
Moon [parecliptic with jawzahar] **[CS]**	**0;3,10,37** *Mul.*, **I.2[4]**	–	3′ J: Ch. 13 (52)	**0;3,10,37**,18,40,26 (Kennedy, 156) or **0;3,10,37**,17,40,26 S-K (374) or approx. 0;3 *Zīj*: Ch. 30 (3:76)	3′ plus a fraction T: II.7[8] (1:150–1)
Moon [inclined orb] **[CS]**	**11;9,7,43** *Mul.*, **I.2[5]**	–	11;9 J: Ch. 13 (52)	11;12 – 0;3 = 11;9 *Zīj*: Ch. 30 (3:76)	11;9 T: II.7[9] (1:152–3)
MOTION IN A NYCHTHEMERON (I.2[8])					
Sun[a]	**0;59,8,20** *Mul.*, **I.2[8]**	0;59,8,17, 13,12,31 *Alm.*, III.2 (143)	ca. 59′ J: Ch. 13 (50)	**0;59,8,20**,46 T: 2:493	0;59 T: II.7[11] (1:152–3)

[a] **Kharaqī's *Muntahā*** [*Bāb* 8, *Faṣl* 2 (180 [57])] has "about 59;8"; **Bīrūnī, *Qānūn*** (2:688) has 0;59,8,40,7,56,33.

MOTIONS OF THE DEFERENT ORBS ABOUT THEIR ECCENTRIC CENTERS IN A NYCHTHEMERON (I.2[9])					
	Jaghmīnī	Ptolemy	Farghānī	Battānī	Ṭūsī
Saturn **[S]**	**0;2,0,35** *Mul.*, **I.2[9]**	0; 2,0,33,31,28,51 *Alm.*, IX.4 (429)	ca. 2′ J: Ch. 14 (59)	**0;2,0,35**,55,48,3 Fatih 3418, f. 44a; or 0;2,0,36,4,43,2,8 (Kennedy, 159)	2′ T:II.9[8] (180–1)
Jupiter **[S]**	**0;4,59,16** *Mul.*, **I.2[9]**	0;4,59,14,26,46,31 *Alm.*, IX.4 (432)	ca. 5′ J: Ch. 14 (59-60)	**0;4,59,16**,54,54,57 Fatih 3418, f. 44a or **0;4,59,16,** 19,53,47,11,20 (Kennedy,159)	5′ T: II.9[8] (1:180–1)
Mars **[S]**	**0;31,26,40** *Mul.*, **I.2[9]**	0;31,26,36,53,51,33 *Alm.*, IX.4 (435)	ca. 31′ J: Ch. 14 (60)	**0;31,26,40**,15,11,13 Fatih 3418, f. 44a; or **0;31,26,39,36,** 34,5,16,50 (Kennedy, 159)	31′ T: II.9[8] (1:180–1)
Venus **[S]**	**0;59,8,20** *Mul.*, **I.2[9]**	0;59,8,17,13,12,31 *Alm.*, IX.4 (438)	Same as Sun J: Ch. 14 (59)	**0;59,8,20**,46 [same as Sun]	Same as Sun T: II.9[8] (1:180–1)

MOTIONS OF THE DEFERENT ORBS ABOUT THEIR ECCENTRIC CENTERS IN A NYCHTHEMERON (I.2[9])

	Jaghmīnī	Ptolemy	Farghānī	Battānī	Ṭūsī
Mercury [S]	**1;58,16,40** *Mul.*, **I.2[9]**	0;59,08,17,13,12,31 [twice this amount] *Alm.*, IX.4 (441)	Twice the Sun J: Ch. 14 (58)	Twice the Sun [see Sun]	Twice the Sun T:II.8[10] (1:168–9)
Moon [S]	**24;22,53,22** or 24;23,53,22 or 24;23 *Mul.*, **I.2[9]**[a]	12;11,26,41,20,17,59 [twice this amount or double elongation] *Alm.*, IX.4 (187)	24;23 J: Ch. 13 (51)	24;23 *Zīj*: Ch. 30 (3:76)	24;23 T:II.7[10] (1:152–3)

[a] **I.2[9]. for the Moon: 24;<u>22</u>,53,22.** This value is in MS F (6a), MS K (f. 7b), and MS L (f. 66b). MS B (p. 16) rounds it to 24;**23**, as does Ṭūsī (II.7[10]; Ragep, *Tadhkira*, 1:152–53). These MSS then also agree with 24;23 found in Battānī's *Zīj*. Only MS S (f. 54b) has 24;**23**,53,22, which is the variant written unambiguously (since he wrote it out) in ʿAbd al-Wājid's commentary (Laleli 2127, f. 26b). Nevertheless, it is more decisive that Jaghmīnī repeats the value 24;**22**,53,22 in **I.5[37]**, in his discussion of "What occurs to the Moon in relation to the Sun."

I.2[10]. ...this is the case for the epicycles of the five vacillating planets: Throughout the *Mulakhkhaṣ* Jaghmīnī uses the word *mutaḥayyira*, which I have translated as vacillating planets, in order to designate the five retrograding planets, namely Saturn, Jupiter, Mars, Mercury, and Venus (see **I.2[10]**, **I.5[5, 9, 12, 17, 29, 32]**). This term does not include the Sun and the Moon, which are also "planets" [i.e., "wandering stars" with respect to the "fixed stars"]; for the general term, Jaghmīnī uses the word *al-sayyāra,* but this only once (**I.3[12]**). Jaghmīnī's distinction between *al-mutaḥayyira* (for the *five* planets) and the more general *al-sayyāra* (for the *seven* planets) is most likely due to his following Kharaqī in his *Muntahā al-idrāk fī taqāsīm al-aflāk*:

...الكواكب السيّارة، أعني الشمس والقمر والكواكب المتحيّرة...

The wandering planets [*al-kawākib al-sayyāra*], i.e., the Sun, Moon, and the vacillating planets [*al-kawākib al-mutaḥayyira*] (See Ghalandari, "A Survey of the Works of '*Hayʾa*' in the Islamic Period," 172 [42]; cf. Berlin, Staatsbibliothek, Landberg MS 33, f. 8b.)

Exactly when this distinction between *sayyāra* and *mutaḥayyira* came into general use is not clear. In the *Planetary Hypotheses*, planets [πλανώμενοι] is translated as *mutaḥayyira* (see Goldstein, "The Arabic Version of Ptolemy's Planetary Hypotheses," 13 [variant in MS L; Heiberg, 70, title], 15 [lines 10 and 15; Heiberg, 76, lines 20 and 29]). For the last case, MS L correctly translates τὰ ἀπόγεια τῶν ε̄ πλανωμένων as وأوجات الكواكب الخمسة المتحيّرة [apogees of the five planets], seeming to imply that there is a special category for five of the planets. But already in Khwārizmī's tenth-century *Mafātīḥ al-ʿulūm* (edition by G. van Vloten [Leiden: E. J. Brill, 1895], 210, 228) and Bīrūnī's eleventh-century *Qānūn* (3:987), a clear distinction is made between *sayyāra* for the seven planets and *mutaḥayyira* for the five (Kunitzsch, "al-Nudjūm," in *EI2* [1995], 8:101a).

MOTIONS OF THE EPICYCLES IN EACH NYCHTHEMERON (I.2[10])

	Jaghmīnī	Ptolemy	Farghānī	Battānī	Ṭūsī
Saturn	**0;57,7,44** *Mul.*, **I.2[10]**	**0;57,7,43,41**, 43,40 *Alm.*, IX.4 (429)	57′ J: Ch. 14 (59)	**0;57,7,44**,48 S-K (374)	='s excess of Sun's mean over planet's mean T: II.9[11] (1:182–3)
Jupiter	**0;54,9,3** *Mul.*, **I.2[10]**	**0;54,9,2,46**, 26,0 *Alm.*, IX.4 (432)	54′ J: Ch. 14 (59)	**0;54,9,3**,52 S-K (374)	='s excess of Sun's mean over planet's mean T: II.9[11] (1:182–3)
Mars	**0;27,41,40** *Mul.*, **I.2[10]**	**0;27,41**, 0,19,20,58 *Alm.*, IX.4 (435)	28′ J: Ch. 14 (60)	**0;27,41,40** S-K (374)	='s excess of Sun's mean over planet's mean T: II.9[11] (1:182–3)
Venus	**0;36,59,29** or 0;37 or 0;37,19,29 *Mul.*, **I.2[10]**[a]	**0;36,59**, 25,53,11,28 *Alm.*, IX.4 (438)	37′ J: Ch. 14 (59)	**0;36,59,29**, 27,42,45 Fatih 3418, f.44a or **0;36,59**, 45,27,42,45 S-K (374) or **0;36,59,29**, 28,42,45 (Kennedy, 156)	0;37 [S in upper half] T: II.9[11] (1:182–3)
Mercury	**3;6,24,7** *Mul.*, **I.2[10]**	**3;6,24,06,59**, 35,50 *Alm.*, IX.4 (441)	3;6 J: Ch. 14 (58)	**3;6,24,7**, 45,53,33 Fatih 3418, f.44a or **3;6,[2]4,7**, 45,53,33 (Kennedy, 156)	3;6 T: II.8[13] (1:170–1)
Moon	**13;3,53,56** *Mul.*, **I.2[10]**	**13;3,53,56**, 17,51,59 *Alm.*, IV.4 (186)	13;4 J: Ch. 14 (51)	**13;3,53,56**, 17,51,59 Fatih 3418, f.44a or 13;4 *Zīj*: Ch. 30(3:77)	13;4 [CS in upper half] T: II.7[13] (1:154–5)

[a] **I.2[10]. The motions of the epicycles in each nychthemeron are…for Venus: 0;36,59,29:** This value is in MS K (f. 7b); however, it has been changed in MS F (f. 6a), MS L (67a), and MS S (f. 55a) to 0;37,19,29. The source of **0;37,19,29** is not clear, but ʿAbd al-Wājid's *Sharḥ al-Mulakhkhaṣ* gives it unambiguously (but without explanation) (Laleli 2127, f. 28b). MS B (p. 17) has **0;37** (and this is the value given by several sources such as Ṭūsī and Farghānī), but obviously this approximation could apply to either 0;36,59,29 or 0;37,19,29. On the other hand, an early *Mulakhkhaṣ* commentary by Alānī (Ahmet III 3308, f.25a) has 0;36,59,29; and so does Battānī, whose values Jaghmīnī seems to rely on throughout.

Chapter 3 of Part I: On the Circles

I.3[12]. The imaginary circles that are traced by the rotation of points in the planetary orbs are either traced on the surfaces of the spheres or else are not traced on the surfaces: It is not clear whether the words "by the rotation of points in the planetary orbs" were in the original version of the text; they are found in MSS K and L. MS S (f. 58) and MS F (f. 7b) added this in their margins (and both had correction marks). MS B (p. 22) had a variant that was an incomplete sentence.

I.3[13]. Only MS B (p. 23) adds an additional sentence at the conclusion of this paragraph, which marks the end of Chapter 3, to the effect that the equant orb is one of "the traced circles": "it being a circle traced by the motion of the line extending from a point that toward which the epicycle's diameter is always directed however it turns." The fact that this variant also states that a further clarifying explanation of this is in the chapter on circles clearly makes it suspect given that this *is* the chapter on circles. Princeton, Garrett MS 373 (p. 343) also includes a modified version of this sentence; however, it states that the clarifying explanation is given in Chapter 5, which is correct.

Chapter 4 of Part I: On the Arcs

I.4[3]. The co-ascension of each arc along the zodiacal orb is that which rises with it along the equinoctial: There is a variant in MS B (p. 24) that is also given as an alternative reading in the margin of MS S (f. 59a) with only minor variations. The variant states: "Among [the arcs] is the co-ascension: when an arc rises along the zodiacal circle, there then necessarily rises with it an arc on the equinoctial, and that arc on the equinoctial is called the co-ascension of that arc along the zodiacal orb."

I.4[5]. The equation of daylight… Let us take an example for this: Fig. C1 is a three-dimensional rendition of Jaghmīnī's passage. Cf. Ṭūsī: III.3[2] (Ragep, *Tadhkira*, 1:260–61; "On the Co-ascensions of the Ecliptic," III.7 [1:282–87]; and Fig. C28 [1:363]). For comparison, ʿAbd al-Wājid's commentary provides a two-dimensional diagram (Laleli 2127, f. 49b).

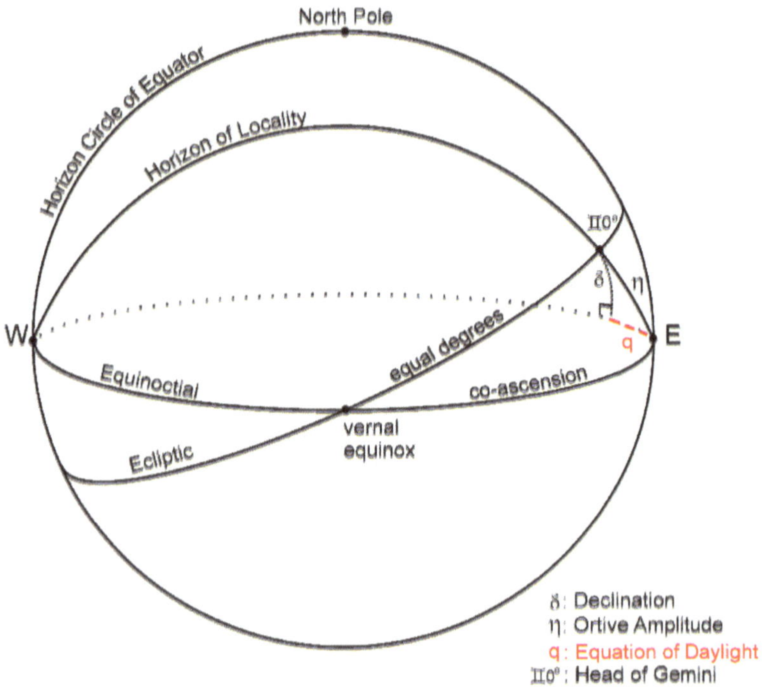

Fig. C1 Illustration of the Equation of Daylight

I.4[5]. When the head of Gemini is found toward the east for a horizon <u>other</u> than the equator: The omission of غير by two key texts (MS F [f. 8a] and MS S [f. 59a]) may be an indication of what was missing in an earlier version of the *Mulakhkhaṣ*, and then added later, perhaps in a revision by Jaghmīnī himself. It is also omitted in ʿUbaydī's commentary (Istanbul, Süleymaniye Library, Laleli 2128, f. 33b). However, it is included in MS B (p. 25), MS K (f. 13a), and MS L (f. 69a). It is also in the commentaries of ʿAbd al-Wājid (Laleli 2127, f. 48b) and several copies of Qāḍīzāde's (UCLA, Caro Minasian 33, p. 246; Tehran, Majlis-i Shūrā 18045 [not foliated]; Isfahan, Abū al-Barakāt 50 [not foliated]; and Qum, Faydiyya 01832, p. 107).

I.4[6]. …the angle of the equation: Cf. Bīrūnī, *Tafhīm*, nos. 172–174 (89–90); and Ragep, *Tadhkira*, II.6[5] (1:148–49).

I.4[9]. …that they called sectors: Cf. Bīrūnī, *Tafhīm,* no. 201 (107–10) and Ṭūsī, *Tadhkira*, II.14[1] (Ragep, 1:240–41, 2:463). See also E. S. Kennedy, "A Survey of Islamic Astronomical Tables," 143; Kennedy, "The Sasanian Astronomical Handbook *Zīj-i Shāh* and the Astrological Doctrine of 'Transit' (*Mamarr*)," *Journal of the American Oriental Society* 78, no. 4 (Oct.-Dec., 1958): 247–53; and Kennedy, *The Planetary Equatorium of Jamshīd Ghiyāth al-Dīn al-Kāshī* (Princeton, NJ: Princeton University Press, 1960), 218–22.

I.4[9]. The epicycle is divided by two lines, one of them extending from the deferent center: For the epicycle orb, Ṭūsī states the line is produced from the World center rather than from the deferent center (*Tadhkira*, II.14[1]; Ragep, 1: 240–41). Also MS S (f. 61b) draws it from the World center. However, the text clearly states it is from the deferent center, and MS B (p. 29) and MS K (f. 16a) have it drawn this way, as do several commentaries I also checked (see ʿAbd al-Wājid, Laleli 2127, f. 56a; ʿUbāydī, Laleli 2128, f. 38b; Qāḍīzāde, UCLA, Caro Minasian 33, p. 270).

I.4[12]. The second declination is an arc along a latitude circle between the two of them, I mean between the equinoctial and the zodiacal circle: Given that this sentence could have been written more succinctly as "The second declination is an arc along a latitude circle between the equinoctial and the zodiacal circle," it does raise the question about the orality of the text, especially since he uses the first person "I mean."

I.4[12]. ...and its amount is 23;35: Jaghmīnī's value for the obliquity is not the Ptolemaic one of 23;51,20 (*Almagest*, I.12 [Toomer, *Ptolemy's Almagest*, 61–63, esp. 63n75]). The 23;35 value is the one that derives from the time of Maʾmūn (r. 813–33). It is found in Farghānī's *Jawāmiʿ ʿilm al-nujūm*, Ch. 5 (*Jawāmiʿ*, 18), who also reports Ptolemy's value 23;51; Battānī's *Zīj*, Ch. 4 (3:18); and Ṭūsī's *Tadhkira* (II.4[1]) as (23+⅓+¼)° (Ragep, 1:120–21, 2:394). However, Ṭūsī "updates" the value to 23;30 in his *Īlkhānī Zīj*, written some four years after the *Tadhkira*. Noteworthy is that ʿAbd al-Wājid uses the value of 23;30 throughout his commentary, informing us (within his comments on Jaghmīnī's 23;35 value) that 23;30 is due to "new observations" (Laleli 2127, ff. 58a–58b). Ibn Sīnā found the value of the obliquity to be 23;33,30, but seemingly he had few followers (see S. P. Ragep, "Ibn Sīnā," in *BEA*, 1:570–72; and F. Jamil Ragep and Sally P. Ragep, "The Astronomical and Cosmological works of Ibn Sīnā," 6, 10).

I.4[15]. Parallax [lit., divergence of sight]: Ṭūsī devotes an entire chapter to this subject. See Ragep, *Tadhkira*, II.12[1–8] (1:222–29, 2:458).

I.4[16]. The ortive amplitude: See Bīrūnī, *Tafhīm*, no. 220 (129) for a clear explanation of this term along with a diagram.

I.4[22]. The measure of each one of these six arcs is similar to its [corresponding arc] along the equinoctial: The six arcs are: (1) the arc of daylight; (2) the arc of night; (3) the planet's arc of daylight; (4) the planet's arc of night; (5) the turning of the orb [daylight]; and (6) the turning of the orb [night]. Note the meaning here of **"these six arcs [are] similar"** is analogous to what is meant by similar triangles.

Chapter 5 of Part I: On What Occurs to the Planets in Their Motions

I.5[2]. The Sun has a single anomaly (Fig. C2):

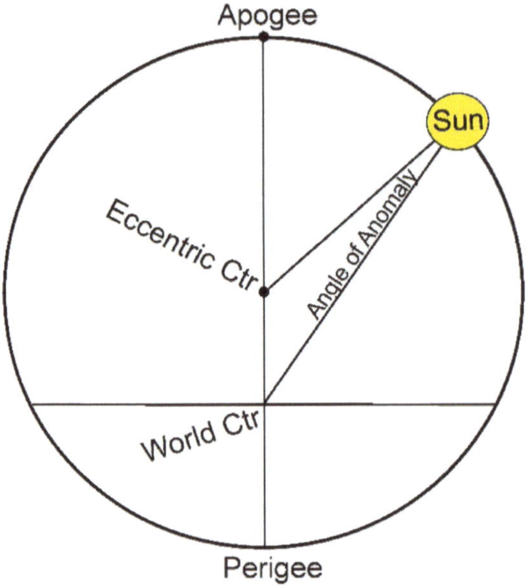

Fig. C2 Illustration of the Sun's Single Anomaly

I.5[3]. As for the remaining planets, they have numerous anomalies in longitude: I translated *ikhtilāf* (lit., difference) as anomaly. Jaghmīnī is using the term in a technical sense and lists them in this chapter. It should be understood as an "irregular, or anomalous, speed, i.e., one that differs from the mean" (see *Tadhkira*, II.7[1]18–21; Ragep, 1:148–49 and 2:417).

I.5[3]. Starting with the parameters in this section, MS B (p. 33) has an interesting way of treating the fractional parts of the parameters, and provides us with an example of late Ottoman mathematical notation. These are given as variants in the apparatus. This is discussed in *II.1d: Parameters.*

THE RADII OF THE EPICYCLES AT THEIR MEAN DISTANCES (I.5[3])[a]

	Jaghmīnī, *Mulakhkhaṣ* [=*Mul.*]	Ptolemy, *Almagest* [=*Alm.*]	Battānī, *Zīj* [=*Zīj*]	Farghānī, *Jawāmiʿ* [=J]	Ṭūsī, *Tadhkira* [=T]	Kharaqī, *Muntahā* [=M] and [*Tabṣira*]
Saturn	**6;30** *Mul.*, **I.5[3]**	6;30 *Alm.*, XI.6 (540)	6;29,50 *Zīj*: Ch. 28 (3:73)	6½ J: Ch. 16 (65)	6½ T:II.9[13] (1:184–5)	6;30 M: *Bāb* 10, *Faṣl* 2 (207 [110])

THE RADII OF THE EPICYCLES AT THEIR MEAN DISTANCES (I.5[3])[a]

	Jaghmīnī, *Mulakhkhaṣ* [=*Mul.*]	Ptolemy, *Almagest* [=*Alm.*]	Battānī, *Zīj* [=*Zīj*]	Farghānī, *Jawāmiʿ* [=J]	Ṭūsī, *Tadhkira* [=T]	Kharaqī, *Muntahā* [=M] and [*Tabṣira*]
Jupiter	**11;30** *Mul.*, **I.5[3]**	11;30 *Alm.*, XII.3 (570)	11;30,5 *Zīj*: Ch. 28 (3:73)	11½ J: Ch. 16 (65)	11½ T:II.9[13] (1:184–5)	11;30 M: *Bāb* 10, *Faṣl* 2 (207 [110])
Mars	**39;30** *Mul.*, **I.5[3]**	39;30 *Alm.*, XII.4 (574)	39;**25**,22 39;55,22 *Zīj*: Ch. 28 (3:73)[b]	39⅙ J: Ch. 16 (65)	39½ T:II.9[13] (1:184–5)	39;30 M: *Bāb* 10, *Faṣl* 2 (207 [110])
Venus	**45;0** or 43⅙ [=43;10] *Mul.*, **I.5[3]**[c]	43;10 *Alm.*, X.2 (472)	**44;9,5** *Zīj*: Ch. 28 (3:73)	43⅙ J: Ch. 16 (65)	43⅙ T:II.9[13] (1:184–5)	43;10 or 43;30 M: *Bāb* 10, *Faṣl* 2 (207 [110]); 43;32[*Tabṣ.*]
Mercury	**25;0** or 22;30 *Mul.*, **I.5[3]**[d]	22;30 *Alm.*, IX.9 (459–60)	22;30,30 *Zīj*: Ch. 28 (3:73)	22½ J: Ch. 16 (65)	22½ T:II.8[13] (1:170–1)	22;30 M: *Bāb* 11, *Faṣl* 2 (218 [133])
Moon	**6;20** or 5;15 *Mul.*, **I.5[3]**[e]	5;14 *Alm.*, IV.6 (202)	5;15 *Zīj*: Ch. 28 (3:73)	6⅓ J: Ch. 16 (65)	5;15 T:II.7[16] (1:156–7)	ca. 5 M: *Bāb* 9, *Faṣl* 3 (197 [87])

[a] For abbreviations and references, see above *"On the Sources"* in **I.2[9]**.

[b] **I.5[3]. The radii of the epicycles at their mean distances are…for Mars: 39;30:** J. B. Delambre gives the Battānī value as 39;55,22. However, the Arabic text, written alphanumerically as كه, is 39;**25**,22. Cf. Nallino, *Al-Battānī Sive Albatenii Opus Astronomicum*, Ch. 28, 3:73 and Delambre, *Histoire de l'astronomie du Moyen Age* (Paris, 1819), Ch. 2: Albategnius, 10–62) [repr. In *Islamic Mathematics and Astronomy*, ed. Fuat Sezgin (Frankfurt am Main, 1998), 68:37].

[c] **I.5[3]. The radii of the epicycles at their mean distances are…for Venus: 45;0:** MS F (f. 10b), MS K (f. 19a), MS L (f. 71b), and MS S (f. 64b) have a value of 45;0, which could be a rounding up of Battānī's parameter. MS B (p. 33) gives Ptolemy's value of 43⅙, which is also found in Farghānī, Kharaqī, and Ṭūsī.

[d] **I.5[3]. The radii of the epicycles at their mean distances are…for Mercury: 25;0:** MS F (f. 11a), MS L (f. 71b), and MS S (f. 64b) all have the 25;0 value; however, MS B (p. 33) and MS K (f. 19a) have the Ptolemaic value 22;30, which is also the value found in the other sources (and clearly closer to Battānī's value of 22;30,30 [*Zīj*, Ch. 28, 3:73]).

[e] **I.5[3]. The radii of the epicycles at their mean distances are…for the Moon: 6;20:** The two values given for the Moon's parameters (6;20 and 5;15) derive from different reference radii: (1) 6;20 is based on the deferent orb having a radius of 60p; and (2) 5;15 is based on an inclined orb having a radius of 60p. Jaghmīnī has opted for the former, as does Farghānī. MSS F, S, K and L have 6;20; only MS B has 5;15. Cf. Ptolemy, *Almagest*, IV.5; and Pedersen, *A Survey of the Almagest,* 159–202 (Ch. 6: The Theories of the Moon), 424.

I.5[5]. The third anomaly [regarding what occurs to epicycle centers]**:** For Ṭūsī's explanation of this anomaly regarding the Moon, see *Tadhkira*, II.7[25] (Ragep, 1:158–61).

I.5[9]. the mean apex: For Ṭūsī's definition of the mean apex, see "On the Orbs and Longitudinal Motions of the Remaining Planets," *Tadhkira*, II.9[11] (Ragep, 1:182–83).

THE DISTANCE OF THE ECCENTRIC CENTER FROM THE WORLD CENTER (I.5[11])[a]

	Jaghmīnī, *Mulakhkhaṣ* [=*Mul.*]	Ptolemy, *Almagest* [=*Alm.*]	Battānī, *Zīj* [=*Zīj*]	Farghānī *Jawāmi'* [=J]	Ṭūsī, *Tadhkira* [=T]	Kharaqī, *Muntahā* [=M] and [*Tabṣira*]
Sun	**2;29,30** *Mul.*, **I.5[11]**	2;29,30 *Alm.*, III.4 (153–5)	2;4,45 *Zīj*: Ch. 28 (3:73; 2:244)	2½ J: Ch. 16 (64–5)	2;5 [also 2;30] T:II.6[4] (1:146–7; 2:416, n6)	ca. 2 M: *Bāb* 8, *Faṣl* 2 (183 [62])
Moon[b]	**10;19** *Mul.*, **I.5[11]**	10;19 *Alm.*, V.4 (226)	10;19 *Zīj*: Ch. 30 (3:82; 2:223)	12½ [=10;19] J: Ch. 16 (64–5)	10;19 T:II.7[18] (1:156–9)	10+⅓ M: *Bāb* 9, *Faṣl* 1 (189 [71])

[a] For abbreviations and references, see above *"On the Sources"* in **I.2[9]**.

[b] The two values given for the Moon's parameters (10;19 and 12;30) derive from different radii (See Pedersen, *A Survey of the Almagest*, 184–85, 424): (1) 12;30 is based on the deferent orb having a radius of 60; and (2) 10;19 is based on the inclined orb having a radius of 60p.

THE DISTANCE OF THE EQUANT CENTER FROM WORLD CENTER (I.5[12])

	Jaghmīnī	Ptolemy	Farghānī	Ṭūsī	Kharaqī
Saturn	**6;50** *Mul.*, **I.5[12]**	6;50 [3;25x2] *Alm.*, XI.5 (537)	3+¼+⅙ [x2] J: Ch. 16 (64–5)	3+¼+⅙ [x2] T: II.9 [9] (1:180–3)	6+½+⅓ M: *Bāb* 10, *Faṣl* 2 (205 [152])
Jupiter	**5;30** *Mul.*, **I.5[12]**	2;45 [x2] *Alm.*, XI.3 (524)	2½+¼ [x2] J: Ch. 16 (64–5)	2¾ [x2] T: II.9[9] (1:180–3)	5+½ M: *Bāb* 10, *Faṣl* 2 (205 [152])
Mars	**12;0** *Mul.*, **I.5[12]**	6;0 [x2] *Alm.*, X.7 (498)	6;0 [x2] J: Ch. 16 (64–5)	6;0 [x2] T : II.9[9] (1:180–3)	12 M: *Bāb* 10, *Faṣl* 2 (205 [152])
Venus	**2;5** *Mul.*, **I.5[12]**[a]	1;15 [x2] *Alm.*, X.2 (472–4)	1;15 [x2] J: Ch. 16 (64–5)	ca. ½ Sun [x2] T: II.9[9] (1:180–3)	2;5 M: *Bāb* 10, *Faṣl* 2 (205 [152])

[a] Note Jaghmīnī's value for Venus is not the Ptolemaic one; rather it is the same as Kharaqī's value. This could be an indication that they are deriving it from new values for the Sun's eccentricity.

DISTANCE BETWEEN EQUANT AND DIRIGENT, AND BETWEEN DIRIGENT AND DEFERENT CENTER (I.5[13])

	Jaghmīnī	Ptolemy	Farghānī	Ṭūsī	Kharaqī
Mercury	**3;10** *Mul.*, **I.5[13]**	3;0 *Alm.*, IX.9 (459)	3;0 J: Ch. 16 (64–5)	Half 6;0 T: II.8[14] (1:170–1)	**3 parts+⅙** M: *Bāb* 11, *Faṣl* 2 (217–18 [133])

LATITUDE OF THE PLANETS (I.5[16])
(=maximum inclination of the inclined orb from the zodiacal orb)

	Jaghmīnī, *Mulakhkhaṣ*	Ptolemy, *Almagest*	Ṭūsī, *Tadhkira*
Saturn	**2;30°** *Mul.*, **I.5[16]**	2½° *Alm.*, XIII.3 (605)	2½° II.10[1] (1:188–9)
Jupiter	**1;30°** *Mul.*, **I.5[16]**	1½° *Alm.*, XIII.3 (605)	1½° II.10[1] (1:188–9)
Mars	**1;0°** *Mul.*, **I.5[16]**	1° *Alm.*, XIII.3 (604)	1° II.10[1] (1:188–9)
Venus	**+0;10°** *Mul.*, **I.5[16]**	+⅙° *Alm.*, XIII.3 (601)	+⅙° II.10[1] (1:188–9)
Mercury	**–0;45°** *Mul.*, **I.5[16]**	–¾° *Alm.*, XIII.3 (601)	–(½+¼)° II.10[1] (1:188–9)
Moon	**5;0°** *Mul.*, **I.5[16]**	5° *Alm.*, V.7, 12 (237, 247)	5° II.7[4] (1:150–1)

EPICYCLE LATITUDE AT MAXIMUM (I.5[17])
(= maximum deviation in either direction of the apex [or epicyclic perigee] from the inclined deferent)[a]

	Jaghmīnī, *Mulakhkhaṣ* *Mul.*,I.5[17]	Ptolemy, *Almagest*	Ṭūsī, *Tadhkira* T: II.10[4] (1:190–3)	Alānī, *Sharḥ* Ahmet III 3308, f. 51a	Kharaqī, *Muntahā* M: *Bāb*15, *Faṣl* 3 (281 [chart])
Saturn	**0;32°**	0;35[b]	0;35	0;32	0;35
Jupiter	**0;38°**	0;38[b]	0;38	0;38	0;38
Mars	**6;16°** or 6;56[c]	6 *Alm.*, XIII.3 (603 4)	6⅒ (=6;6)	6;16	6;7
Venus	**1;2°**	1;2 *Alm.*, XIII.3 (602)	1;2	1;2	1;2
Mercury	**1;45°**	1;45 *Alm.*, XIII.3 (602)	1;45	1;45	1;45

[a] Jaghmīnī's values for Saturn, Jupiter, and Mars are with respect to the maximum deviation of the perigee; however, for Venus and Mercury he gives the values for the maximum deviation of the apex.

[b] Values derivable from latitude tables *in Almagest*, XIII.5.

[c] Value found in MS S (f. 66b) and MS F (f. 8a).

I.5[17–18]. F. J. Ragep provides a table listing values for the deviation and slant of planet epicycles for Ptolemy and Ṭūsī (*Tadhkira*, II.10[4] and [5], 2:424–25) as well as a diagram of the deviation for the upper planets (Fig. C6a, 1:347). See also Swerdlow, "Ptolemy's Theories of the Latitude of the Planets in the *Almagest*, *Handy Tables*, and *Planetary Hypotheses*," 41–71, esp. 63.

I.5[18]. It is called the latitude of the slope [*wirāb*], the slant [*inḥirāf*], and the twist [*iltiwāʾ*]. Its maximum for both [i.e., Mercury and Venus] is 2;30: These are Ptolemy's values for Mercury and Venus (*Almagest*, XIII.5 [Toomer, *Ptolemy's Almagest*, 633–34]). Cf., Ṭūsī, *Tadhkira*, II.10[5] (Ragep, 1:192–95, 2:424–26).

I.5[26]. As for the position of the apogees, they are for the beginning of the year 1517 (غثيز) of Dhū al-Qarnayn: The term "Dhū al-Qarnayn" (the two-horned) is in reference to the era of Alexander the Great; see W. Montgomery Watt, "al-Iskandar," in *EI2* (1978), 4:127. A variant reading of this number is 1317 (غشيز) due to a mistake in reading ش (300) instead of ث (500); for an example of this, see ʿAbd al-Wājid's commentary (Laleli 2127, f. 85a). However, some commentators removed all doubt by spelling the year out in addition to writing the number alphanumerically; two prominent examples are Alānī (Ahmet III 3308, f. 55b) and Qāḍīzāde (Ayasofya 2662, f. 42b). The date 1517 is also found in at least two fifteenth-century Persian commentaries (Muḥammad b. ʿUmar al-Andiqānī [Ayasofya 2592, f. 26a] and Ḥamza b. Ḥājjī Sulaymān [Ayasofya 2593, f. 121a]). It is also noted in the articles of Ghalandari ("Chaghmīnī," *The Great Islamic Encyclopedia*) and Qāsimlū ("Chaghmīnī," *Encyclopaedia of the World of Islam*). Rudloff and Hochheim omitted the year entirely in their German translation ("Die Astronomie des Maḥmûd ibn Muḥammed ibn ʿOmar al-Ǵagmînî," 253); had they included it much of the controversy regarding Jaghmīnī's dates may have been avoided.

The date 1517 and Jaghmīnī's parameters for the apogee and nodes are important in establishing that he was alive in 1205 CE (=602 H). An even more precise calculation of the date is 1516 years from Monday, 1 October 312 B.C.=1 October 1205 (=16 Ṣafar 602 H [±2 days]). This is because the beginning of year one of the Alexander epoch is calculated starting with Monday, 1 October 312 B.C. in the Julian calendar (see Benno van Dalen, "Taʾrīkh," section 2.a. "Calendars and eras," TABLE 2 in *EI2* (2000), 10:264–71; how to convert dates is also provided within the article); cf. Bīrūnī's listing of eras (*Tafhīm*, no. 282 [174]).

JAGHMĪNĪ'S PARAMETERS FOR THE POSITIONS OF THE APOGEES (I.5[26]) AND NODES (I.5[27])				
These values are found in all 5 MSS and also the commentaries of Alānī (Ahmet III 3308, f. 55b) and Qāḍīzāde (Ayasofya 2662, ff. 42b–43a). Furthermore, since the planetary nodes are fixed with respect to the apogees (i.e., being 90 degrees apart), they provided additional confirmation that Jaghmīnī was internally consistent within the *Mulakhkhaṣ*.				
	Apogee Position	Head Node Position	Tail Node Position	Midpoint (between head & tail)
Saturn [**50**° beyond midpoint]	**Sagittarius 9;23,33** =249;23,33 [=199;23;33+**50**]	**Cancer 19;23,33** =109;23,33	289;23,33	199;23;33
Jupiter [**20**° in advance of midpoint]	**Virgo 19;23,33** =169;23,33 [=189;23,33–**20**]	**Cancer 9;23,33** =99;23,33	279;23,33	189;23,33
Mars [= midpoint]	**Leo 11;53,46** =131;53,46	**Taurus 11;53,46** =41;53,46	221;53,46	131;53,46
Venus [= midpoint]	**Gemini 27;10,33** =87;10,33	**Pisces 27;10,33** =357;10,33	177;10,33	87;10,33
Mercury [= midpoint]	**Libra 26;23,33** =206;23,33	**Capricorn 26;23,33** =296;23,33	116;23,33	206;23,33

Most likely, Jaghmīnī relied on Battānī's values for his computations **(see chart below)**. My conclusion was facilitated by the fact that both Jaghmīnī and Battānī give their values in the era of Dhū al-Qarnayn (in contrast to Bīrūnī who uses the Yazdigird calendar), and both agree that the apogee moves 1° per 66 years. Based on this, I calculated a constant value of **4;55,33** that the apogee would have moved between their two dates (1517 Dhū al-Qarnayn for Jaghmīnī; 1191 Dhū al-Qarnayn for Battānī. However, note that some tweaking was necessary since a 325-year (rather than 326) difference was necessary to make the calculation work out; and alternative readings for the alphanumerical values of Mars and Venus are suggested.

POSITION OF THE APOGEES			
	Jaghmīnī, *Mulakhkhaṣ* **I.5[26]**	Battānī, *Zīj* (Ch. 45, 3:172–3; Ch. 28, 3:67 [Sun])	Value of Jaghmīnī minus value of Battānī
Year	1517 Dhū al-Qarnayn	1191 Dhū al-Qarnayn	
Sun	**Gemini 27;10,33** =87;10,33	[same as Venus]	
Saturn	**Sagittarius 9;23,33** =249;23,33	244;28	249;23,33 244;28,00 **4;55,33**
Jupiter	**Virgo 19;23,33** =169;23,33	164; 28	169;23,33 164;28,00 **4;55,33**

POSITION OF THE APOGEES			
	Jaghmīnī, *Mulakhkhaṣ* **I.5[26]**	Battānī, *Zīj* (Ch. 45, 3:172–3; Ch. 28, 3:67 [Sun])	Value of Jaghmīnī minus value of Battānī
Mars	**Leo 11;53,46**[a] =131;53,46	126;18	131;53,46 126;**18**;00 5;35,46
Misprint in Battānī?		126;**58** [58 (نح) not 18 (يح) ?]	131;53,46 126;**58**;00 **4;55**,46
Venus	**Gemini 27;10,33** =87;10,33	82;14	87;10,33 82;14,00 **4;56;33**
Misprint in Battānī?		82;**15** [15 (يه) not (يد) ?]	87;10,33 82;**15**,00 **4;55;33**
Mercury	**Libra 26; 23,33** =206;23,33	201;28	206;23,33 201;28,00 **4;55,33**

[a] Note that Jaghmīnī repeats 33 seconds for all the apogee positions, except Mars has 46 seconds (with no variants on this to date); 46 is a mystery, seemingly a mistake introduced at an early time and repeated. The repetition of 33 seconds though I assume was based on a *zīj*. See Kennedy ("A Survey of Islamic Astronomical Tables," 169) for his take on newly observed parameters versus calculated ones.

I find it puzzling that Jaghmīnī does not rely on Bīrūnī's often more accurate parameters (which clearly do not correspond with Jaghmīnī's); they were fellow Khwārizmians, and Bīrūnī flourished a century after Battānī. Perhaps Jaghmīnī found working in the Alexandrian calendar more congenial and thus followed Battānī, but it is far more likely the authority of Battānī's *Zīj* still held sway despite the intervening centuries and the availability of better parameters.

I.5[28]. Then for every year, one adds to their positions what the orb of the fixed stars moves in the year: Jaghmīnī has accepted the fact that precession is a fixed constant of 1 degree per 66 years, thus rejecting the notion of variable precession. This is in contrast to what one finds in the western Islamic world and the West.

I.5[29]. What occurs to the vacillating planets regarding retrogradation, direct motion, and stations: Ṭūsī uses *wuqūf* not *maqām* for station (*Tadhkira*, II.5[8]24 [Ragep, 1:136–37, 2:414]). Cf., *Almagest*, IX.2 (Toomer, *Ptolemy's Almagest*, 420–21).

I.5[32]. What occurs to [the vacillating planets] in relation to the Sun: Cf. Ṭūsī, *Tadhkira*, II.9[12] and [14] (Ragep, 1:182–85) and IV.6[3] (2:525). For Bīrūnī's definition of a planet in combust, see *Tafhīm*, no. 153 (64–65).

I.5[35]. What occurs to the Moon in relation to the Sun: The new Moon [*muḥāq*], waxing, full Moon, waning, its eclipsing of the Sun, and lunar eclipses: See **Figs. C3**, **C4**, and **C5**.

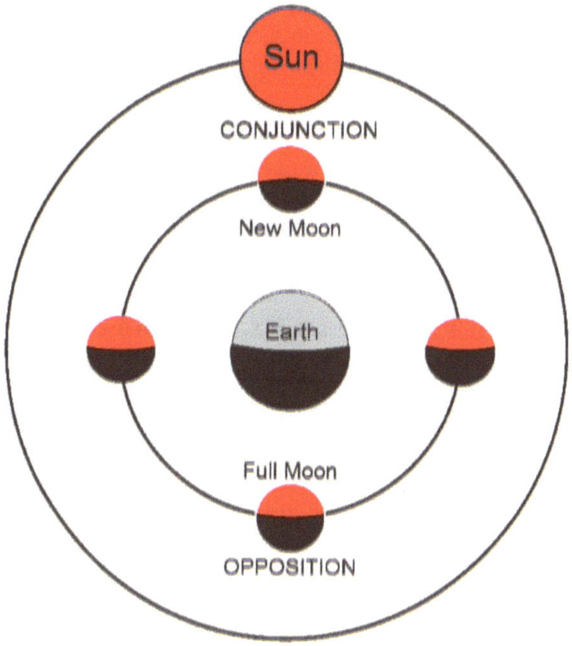

Fig. C3 Illumination of the Moon in Relation to the Sun

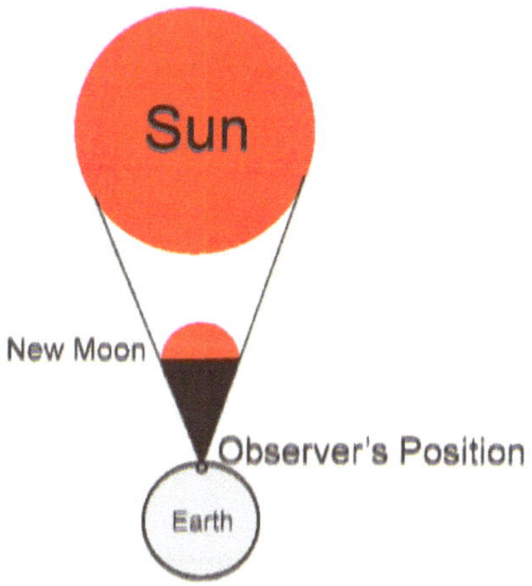

Fig. C4

Fig. C5

I.5[35]. What occurs to the Moon in relation to the Sun: The new Moon [*muḥāq*]: According to Bīrūnī, this term technically applies to the Moon's setting or disappearance two days prior to appearing new in the west (*Tafhīm*, no. 154 [64–66]).

I.5[37]. Among what occurs to the Moon is that the Sun is always in the middle between the [Moon's] apogee and the center of its epicycle: For an indication that the *Mulakhkhaṣ* was studied (and not merely copied), the mathematics of this is worked out in the margin of Princeton, Garrett 373 (pp. 358, 358A).

The Second Part: On an Explanation of the Earth and What Pertains to It in Three Chapters

Chapter 1 of Part II: On the Inhabited Part of the Earth and Its Latitude, Its Longitude, and Its Division into the Climes

II.1[1]. the cupola of the Earth: Cf. Bīrūnī, *Tafhīm*, no. 239 (140); and Ṭūsī, *Tadhkira*, III.1[7] (Ragep, 1:250–51).

II.1[2]. Ptolemy, after writing the *Almagest*, claimed that he found habitation below the equator to a distance of 16;25: Jaghmīnī is here referring to Ptolemy's *Geography*, which Ptolemy wrote after his *Almagest*. Regarding habitation beneath the equator, Ptolemy states in the *Almagest*: "For those who live beneath the equator…It is said that the regions beneath the equator could be inhabited, since the climate must be quite temperate…But what these inhabited regions are we have no reliable grounds for saying. For up to now they are unexplored by men from our part of the inhabited world, and what people say about them must be considered guesswork rather than report" (Toomer, *Ptolemy's Almagest*, 83 [II.6]). In comparison, in his *Geography*, Ptolemy gives the southern limit of habitation as $16\frac{5}{12}^\circ$ south of the equator (Berggren and Jones, *Ptolemy's Geography*, 85 (Bk 1.23[4], [23])). Note that the *Mulakhkhaṣ* commentators are well aware that Jaghmīnī is referring to Ptolemy's *Geography* here (for examples, see Alānī, Ahmet III 3308, f. 65a; ʿAbd al-Wājid, Laleli 2127, f. 101a; and Qāḍīzāde, Ayasofya 2662, f. 48b).

II.1[2]. The longitude of the inhabited part is <u>180;0</u>, and its beginning is from the west; however, some of them take it to be from the coast of the enclosing ocean and some of them from islands well into this ocean, their distance from its coast being <u>10;0</u>: Jaghmīnī is here referring to the Eternal Islands (*al-khālidāt*), also called the Fortunate Islands (*suʿadāʾ*) [also referred to as the Isles of the Blest, and usually thought to be the Canary Islands]. See *Tadhkira*, III.1[7]10 (Ragep, 1:250–51 and 2:468).

II.1[3]. This inhabited part was then divided into seven longitudinal sections parallel to the equator: The original versions of Jaghmīnī's *al-Mulakhkhaṣ* contain the Ptolemaic values found in the *Almagest,* II.6 up to Clime VII for the maximum daylight of 16 hours (Toomer, *Ptolemy's Almagest*, 82–87). Kharaqī's *Tabṣira* (*Maqāla* 2, *Bāb* 2 of *hayʾat al-arḍ*) and *Muntahā* (*Maqāla* 2, *Bāb* 4) also contain

Ptolemaic values, and Kharaqī cites Ptolemy and the *Almagest* in both his works. But there are differences, for which see **II.1[8]**.

As discussed in the Introduction (**§ I.1**), an enormous number of extant *Mulakhkhaṣ* witnesses have been modified with parameters found in Ṭūsī's *Tadhkira fī ʿilm al-hayʾa,* composed over fifty years after the *Mulakhkhaṣ*. For more extensive comparative charts of the values of the maximum daylight and latitudes of climes, see F. J. Ragep, "On Dating Jaghmīnī and His *Mulakhkhaṣ,*" 463–64; and Ragep, *Tadhkira*, Commentary III.1[8] (2:469–71), esp. Table 7 (2:470). In addition to Ptolemy, Jaghmīnī, and Ṭūsī, these charts list the parameters found in Bīrūnī, Qāḍīzāde's *Sharḥ al-Mulakhkhaṣ*, Rudloff and Hochheim's German *Mulakhkhaṣ* translation, and modern computations.

Below is a comparative chart of the values of the maximum daylight and latitudes of climes for Ptolemy's *Almagest,* Jaghmīnī's *Mulakhkhaṣ,* Ṭūsī's *Tadhkira*, and Kharaqī's *Muntahā* and *Tabṣira.*

MAXIMUM DAYLIGHT AND LATITUDES OF THE CLIMES (II.1[3–10])

	Maximum Daylight (Hours)	Latitudes Jaghmīnī, *Mulakhkhaṣ* **(II.1[3–10])**[a]	Ptolemy, *Almagest* II.6(82–7)	Ṭūsī, *Tadhkira* III.1[8] (1:250–3; 2:469–71)	Kharaqī, *Muntahā* and *Tabṣira*[b]
I	12	0°			
	12¾ [12;45]	12;30° (II.1[3])	12;30°	12;40° [12+⅔]	12;30°
	13 [13;0]	16;27° (II.1[3])	16;27°	16;37,30° [16+½+⅛]	16;27°
II	13¼ [13;15]	20;14° (II.1[4])	20;14°	20;27° [20+¼+⅕]	20;14°
	13½ [13;30]	**23;51° (II.1[4])**	**23;51°**	24;5° [24+(½ *of* ⅙)] **24;40°** [24+(½+⅙)] copyist error	**23;51°**
III	13¾ [13;45]	27;12° (II.1[5]) 29;12° copyist error	27;12°	27;30° [27+½]	27;12°
	14 [14;0]	30;22° (II.1[5])	30;22°	30;40° [30+⅔]	30;22°
IV	14¼ [14;15]	33;18° (II.1[6])	33;18°	33;37,30° [33+½+⅛]	33;18°
	14½ [14;30]	36;0° (I.1[6])	36;0°	36;22° [36+⅕+⅙]	36;0°
V	14¾ [14;45]	38;35° (II.1[7])	38;35°	38;54° [39 minus ⅒]	38;35°
	15 [15;0]	40;56° (II.1[7])	40;56°	41;15° [41+¼]	40;56°
VI	15¼ [15;15]	43;51° (II.1[8])	43;1° 43;15°	43;22,30° [43+¼+⅛]	43;1° 43;15°
	15½ [15;30]	45;1° (I.1[8])	45;1°	45;21° [45+¼+⅒]	45;1°
VII	15¾ [15;45]	46;51° (II.1[9])	46;51°	47;12° [47+⅕]	46;51°
	16 [16;0]	48;32° (II.1[9])	48;32°	48;52,30° [48+½+¼+⅛]	48;32°
	16¼ [16;15]	50;25° (II.1[10])	50;4°	50;20°	–

[a] See Text Apparatus for parameter variants found in the manuscripts.

[b] *Tabṣira*: Ayasofya 2581, ff. 121a–124a and Laleli 2141, ff. 45a–46b; *Muntahā*: Ghalandari, 327–34 [308–25]. In both *Tabṣira* witnesses, the numbers are written in words.

II.1[3]. The first clime: All 5 witnesses give the value **12 hours** for maximum daylight at the equator, and **12;45 hours** for the maximum daylight at the alternative beginning of Clime I. MSS F, K, L, S have **12;30** degrees latitude for this alternative, but MS B provides no value. In fact, starting here MS B leaves blank spaces for the clime values. For a maximum daylight of **13;0** hours, MSS F, L, S give **16;27** degrees latitude. Alānī's *Sharḥ* (Ahmet III 3308, f. 66a) also gives this Ptolemaic value. But MS K (f. 28a) gives 16;37; this could be due to the obvious scribal error of not changing the ل (30) into a ك (20). Another possibility is that the copyist is using Qāḍīzāde's parameter of 16;37 (Ayasofya 2662, f. 49a) or perhaps abridging Ṭūsī's parameter of 16;37,30 [16+½+⅛] degrees (*Tadhkira*, III.1[8] [Ragep, 1:250-51]). Note that Ṭūsī, unlike Jaghmīnī, begins the first clime at 12;45 [12+½+¼] hours with a latitude of 12;40 (12⅔ degrees).

II.1[4]. the second clime: MSS F, K, L, S begin the second clime where maximum daylight is **13;15** [hours] and the value in MS B is blank. The Ptolemaic value of **20;14** degrees latitude is given in MS L (f. 75b) and MS S (f.72a), though in MS S someone has modified 20;14 to 20;27, which is the value found both in Ṭūsī (20;27=20+¼+⅕ [*Tadhkira*, III.1[8] (Ragep, 1:250–51)]) and in Qāḍīzāde, *Sharḥ Mulakhkhaṣ* (Ayasofya 2662, f. 49b). In other words, someone has tried to "update" the Ptolemaic 14 minutes to Ṭūsī's 27. MS B provides no value. MS F (f. 15a) has 24;0 and MS K (f. 28a) has 24;15. For MSS F and K, 20;14 (ك يد) was most likely misread from a copy that was missing the dots on the ي , leading to the combining of د and ك to form 24. For MS K, the 24;15 is most likely a copyist error. Note that Alānī's commentary (Ahmet III 3308, f. 66b) also has the Ptolemaic value.

<u>**For 23;51 degrees latitude**</u>: 4 out of 5 of my main manuscripts have Ptolemy's value for latitude of 23;51 degrees; MS B provides no values. This specific Ptolemaic value (also used by Kharaqī) was an important factor in selecting manuscripts for this edition, since I chose those witnesses that contain the original Ptolemaic values. Though Alānī's commentary also has the Ptolemaic value of 23;51 degrees (f. 66b), the vast majority of *Mulakhkhaṣ* manuscripts and commentaries have changed the text here to **24;40**, which is found in many copies of the *Tadhkira*. (Note that someone has written 24;40 in the margin of MS S.) This value is itself the result of a copyist's error whereby Ṭūsī's correct value of **24;5** degrees [written as 24+(½ **of** ⅙)] was misread as [24+(½ **and** ⅙) *or* 24⅔], which only involves the addition of a < و >. Because the 24;40 value could only have been transmitted after the *Tadhkira* was copied, and it is the predominant value in most copies and commentaries of the *Mulakhkhaṣ*, it was assumed by Bīrjandī (followed by F. J. Ragep in his commentary on the *Tadhkira*) that the *Mulakhkhaṣ* must postdate the *Tadhkira*. For more details on the significance of this scribal error, see *Tadhkira*, III.1[8] (1:250–51 and 2:471); for J. Ragep's revision of his original assumption, see his "On Dating Jaghmīnī and His *Mulakhkhaṣ*."

II.1[5]. the third [clime]: for **27;12 [degrees latitude]:** MS B provides no value; the remaining 4 MSS all have the value 29;12 (i.e, كط). However, in MS S (f. 72a), someone has attempted to change 29 to 27 in the main text; and also in MS L (f. 75b), someone has written 27 (كز) in the margin. The value 29 remains a mystery. For comparison, 27;12 is in Alānī's commentary (Ahmet III 3308, f. 66b) and in both Kharaqī's *Tabṣira* and *Muntahā*; whereas Ṭūsī (*Tadhkira*, III.1[8] [Ragep, 1:250–51]), Qāḍīzāde (Ayasofya 2662, f. 49b), and ʿAbd al-Wājid (Laleli 2127, f. 102b) all give 27;30. However, al-Andiqānī's fifteenth-century Persian translation of the *Mulakhkhaṣ* has 29;12 (Ayasofya 2592, f. 21b).

For **14** hours, and **30;22** degrees latitude: MS B provides no values; the remaining 4 MSS all have the Ptolemaic value of 30;22. For comparisons: Alānī's commentary also has 30;22 (f. 66b); however, Ṭūsī gives 30⅔ (*Tadhkira*, III.1[8] [Ragep, 1:250–51]) and this is equivalent to Qāḍīzāde's 30;40 (f. 49b). ʿAbd al-Wājid (ff. 102b–103a) gives 30;40 but in addition mentions that some versions have 30;22.

II.1[6]. the fourth [clime]: The entire section on the fourth clime is omitted by MS B. MSS F, L, S have the Ptolemaic value of **33;18** for **14;15** hours. MS K (f. 28a) has the variant (لج لج), which can be read 33;38 or 33;33; however, 33;38 could be a rounding up of Ṭūsī's value of 33;37,30 (see chart and *Tadhkira*, III.1[8] [Ragep, 1:250–51]). This assumption is supported by ʿAbd al-Wājid, who in his commentary (Laleli 2127, f. 103a) provides both the alphanumeric لج لج and Ṭūsī's value of 33+½+⅛ (=33;37,30). For other comparisons: Qāḍīzāde gives 33;37 (Ayasofya 2662, f. 49b); and both Kharaqī's *Tabṣira* and Alānī give the Ptolemaic value of 33;18 (Ahmet III 3308, f. 66b).

For **14;30** hours, MSS F, L, S give the Ptolemaic value of **36;0** degrees latitude (also found in Alānī [f. 66b] and Kharaqī's *Tabṣira*). However, MS K (f. 28a) has 36;22, which is found in Qāḍīzāde (f. 49b), and is equivalent to the 36+⅕+⅙ put forth by Ṭūsī (*Tadhkira*, III.1[8] [Ragep, 1:252–53]) and also ʿAbd al-Wājid (f. 103a), who attributes his value to Ṭūsī.

II.1[7]. the fifth [clime]: MS B provides no values for this clime. MSS F, K, L all have **14;45** hours, and MS S has 14;0 with 45 written beneath the 0. MSS F, K, L, S all have the Ptolemaic value of **38;35** for latitude. Alānī (Ahmet III 3308, f. 66b) also gives 38;35; however, both Ṭūsī (*Tadhkira*, III.1[8] [Ragep, 1:252–53]) and ʿAbd al-Wājid (Laleli 2127, f. 103a) give the value as 39 minus ⅒ which is equivalent to Qāḍīzāde's parameter of 38;54 degrees (Ayasofya 2662, f. 50a).

For **15** hours, MS L (f. 75b) has 44;56, but the 44 has been corrected to 40 in the margin. MS S has been corrected from 41;56 to the Ptolemaic value **40;56**. MS F (f. 15a) has 41;56 (unaltered). MS K (f. 28a) has 41;55; this could be a case of a copyist mistaking نه [55] for يه [15], since 41;15 is the value given by Ṭūsī (41¼) (Ragep, *Tadhkira*, 1:252–53, III.1[8]), Qāḍīzāde (f. 50a), and ʿAbd al-Wājid (f. 103a). Furthermore, in the margin of MS L, someone has written 41;15 in a different hand from the 40 in the margin mentioned above. The Ptolemaic 40;56 degrees is also found in Alānī (f. 66b) and Kharaqī's *Tabṣira*.

II.1[8]. the sixth [clime]: There are a number of variants for the 43;51. Toomer, *Ptolemy's Almagest,* 86 and 86n43 states: "Although not supported by any ms. reading (Ar has 43¼), 43;1 is confirmed by the values for the shadow lengths." However, the Kharaqī witnesses contain both readings: the *Tabṣira* has 43;1 (Ayasofya 2581, f. 123a [copied 885/1480, f. 155a]), and the earlier Laleli 2141 (also containing the earliest dated *Mulakhkhaṣ*), f. 46a has 43;15. The parameters are written out in both, thus removing any ambiguity connected with witnesses using an alphanumerical notation. The *Muntahā* has 43¼ (Ghalandari, 331 [319]).

MS B provides no values for this clime. MS F (f. 15a), MS K (f. 28a), MS S (f. 72a), MS L (f. 75b) all have **15;15** hours with latitude **43;51** (clearly marked), as opposed to the Ptolemaic value of 43;15 (or 43¼). There are attempts to change the 51 to 15 in different hands: in MS S, 15 is added beneath 51; and in MS L, 2 dots are added beneath the ن with يه (15) added in the margin. Also in the main text of MS L, in another hand, someone has written **22** beneath the 51. The value **43;22** is found in Qāḍīzāde (Ayasofya 2662, f. 50a) and in Ṭūsī's *Tadhkira* (III.1[8] [Ragep, 1:252–53]), written as 43+¼+⅛. ʿAbd al-Wājid gives both 43;22 and forty-three parts and a quarter [=43;15] (Laleli 2127, f. 103a). Alānī also gives 43;15 degrees (Ahmet III 3308, ff. 66b–67a). The original source(s) of 43;51 remains a mystery to me (to date); however, it is contained in the published fifteenth-century Persian translation of the *Mulakhkhaṣ* (Andiqānī, 903). Unfortunately, Andiqānī's value was unreadable in Ayasofya 2592 (f. 22a), the only witness I was able to consult.

MSS F, L, S have **15;30** hours and **45;1** degrees latitude. In MS K one finds the odd value of 15;32 hours and an ambiguous value for the latitude that might be read as 44;21 or 44;51, or even 0;21 or 0;51, since his 44 (مد) often is used to represent 0. For comparison, Qāḍīzāde (f. 50a) has 45;21 as does Ṭūsī (=45+¼+⅒) (*Tadhkira*, III.1[8] [Ragep, 1:252–53]) and ʿAbd al-Wājid (f. 103a). Alānī has 45;1 (f. 67a), which is also found in Kharaqī's *Tabṣira*.

II.1[9]. the seventh [clime]: MSS F, L, S have the Ptolemaic value of **46;51** degrees (also found in Alānī [Ahmet III 3308, f. 67a] and Kharaqī's *Tabṣira*); MS B has no value; and MS K has 46;52 (clearly marked). In comparison: Ṭūsī's value is 47⅕ (*Tadhkira*, III.1[8] [Ragep, 1:252–53]), which is also found in ʿAbd al-Wājid (Laleli 2127, f. 103b), and is equivalent to Qāḍīzāde's value of 47;12 degrees (Ayasofya 2662, f. 50b).

MSS F, L, S all have **48;32** degrees latitude for **16** hours; these values are missing in both MS B and MS K. Kharaqī's *Tabṣira* and *Muntahā*, and Alānī's *Sharḥ,* also have these Ptolemaic values (f. 67a). Ṭūsī gives this value as 48+½+¼+⅛ (*Tadhkira*, III.1[8] [Ragep, 1:252–53]); it is also found in ʿAbd al-Wājid (f. 103a), which is equivalent to Qāḍīzāde's 48;52 (f. 50b).

II.1[10]. According to some of them, its end is at the end of the inhabitable land; according to others, it is up to where the latitude is 50;25 [degrees]: In the *Almagest*, Ptolemy gives 16;25 hours for the end of the seventh clime at 50;4 degrees, this purportedly going through the middle of the Maiotic Lake (modern Sea of Azov); see *Almagest*, II.6[18] (Toomer, *Ptolemy's Almagest*, 87, 87n51). MS B has no value, and MSS F, S, L give **50;25** degrees. MS K (f. 28b) has 50;35. Bīrūnī seems to be the origin of 50;25, which is what one finds in his *Tafhīm,* no. 236 (138). For other comparisons: Kharaqī gives no value in his *Tabṣira and Muntahā*; and Ṭūsī gives 50⅓ (*Tadhkira*, III.1[8] [Ragep, 1:252–53]), equivalent to 50;20 (found also in Qāḍīzāde [Ayasofya 2662, f. 50b] and Alānī [Ahmet III 3308, f. 67a]). ʿAbd al-Wājid (Laleli 2127, f. 103b) gives the otherwise unattested 55;20.

II.1[10]. …they do not count that part of the habitable land below the equator as part of the climes: ʿAbd al-Wājid reminds us that this would be that part of the habitable land below the equator in Ptolemy's *Geography* (Laleli 2127, f. 103b).

II.1[10]. …some of them do not count what is between the equator and latitude <u>12;30</u> nor what is between latitude <u>50:25</u> to the end of the habitable land: MSS F, K, L, S all give a latitude of 50;25; MS B has no value. ʿAbd al-Wājid gives 55;25 (=55+¼+⅙) (Laleli 2127, f. 104a), which is close but not exactly the same as the 55;20 cited previously.

II.1[10]. According to what they have claimed, in latitude <u>63</u> is an inhabited island whose residents live in bath-houses due to the severity of the cold: Jaghmīnī is probably referring to the island Thule, usually thought to be the Shetland Islands; *Almagest*, II.6[29] (Toomer, *Ptolemy's Almagest*, 89, 89n66) and *Ptolemy's Geography*, Book I.7 and II.3[32] (Berggren and Jones, 64–65, 180). Kharaqī explicitly mentions the island of Thule in the *Tabṣira* (Laleli 2141, f. 45a).

II.1[10]. in latitude <u>64</u> is a habitation whose residents are an unknown Slavic people: The unknown Slavic people (the *Ṣaqāliba*) could be a reference to Ptolemy's "unknown Scythian peoples" at 64;30 degrees (*Almagest*, II.6[30] [Toomer, *Ptolemy's Almagest*, 89]). It is not clear whether Jaghmīnī is aware of Ibn Faḍlān's tenth-century account of various peoples in the northern latitudes. For a recent study, see James E. Montgomery, "Ibn Faḍlān and the Rūsiyyah," *Journal of Arabic and Islamic Studies* 3 (2000): 1–25.

At the end of this chapter, MS B (p. 48) includes an incomplete and unlabeled illustration for the climes. MSS F, L, K, S do not have a diagram, and there is no indication that one was ever intended here. However, one does find an illustration of the climes in Kharaqī's *Tabṣira* as well as in all the *Mulakhkhaṣ* commentaries I have checked.

Chapter 2 of Part II: On the Characteristics of the Equator and Locations Having Latitude

II.2[1]. The turning of the orb there is wheel-like, I mean similar to the buckets of waterwheels emerging from the surface of the water at right angles: Jaghmīnī here uses the word *ʿaṣāmīr* (plural of *ʿuṣmār*), which is defined as a waterwheel with a bucket; see F. Steingass, *Arabic-English Dictionary* (London, 1884), 701. Cf. *Tadhkira*, III.2[1]23 (Ragep, 2:472), for a definition of *dūlābiyy*[an] (wheel-like).

II.2[2]. …when the Sun reaches the two equinox points, this being the days of Nayrūz and Mihrjān: For Bīrūnī's discussion of Nayrūz and Mihrjān, see *Tafhīm*, nos. 302 and 304, respectively (180–82). Jaghmīnī uses an "Arabized" spelling of

Nayrūz <نيروز>, rather than <نوروز>; and I found no variant spellings in my main manuscripts or in various commentaries I checked (ʿAbd al-Wājid [Laleli 2127, f. 106b]; Alānī [Ahmet III 3308, f. 69a]). Jaghmīnī clearly connects the two holidays with the equinox points (Nayrūz: the vernal equinox; Mihrjān: the autumnal equinox). However, according to Bīrūnī, Mihrjān falls on the 16th day of the month Mihr-māh, which does not necessary occur at the autumnal equinox.

II.2[7]. Among them are [the locations] whose latitude exceeds the complement of the obliquity, i.e., over 66;25: Jaghmīnī's statements in this section are often obscure and seemingly contradictory, and certainly not "obvious" as he claims several times. His attempt to facilitate his points with an example further complicates this. This is due to Jaghmīnī not distinguishing clearly between what is occurring on the northern parts of the ecliptic and the southern parts; he begins the discussion with the former, but his example relates to the latter. In addition, he does not clearly distinguish between those stars that are permanently visible or permanently invisible, and those that are temporarily visible/invisible (i.e., those that rise and set). I have therefore added clarifying phrases in brackets and also footnotes for his passages on these locations. Fortunately, I was greatly assisted in comprehending this section by using ʿAbd al-Wājid's commentary. ʿAbd al-Wājid specifically provides a worked out example of a location of 70 degrees latitude (20 degrees colatitude) in Laleli 2127, ff. 112b–114b.

II.2[8]. …and its altitude is in the amount of the difference of the latitude from 90 degrees, which is the colatitude, I mean [the latitude's] "completion," and it is known as the complement of the arc.: Jaghmīnī first introduces the meaning of the complement of the arc in **I.4[1]: On the Arcs**. However, here he also terms the latitude's complement as its "completion." Cf. ʿAbd al-Wājid, Laleli 2127, f. 113b.

II.2[11]. Since that which sets faces that which rises, then that facing what rises in reverse order will set in reverse order, and vice versa: According to ʿAbd al-Wājid, "vice versa" here means "that facing what rises in regular order will set in regular order" (Laleli 2127, ff. 118b–119a).

II.2[11]. And since the rising in one of the two halves of the [zodiacal] orb in terms of order is contrary to the rising in the second [half] but matches the setting, …:

		RISING	SETTING
Half 1	Capricorn 0° => Gemini 30°	reverse	regular
Half 2	Cancer 0° => Sagittarius 30°	regular	reverse

II.2[11]. …it follows that the rising of each half will be contrary to its setting, so what rises in reverse order will set in regular order, and vice versa: Vice versa here means what rises in regular order sets in reverse order.

II.2[14]. Now then, instruction of the above is sufficient for understanding this [topic]: The abrupt tone of this closing statement suggests that Jaghmīnī's target audience is a student, not a patron.

Chapter 3 of Part II: Miscellaneous Items

A folio is missing in **MS F** which corresponds to omitted text between **II.3[1]** and **II.3[4]**. I have marked the beginning and end of this omission between two asterisks ([*]...[*]) in both the Arabic edition and corresponding English translation, and it is also noted in the critical apparatus. The missing part starts within the passage on the ascendant and returns within the section on determining the *qibla* bearing, so **Fig. 8** (the illustration of the Indian circle) is also missing.

II.3[1]. Then if the star is [aligned] with one of the two solstice points or it has no latitude, its degree, i.e., [the star's projected] place on the zodiacal orb, is its degree of transit: Regarding the definition of the degree of transit of the star, it seems redundant that Jaghmīnī dichotomizes between a star *at* the solstice point with a star with no latitude, since a star at a solstice point is on the ecliptic and thus would have no latitude. However, ʿAbd al-Wājid comments that Jaghmīnī really meant that a star *aligned* with the solstice points on the solstitial colure will have the same degree of transit as those points. See ʿAbd al-Wājid, Laleli 2127, f.121a.

SUMMARY OF STAR TRANSIT		
	NORTH latitude	**SOUTH latitude**
Half 1 of zodiacal orb: Cancer 0° => Sagittarius 30°	star reaches meridian after its degree	star reaches meridian before its degree
Half 2 of zodiacal orb Capricorn 0° => Gemini 30°	star reaches meridian before its degree	star reaches meridian after its degree

For discussions of the star's degree, see Bīrūnī, *Tafhīm*, no. 243 (147–48) and Ṭūsī's chapter entitled "On the Degrees of Transit of the Stars on the Meridian and on Their [Degrees of] Rising and Setting" (*Tadhkira*, [III.11[1–3]; Ragep, 1:302–5, 370 [Fig. C35] and 2:495–96).

II.3[1]. As for the right orb, the rules for this are exactly the same. As for the inclined orbs, one needs to take into account the horizons: The various latitudes for these cases are discussed by ʿAbd al-Wājid (Laleli 2127, f. 123a–123b).

II.3[2]. Jaghmīnī refers to 2 kinds of shadows: the "first shadow" is produced by a horizontal gnomon erected parallel to the horizon plane and is called *maʿkūs* [*umbra versa*] because it produces a reversed shadow directed toward the ground. This shadow is also called *muntaṣib* (erect) since it is perpendicular to both the gnomon and the horizon plane. Jaghmīnī's "second shadow" is produced by a vertical gnomon perpendicular to the horizon plane. It is called *mustawī* [*umbra recta*] because this planar shadow is level with the horizon plane (see **Fig. C6**).

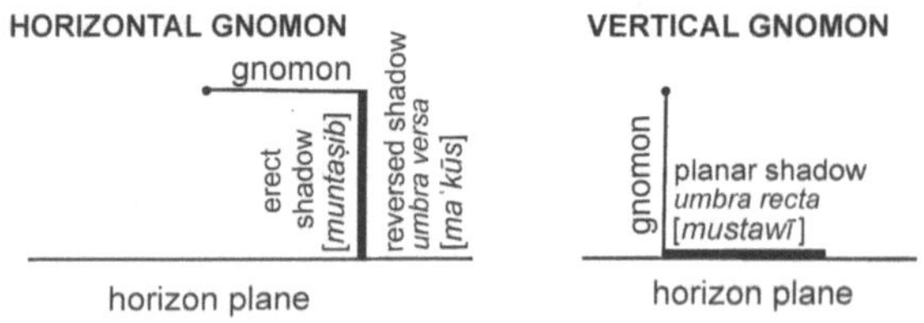

Fig. C6

Jaghmīnī informs us here how to determine the start-time for the afternoon [*ʿaṣr*] prayer using gnomon shadows according to both the rulings of al-Shāfiʿī and Abū Ḥanīfa. It is noteworthy that the attribution to al-Shāfiʿī has been added in the margins of both MS S (f. 78a) and MS L (f. 78b); since these manuscripts are witnesses to the earliest version of the text, this could indicate that in the original version Jaghmīnī did not think there was a need to cite al-Shāfiʿī explicitly, since, as a Shāfiʿī presumably teaching in a Shāfiʿī madrasa, Jaghmīnī probably assumed that Shāfiʿī's opinion on prayer would have been common knowledge. On the other hand, he did need to cite the source of the second opinion, i.e., that of Abū Ḥanīfa. In any event, someone writing after the first edition of the *Mulakhkhaṣ* felt the need to reference both.

For Bīrūnī on the various divisions of the gnomon and kinds of shadow, see *Tafhīm*, nos. 227, 228, and 229 (133–34); and E. S. Kennedy, "Al-Bīrūnī on the Muslim Times of Prayer," in *The Scholar and the Saint: Studies in Commemoration of Abū'l-Rayḥān al-Bīrūnī and Jalāl al-Dīn al-Rūmī* (New York: New York University Press, 1975), 88; repr. in E. S. Kennedy, et al. *Studies in the Islamic Exact Sciences,* ed. D. A. King and M. H. Kennedy (Beirut, 1983), 304. See also E. S. Kennedy, *The Exhaustive Treatise on Shadows by Abu al-Rayḥān Muḥammad b. Aḥmad al-Bīrūnī, Translation & Commentary.* Vol. 1, *Translation* (Aleppo: Institute for the History of Arabic Science, 1976), 62–67 (Ch. 6: On the Method by Which the Use of the Shadow and the Gnomon is Arranged); 68–80 (Ch. 7: On the Divisions into Which Gnomons are Divided); 210–30 at 219 (Ch. 25: On the Recital of the Opinions of the Imāms Regarding the Times of Prayer, and What is Resorted to in Determining Them). Cf. David A. King, "On the Role of the Muezzin and the Muwaqqit in Medieval Islamic Society," in *Tradition, Transmission, Transformation,* especially the sections "On the Times of Prayer in Islam," 289 and "Simple Techniques for Time-Keeping by Day and Night," 296.

II.3[3]. On determining the meridian line: On finding the meridian line and defining the **Indian circle**, cf. Ṭūsī, *Tadhkira*, III.12[2–3] (Ragep, 1:306–7, 2:496–97); and on how to determine the Indian circle, see Bīrūnī, *Tafhīm*, no. 131 (49–52). There are some minor discrepancies between Jaghmīnī, Ṭūsī, and Bīrūnī: Both Bīrūnī and Ṭūsī prefer to define the gnomon length as half the radius, whereas Jaghmīnī uses the equivalent ¼ the diameter (cf. *Tafhīm,* 49; *Tadhkira*, 1:306–7); and whereas Jaghmīnī gives two methods to determine that the gnomon is perpendicular, Bīrūnī uses the plumb-line option only and Ṭūsī is silent on this matter. *Al-shāqūl* (the plumb-line or plummet) is a suspended string with an attached weight that points towards the Earth's center of gravity.

II.3[4]. Since the longitude and latitude of Mecca are less than the longitude and latitude of our locality: We have here an example of Jaghmīnī personalizing the exercise by referring to his hometown. Unfortunately, he does not mention the locale specifically (presumably the students knew where they lived), and the commentators I checked either omitted this information (see Alānī, Ahmet III 3308, f. 83; and ʿAbd al-Wājid, Laleli 2127, f. 128a) or only cited the general region; for example, Qāḍīzāde just gives Khwārizm (Ayasofya 2662, f. 62b).

II.3[4–5]. In these passages, Jaghmīnī instructs us how to determine the *qibla* bearing for:

1. **Locations whose longitude and latitude are greater or less than those of Mecca** (by constructing **Fig. 9**) using an Indian circle (see **Fig. 8**). Cf. ʿAbd al-Wājid, who discusses eight different possibilities using various combinations of greater, equal to, and less than for the latitudes and longitudes (ff. 128b–129a);
2. **Locations whose longitude are the same as that of Mecca**;
3. **Locations whose latitude equals Mecca's latitude**. This is not a simple determination based on facing due east or west. (ʿAbd al-Wājid points out Kūshyār ibn Labbān made this error [f. 129a].) Jaghmīnī provides detailed instructions here on determining this *qibla* bearing using an astrolabe, for the two specific times of the year (Gemini 7;21 and Cancer 22;39) when the Sun would be directly overhead in Mecca.

Three points: [1] Gemini 7;21 and Cancer 22;39 are derived by using a latitude of 21;40 for Mecca and an obliquity of 23;35; [2] Jaghmīnī uses the term *khaṭṭ wasaṭ al-samāʾ* here, literally "mid-heaven line" (*linea medii coeli*) for the meridian; and [3] Jaghmīnī assumes the reader is already familiar with how to use the astrolabe, its parts, and its various functions. Alternatively, we would have to speculate that he was providing basic definitions of its parts, its use, and applications while teaching the exercise.

II.3[5]. So wherever the [chosen zodiacal] degree lands on the altitude almucantars, you will observe the Sun when it reaches that altitude and erect a gnomon; then its shadow at that time is the bearing for the *qibla*: Note that this process of determining the Sun's altitude for the location (using the astrolabe), then observing the Sun at that altitude in the sky, and then erecting a gnomon to observe the cast shadow can only be done twice a year (Gemini 7;21 and Cancer 22;39), namely when the Sun is directly overhead at Mecca at noon and will cast its shadow in a direct line to the location. Cf. Ṭūsī, who was able to use the idea behind Jaghmīnī's technique for latitudes equal to that of Mecca and generalize to all locations (*Tadhkira*, III.12[3–4]; Ragep, 1:306–9, 2:497). Also see David A. King, who points out that Battānī and Jaghmīnī both used methodological procedures that were cartographic (*World-Maps for Finding the Direction and Distance to Mecca* [London: Al-Furqān Islamic Heritage Foundation; Leiden: E. J. Brill, 1999], 59, 59n25, no. 1 [Ch. 2: The Determination of the Sacred Direction in Islam]).

Jaghmīnī's section on determining the *qibla* was most likely taken from Kharaqī's *Tabṣira,* given that Kharaqī has the exact same astrolabe/gnomon exercise (Laleli 2141, *Bāb* 12, ff. 55a–56b).

II.3[7]. This is recorded in the *zīj*es [astronomical handbooks]: Jaghmīnī mentions *zīj*es twice in the *Mulakhkhaṣ,* here and in **I.2[10]**. According to E. S. Kennedy, these astronomical handbooks with tables were used by "the practicing astronomer, or astrologer, to solve all the standard problems of his profession" ("A Survey of Islamic Astronomical Tables," 123). See D. A. King and J. Samsó for a supplement to Kennedy's survey with additional tables that are not contained within *zīj*es ("Astronomical Handbooks and Tables from the Islamic World [750–1900]: an Interim Report," *Suhyal* 2 [2001]: 9–105).

II.3[8]. The duration of daytime: For this section, cf. Ṭūsī, *Tadhkira,* III.10 (Ragep, 1:298–303, 2:489–95).

II.3[8]. It has thus become clear that regularized hours are those whose number varies according to the length and shortness of daytime, but their units of time do not vary; seasonal hours are those whose units of time vary, but their number does not vary:

In sum: in **case 1** the number of regularized hours during daytime can vary, say, between a short winter and a long summer, but one winter hour would still equal one summer hour; in **case 2,** 1 of the 12 summer hours would be longer than 1 of the 12 winter hours, but the 12 for daytime and night remains constant throughout the year.

II.3[9]. For some of them said 365¼ days; according to Ptolemy, 365¼ days less 1/300 part of a day; and according to Battānī 365¼ days less 3 parts 24 minutes out of 360 parts of a day: F. J. Ragep provides comparative charts that summarize these reported values (see "Al-Battānī, Cosmology, and the Early History of Trepidation," 285; and *Tadhkira,* 2:493). See also *Ptolemy's Almagest,* III.1 (Toomer, 140); and Battānī's *Zīj,* Ch. 27 (3:61–62 and 1:40–41).

II.3[10]. ...they take away the Sun's mean [motion] from the Moon's mean, and they divide the remainder by the rotation of the orb, namely 360 degrees, thus resulting in 29;31,50,8 days, which is the amount of a month. They then multiplied that by 12, obtaining the days in a lunar year: 354+⅕+⅙ days. This year is less than the solar year by approximately 10 days and 20½ hours:

This passage, including *all* Jaghmīnī's parameters, is found in Kharaqī's *Tabṣira* (Laleli 2141, *Bāb* 14 of *hay'at al-arḍ,* esp. f. 58b). According to the parameters of Jaghmīnī (see *Mulakhkhaṣ,* **I.2[3]**, **I.2[8]**, **I.5[37]**)—and Kharaqī—one would subtract 0;59,8,20 [Sun's mean] from 13;10,35,2 [Moon's mean] and then divide 360 by the remainder of 12;11,26,42 for a result of **29;31,50,8 days**. The value 29;31,50,8,20 days is the mean length Ptolemy claims to derive for the synodic month; however, as Toomer points out, this value is not actually what one obtains by Ptolemy's calculation (nor by Jaghmīnī's) but instead is Hipparchus's value which he took from Babylonian sources (see *Almagest* IV.2 [*Ptolemy's Almagest,* 176, esp. n10]).

On the machinations of calculating the lunar calendar, and the value of **354+⅕+⅙ days** (or $354^{11}/_{30}$ days), see Ragep's commentary on Ṭūsī's *Tadhkira*, III.10[2] (2:491–92). As for the lunar year being less than the solar year by approximately **10 days and 20½ hours:** According to Qāḍīzāde, it would have been more correct for Jaghmīnī (and presumably Kharaqī) to have stated that the difference was approximately **10 days and 21 hours** [Ayasofya 2662, f. 69a], which is closer to what I calculated.

II.3[11]. This is as much as allowed by [my] ignoble character, a tormented mind, thought befuddled by preoccupations beyond counting, and concerns [so overwhelming] they would make a mother neglect her child:

«لا يُنادى وليدُها» This is literally "her child will not be called out to." The idiom, which signifies difficulty or distress, is from a proverb whose original meaning implies that the distress is so overwhelming that a mother would forget her child and not call out to him (see Edward W. Lane, *An Arabic-English Lexicon* [Beirut, 1968]: 8:2967).

Appendices

Appendix I: Jaghmīnī's Works

A. Astronomy

	Title	*Description*
1.	الملخّص في علم الهيئة البسيطة *al-Mulakhkhaṣ fī ʿilm al-hayʾa al-basīṭa*	An introductory work on the discipline of *hayʾa basīṭa*, dedicated to Badr al-Dīn al-Qalānisī, composed 602-3/1205-6 [extant]
	*GAL*1: 473; *GAL* suppl. 1:865; Ghalandari, "Chaghmīnī," no. 1; Kaḥḥāla, 12:198; Qāsimlū, 12:62 (no. 8); Ziriklī, 7:181.	
2.	رسالة في أقدار أجرام الكواكب وأبعادها *Risāla fī aqdār ajrām al-kawākib wa-abʿādihā*	A treatise on planetary distances and sizes, dedicated to Badr al-Dīn al-Qalānisī [extant]
	King, *Survey*, 150 (G17, 1.27); King, *Catalogue*, 2:21 [2] [al-Qalānisī is misread as "al-Falāsitī (?)"]. See also Bratislava, Bratislava Library, TG 15, Ordinal Number 291.	
3.	تحرير القواعد لتحليل أستار الفرائد *Taḥrīr al-qawāʿid li-taḥlīl astār al-farāʾid*	A treatise on rules for clarifying various miscellaneous items in astronomy [extant]
	*GAL*1:625; de Slane, 516, no. 2865.	

B. Astrology

4.	الكتاب في قوى الكواكب وضعفها *al-Kitāb fī quwā al-kawākib wa-ḍaʿfihā*	A work on the strengths and weaknesses of the planets, dedicated to a Shihāb al-Dīn. Based on a date given for the planetary apogees, this treatise was composed one year earlier than the *Mulakhkhaṣ*, i.e., ca. 600/1204 [extant]

S.P. Ragep, *Jaghmīnī's Mulakhkhaṣ*, Sources and Studies in the History of Mathematics and Physical Sciences, DOI 10.1007/978-3-319-31993-3

(continued)

<table>
<tr><td>4.</td><td>Paris, BnF, MS ar. 2589, f. 174b contains the abbreviated name of the dedicatee; this is missing in Jaghmīnī, Talkhīṣ kitāb Ūqlīdis, 249. On the identification of Shihāb al-Dīn, see supra I.2.3c. See also de Slane, 468, no. 2589; Ghalandari, "Chaghmīnī," no. 4; Kaḥḥāla, 12:198; Qāsimlū, 12:61 (no. 1); Ziriklī, 7:182.</td></tr>
</table>

C. Mathematics

<table>
<tr><th></th><th>Title</th><th>Description</th></tr>
<tr><td>5.</td><td>تلخيص كتاب أوقليدس
Talkhīṣ kitāb Ūqlīdis</td><td>The Epitome of Euclid's Elements, composed at the request of Shihāb al-Dīn Abū Sa'd ibn 'Imrān al-Khwārizmī al-Khīwaqī, and completed Sunday, 22 Ṣafar 615 H (= Saturday-Sunday, 19-20 May 1218 CE)</td></tr>
<tr><td></td><td colspan="2">GAS 5:115 (no. 56); Ghalandari, "Chaghmīnī," no. 3; Qāsimlū, 12:62 (no. 7), 63; Talkhīṣ kitāb Ūqlīdis, 15–246 (the dedicatee is given on 16, followed by seven verses of poetry).</td></tr>
<tr><td>6.</td><td>الموجز في الحساب [= الضرب ؟]
al-Mūjaz fī al-ḥisāb</td><td>A summary on arithmetic that includes a discussion of multiplication</td></tr>
<tr><td></td><td colspan="2">Ghalandari, "Chaghmīnī," no. 6; Qāsimlū, 12:61 (lists this as two separate works: "al-ḍarb" [2] and "ṣuwar al-hisāb" [3]); Talkhīṣ kitāb Ūqlīdis, 254–55.</td></tr>
<tr><td>7.</td><td>رسالة صوّر الحساب التسع (=رسالة لطيفة في حساب ؟= رسالة في الحساب ؟)
Risālat ṣuwar al-ḥisāb al-tis' (=Risāla laṭīfa fī ḥisāb?=Risāla fī al-ḥisāb?)</td><td>A treatise on nine types of arithmetic [extant]</td></tr>
<tr><td></td><td colspan="2">GAL suppl. 1:865; Ghalandari, "Chaghmīnī," no. 5; Hitti, 324 (no. 1032 = Princeton University, Garrett no. 502H); Kaḥḥāla, 12:198; Qāsimlū, 12:61 (no. 4); Ziriklī, 7:182. See Tehran, Central Library of the University of Tehran, MS 6911. King ambiguously mentions a "R. mukhtaṣara fī l-Ḥisāb, on simple arithmetic" (King, Survey, 150 [G17, 6.3.11]); MAMS2 follows King, 198 (no. 547), M1.</td></tr>
<tr><td>8.</td><td>شرح طرق الحساب في مسائل الوصايا
Sharḥ Ṭuruq al-ḥisāb fī masā'il al-waṣāyā</td><td>A Commentary on using arithmetic in questions related to inheritance [extant]</td></tr>
<tr><td></td><td colspan="2">GAL suppl. 1:865; Ghalandari; "Chaghmīnī"; Qāsimlū, 12:62 (no. 6); Kaḥḥāla, 12:198; Ziriklī, 7:182.</td></tr>
</table>

	Title	*Description*
9.	منظومة في الجبر والمقابلة *Manẓūma fī al-jabr wa-l-muqābala*	A treatise in rhyme on algebra: a poem in 25 verses on problems about algebraic equations
	Qāsimlū, 12:61–62 (no. 5).	

D. Medicine

10.	قانونچه *Qānūnča*	The "little *Qānūn,*" an abridged treatise of Ibn Sīnā's medical text *al-Qānūn fī al-ṭibb* [extant]
	*GAL*2:213; *GAL* suppl. 1:826, 865; Ghalandari, "Chaghmīnī," no. 2; Qāsimlū, 12:62–63 (no. 9); earliest copy dated 12 Ramaḍān 601 H (= 3 May 1205 CE) (Ayasofya MS 3735); see supra *I.2.3a.*	

E. Other

11.	قصيدة	A poem (*qaṣīda*) [extant]
	Talkhīṣ kitāb Ūqlīdis, 247–49; Qāsimlū, 12:63.	
12.	?	A small fragment of a mathematical work attributed to Jaghmīnī
	Witkam, *Inventories,* 88: Leiden, Leiden University, Or. 204 (2), f. 30a.	

F. Misattributions

1.	الكتاب القوامي في الحساب *al-Kitāb al-Qiwāmī fī al-ḥisāb*	Arithmetical treatise on extracting roots and operations with decimal fractions
	GAL suppl. 1:865; Ghalandari, "Chaghmīnī"; Qāsimlū (12:63). All three raise the possibility that this is a misattribution. Qāsimlū states it may be a treatise by Abū Naṣr Samaw'al ibn Yaḥyā al-Maghribī [composed in 1173] that bears this name. Cf. *MAMS*2, 185 (no. 487), M3; and Rashed, *Entre arithmétique,* 140–45 (for Samaw'al's text).	
2.	الرسالة الموسومة بطبّ النبي *al-Risāla al-mawsūma bi-Ṭibb al-Nabī*	A treatise on medicine of the Prophet
	There are several lithographs (Tehran, late nineteenth century) that attribute the authorship of this work to Jaghmīnī; this was also Elgood's conclusion (43, 186–92). However, the Majlis-i Shūrā and Millī Libraries (Tehran) have correctly identified the author as Jaʿfar ibn Muḥammad al-Mustaghfirī (d. 432/1040) (*GAL* suppl. 1:617). The misattribution may have stemmed from al-Mustaghfirī's work having been copied together with Jaghmīnī's *Qānūnča* within one or more codices.	

Appendix II: Works Derivative from the *Mulakhkhaṣ*

A. Commentaries, Supercommentaries, and Glosses (Arabic)

	Author	*Title and Description*	
1.	Muḥammad ibn Mubārak-shāh Mīrak al-Bukhārī (d. 741/1341)	*Sharḥ al-Mulakhkhaṣ;* composed last part of Rabīʿ II, 727 (March 1327)	[1]
	*MAMS*2, 256 (no. 753), A1 and (no. 694), A4; *OALT,* 1:lxxviii; *Riyazî ilimler*, 1:389, e.1.		
2.	Yūsuf ibn Mubārak al-Alānī (ca. 735/1334)	*Sharḥ al-Mulakhkhaṣ;* composed Sunday, 19 Ramaḍān 735 [13-14 May 1335] and dedicated to Jānī Beg Khān (r. 1341–57) of the Golden Horde of the Mongol Empire	[2]
	Riyazî ilimler, 1:389, e.2.		
3.	Faḍl Allāh al-ʿUbaydī (d. 751/1350)	*Sharḥ al-Mulakhkhaṣ;* composed in three days at the request of professors and students; ʿUbaydī was Quṭb al-Dīn al-Shīrāzī's student	[3]
	Fazlıoğlu, "ʿUbaydī," *BEA*, 2:1157; *KZ,* 2:col. 1819; Flügel, 6:113; *OALT,* 1:lxxviii; *Riyazî ilimler*, 1:389, e.3; Ṭāshkubrīzāde, *Miftāḥ al-saʿāda,* 349.		
4.	Saʿd al-Dīn Ḥamza ibn ʿAlī al-Bayhaqī (early 8th/14th c.)	*Sharḥ al-Mulakhkhaṣ*	[4]
	*MAMS*2, 248 (no. 723); *Riyazî ilimler*, 1:391, e.11.		
5.	Kamāl al-Dīn al-Turkmānī: Muḥammad ibn Aḥmad al-Ḥanafī (d. 758/1357)	*Sharḥ al-Mulakhkhaṣ;* composed in 755/1354 in Gülistan/Saray, the capital city of the Golden Horde State, and offered to Jānī Beg	[5]
	Fazlıoğlu "Kamāl al-Dīn al-Turkmānī," *BEA*, 1:609; *KZ,* 2:col. 1819; Flügel, 6:113; *OALT,* 1:lxxix; *MAMS*2, 252 (no. 738), A1; *Riyazî ilimler*, 1:389–90, e.4; Ṭāshkubrīzāde, *Miftāḥ al-saʿāda,* 349.		
5. (a)	Faṣīḥ al-Dīn Muḥammad ibn ʿAbd al-Karīm Niẓāmī al-Kūhistānī (d. 1530)	*Ḥāshiya* on Kamāl al-Dīn al-Turkmānī's *Sharḥ al-Mulakhkhaṣ*	[6]
	Flügel, 6:114(?); *MAMS*2, 309 (no. 914), A6.		

	Author	Title and Description	
6.	Muḥammad ibn al-Ḥusayn ibn al-Rashīd al-Mashhadī al-Khwārizmī (8th/14th c.) *KZ,* 2:col. 1820; Flügel, 6:114; *Riyazî ilimler*, 1:390, e5.	*Sharḥ al-Mulakhkhaṣ*; its only extant copy bears a date of 774/1372-73	[7]
7.	Anonymous *Riyazî ilimler*, 1:390, e.7.	*Sharḥ al-Mulakhkhaṣ*; parts missing	[8]
8.	Kamāl al-Dīn ʿAbd al-Raḥmān ibn Muḥammad ibn Ibrāhīm al-ʿAtāʾiqī (d. 790/1388) The only known witness is Isfahan, Maktabat al-Zahrā 144.	*Sharḥ al-Mulakhkhaṣ;* composed in 770/1368-69	[9]
9.	Humām al-Ṭabīb: Muḥammad ibn Muḥammad ibn Abī Ṭālib (d. after 813/1410) *KZ,* 2:col. 1820; Flügel, 6:114; *MAMS*2, 267 (no. 794), A1; *Riyazî ilimler*, 1:390, e.8.	*Sharḥ al-Mulakhkhaṣ*	[10]
10.	al-Sayyid al-Sharīf al-Jurjānī (d. 816/1413) *KZ,* 2:col. 1819; Flügel, 6:113–14; *MAMS*2, 266 (no. 788), A2; *Riyazî ilimler*, 1:390–91, e.9; Fazlıoğlu, "The Samarqand Mathematical-Astronomical School," 34–36; Ṭāshkubrīzāde, *Miftāḥ al-saʿāda,* 349.	*Sharḥ al-Mulakhkhaṣ*	[11]
10. (a)	Anonymous The only known witness is Ayasofya 2608.	*al-Tawḍīḥ al-Ḥusaynī li-Sharḥ Mulakhkhaṣ al-Jaghmīnī;* gloss on 10. with a dedication to Mehmed II (r. 1444–46, 1451–81)	[12]
10. (b)	Anonymous *Riyazî ilimler*, 1:390–91, e.9.	Gloss on al-Jurjānī's *Sharḥ al-Mulakhkhaṣ;* only extant witness copied in 880 H. by Muṣṭafā ibn ʿAbd Allāh	[13]
11.	ʿAbd al-Wājid (wrongly: Wāḥid) ibn Muḥammad ibn Muḥammad al-Ḥanafī al-Kutāhī (d. 838/1435) *KZ,* 2:col. 1820; Flügel, 6:114; *MAMS*2, 267 (no. 791), A3; *OALT*, 1:24 (no. 2); Ragep, "Astronomy in the Fanārī-Circle"; *Riyazî ilimler*, 1:388, d.2; Topdemir, "ʿAbd al-Wājid".	*Sharḥ al-Mulakhkhaṣ;* it was presented to Sultan Murād II (r. 1421–51)	[14]

	Author	*Title and Description*	
12.	Qāḍīzāde al-Rūmī (d. ca. 835/ 1440)	*Sharḥ al-Mulakhkhaṣ;* composed in 814/1412 and dedicated to Ulugh Beg	[15]
	KZ, 2:col. 1819; Flügel, 6:113; *MAMS*2, 273–74 (no. 808), A1; *OALT,* 1:8–21 (no. 3); *Riyazî ilimler*, 1:372–73; Fazlıoğlu, "The Samarqand Mathematical-Astronomical School"; Ragep, "Qāḍīzāde," *BEA*, 2:942; Ṭāshkubrīzāde, *Miftāḥ al-saʿāda,* 349.		
12. (a)	Fatḥ Allāh al-Shīrwānī (d. 891/ 1486)	*Ḥāshiya ʿalā Sharḥ al-Mulakhkhaṣ;* gloss on 12. by Qāḍīzāde's student; presented to Mehmed II in 878/1473	[16]
	İhsanoğlu, *History*, 2:533, 535–36; *KZ,* 2:col. 1819; Flügel, 6:114; *MAMS*2, 292 (no. 868), A2; *OALT,* 1:43–44 (no. 1); *Riyazî ilimler*, 1:385, ç.1; Fazlıoğlu, "Shirwānī," *BEA*, 2:1055–56; Fazlıoğlu, "The Samarqand Mathematical-Astronomical School".		
12. (b)	Sinān Pāshā (d. 890/1486): Sinān al-Dīn Yūsuf ibn Khiḍr Beg ibn Jalāl al-Dīn ʿĀrif	*Ḥāshiya ʿalā Sharḥ al-Mulakhkhaṣ;* gloss on 12. by vizier of Mehmed II; dedicated to Bāyazīd II	[17]
	İhsanoğlu, *History*, 2:534–35; *KZ,* 2:col. 1819; Flügel, 6:114; *MAMS*2, 290 (no. 858), A2; *OALT,* 1:47 (no. 1); *Riyazî ilimler*, 1:385, ç.3.		
12. (c)	Fakhr al-Dīn al-ʿAjamī (9th/15th c.)	*Ḥāshiya ʿalā Sharḥ al-Mulakhkhaṣ;* gloss on 12. by a student of ʿAlī Qushjī	[18]
	OALT, 1:54; *Riyazî ilimler*, 1:385, ç.4.		
12. (d)	Niksārī: Muḥyī al-Dīn Muḥammad ibn Ibrāhīm ibn Ḥasan al-Niksārī al-Rūmī (d. 901/1495)	*Ḥāshiya ʿalā Sharḥ al-Mulakhkhaṣ;* gloss on 12. by a student of Shīrwānī; dedicated to Bāyazīd II	[19]
	*MAMS*2, 293 (no. 871), A1; *OALT,* 1:62; *Riyazî ilimler*, 1:385, ç.5.		
12. (e)	Kubnawī: al-Ḥaqq ibn Abī Isḥāq Kubnawī (late 9th/15th c.)	*Ḥāshiya ʿalā Sharḥ al-Mulakhkhaṣ;* gloss on 12.; Kubnawī worked at the Diyarbakir court of Aq Qoyunlu Sultan Yaʿqūb Bahādur-Khān (r. 1478–90)	[20]
	*MAMS*2, 282 (no. 833), A2.		
12. (f)	Dellākoğlu: Ḥusām ibn Shams al-Dīn al-Khattābī al-Lāhijānī al-Jīlānī (d. 901/1495)	*Ḥāshiya ʿalā Sharḥ al-Mulakhkhaṣ;* gloss on 12.	[21]
	OALT, 1:20, 63–64.		

	Author	*Title and Description*	
12. (g)	Akhawayn: Muḥyī al-Dīn Muḥammad ibn Qāsim (d. 904/ 1499)	*Ḥāshiya ʿalā Sharḥ al-Mulakhkhaṣ;* gloss on 12.	[22]
	*MAMS*2, 303 (no. 893), A4; *OALT,* 1:20, 65–66 (no. 4); *Riyazî ilimler*, 1:385, ç.6.		
12. (h)	Mīrim Çelebī (d. 931/1525)	*Ḥāshiya ʿalā Sharḥ al-Mulakhkhaṣ;* gloss on 12.	[23]
	OALT, 1:100–1 (no. 4).		
12. (i)	ʿAbd al-ʿAlī al-Bīrjandī (d. 935/1528)	*Ḥāshiya ʿalā Sharḥ al-Mulakhkhaṣ;* gloss on 12.	[24]
	İhsanoğlu, *History*, 2:548; *KZ,* 2:col. 1820; Flügel, 6:114; *MAMS*2, 315–16 (no. 938), A11; *OALT,* 1:101–4 (no. 1); *Riyazî ilimler*, 1:381–82, d.1.		
12. (i.1)	ʿAbd al-Raḥmān ibn Ibrāhīm al-Suhrānī al-Shāfiʿī (d. 1066/1656)	Gloss on 12.(i) entitled *Taʿlīqāt ʿalā Ḥāshiya ʿalā Sharḥ al-Mulakhkhaṣ*	[25]
	OALT, 1:294; *Riyazî ilimler*, 1:384, c.1.		
12. (i.2)	Aḥmad al-ʿImādī: Mawlānā Aḥmad ibn Sayyid Aḥmad al-ʿImādī (11th/17th c.)	Gloss on 12.(i)	[26]
	OALT, 1:331 (no. 2); *Riyazî ilimler*, 1:384, c.2.		
12. (i.3)	See infra, C. Translation (Turkish), no. 1.		
12. (j)	Faṣīḥ al-Dīn Muḥammad ibn ʿAbd al-Karīm Niẓāmī al-Kūhistānī (d. 1530)	*Ḥāshiya ʿalā Sharḥ al-Mulakhkhaṣ;* gloss on 12.	[27]
	Flügel, 6:114(?); *MAMS*2, 309 (no. 914), A7.		
12. (k)	Manṣūr al-Dashtakī: Ghiyāth al-Dīn Manṣūr ibn Muḥammad al-Ḥusaynī al-Dashtakī al-Shīrāzī (d. 948/1541)	*Ḥāshiya ʿalā Sharḥ al-Mulakhkhaṣ;* gloss on 12.	[28]
	Rūḥullāhī, "Dashtakī," title no. 7.		
12. (l)	Burhān al-Dīn Ibrāhīm ibn Muḥammad ibn Ibrāhīm al-Ḥalabī (d. 956/1549)	*Ḥāshiya ʿalā Sharḥ al-Mulakhkhaṣ;* gloss on 12.	[29]
	*MAMS*2, 321 (no. 959), A1.		

	Author	Title and Description	
12. (m)	Saçli Emir (d. 963/1555): Shams al-Dīn Muḥammad ibn ʿAbd al-Awwal ibn Ḥusayn ibn Ḥasan al-Qamarī al-Ḥusaynī al-Tabrīzī al-Ḥanafī (d. 963/ 1555)	*Risāla fī kashf mā dāra ʿalā istidārat al-arḍ wa-kurawiyyatihā;* gloss on the Earth's sphericity (and its relation with prayers as discussed in 12.); composed in 940/1533-34 and presented to the Ottoman vizier Ibrāhīm Pasha	[30]
	OALT, 1:135–36; *Riyazî ilimler,* 1:388, d.1.		
12. (n)	Qāḍī Ḥasan al-Makkī: Ḥasan ibn Muḥammad al-Faṣīḥī al-Makkī (fl. 1014/1605)	*Ḥāshiya ʿalā Sharḥ al-Mulakhkhaṣ;* gloss on 12.	[31]
	OALT, 1:249.		
12. (o)	Bahāʾ al-Dīn Muḥammad ibn Ḥusayn al-ʿĀmilī (d. 1621)	*Ḥāshiya ʿalā Sharḥ al-Mulakhkhaṣ;* gloss on 12.	[32]
	*MAMS*2, 350 (no. 1058), A16.		
12. (p)	Aḥmad al-ʿImādī: Mawlānā Aḥmad ibn Sayyid Aḥmad al-ʿImādī (11th/17th c.)	*Ḥāshiya ʿalā baḥth al-shaʿīrāt fī Sharḥ al-Mulakhkhaṣ li-Qāḍīzāde;* gloss on the study of the standard measure of barleycorns in 12.	[33]
	OALT, 1:330 (no. 1).		
12. (q)	ʿImād al-Dīn Ḥusayn al-Riyāḍī ibn Luṭf Allāh al-Lāhūrī (d. 1732)	*Ḥāshiya ʿalā Sharḥ al-Mulakhkhaṣ;* gloss on 12.	[34]
	*MAMS*2, 374 (no. 1179), A2.		
12. (r)	Walī al-Dīn Jār Allāh Yanī Shahrī (d. 1154/1738)	*Ḥāshiya ʿalā Sharḥ al-Mulakhkhaṣ;* gloss on 12.	[35]
	OALT, 1:403–4; *Riyazî ilimler,* 1:385, ç.7.		
12. (s)	Kasīrī-zāde: Muḥammad Amīn ibn al-Shaykh Muḥammad al-Uskadārī al-Ḥanafī al-Mudarris (d. 1151/1738)	Gloss on 12. dealing with the standard measure of a barleycorn, entitled *Taqrīrāt wāfiya wa-taḥrīrāt kāfiya li-ḥall al-masʾala al-mashhūra bi-l-masʾala al-shaʿīriyya fī sharḥ risālat al-Jaghmīnī li-l-shāriḥ al-mashhūr bi-Qāḍīzāde al-Rūmī*	[36]
	OALT, 1:405–6.		

	Author	*Title and Description*	
12. (t)	Ḥasan al-Jabartī: Badr al-Dīn Ḥasan ibn Burhān al-Dīn Ibrāhīm al-Jabartī (d. 1188/ 1774)	*Ḥāshiya ʿalā Sharḥ al-Mulakhkhaṣ;* gloss on 12. reported by Ḥasan al-Jabartī's son ʿAbd al-Raḥmān al-Jabartī; extant?	[37]
	İhsanoğlu, *History*, 2:586–87; *MAMS2*, 410 (no. 1367); *OALT*, 2:479 (no. 19); *Riyazî ilimler*, 1:386, ç.8.		
12. (u)	Fakhrī-zāde al-Mawṣilī (d. 1188/1774)	*Ḥāshiya ʿalā Sharḥ al-Mulakhkhaṣ;* gloss on 12.	[38]
	MAMS2, 411 (no. 1369); *OALT*, 2:482.		
12. (v)	Anonymous	*Ḥāshiya ʿalā Sharḥ al-Mulakhkhaṣ;* gloss on 12.	[39]
	OALT, 2:744–45.		
12. (w)	Anonymous	*Ḥāshiya ʿalā Sharḥ al-Mulakhkhaṣ;* gloss on 12.	[40]
	OALT, 2:745.		
12. (x)	See infra, B. Translations, Commentaries, Supercommentaries, and Glosses (Persian), no. 4.		
13.	Kāfiyājī: Muḥyī al-Dīn Abū ʿAbd Allāh Muḥammad ibn Sulaymān al-Bargamawī (d. 879/1474)	*Sharḥ al-Mulakhkhaṣ*	[41]
	MAMS2, 291 (no. 863), A2; *OALT*, 1:27 (no. 2); *Riyazî ilimler*, 1:388, d.3.		
14.	Qarā Sinān: Sinān al-Dīn Yūsuf ibn ʿAbd al-Malik ibn Bakhshāyish (d. ca. 885/1480-81)	*Sharḥ al-Mulakhkhaṣ;* dedicated to Bāyazīd II	[42]
	İhsanoğlu, *History*, 2:535–36; *KZ*, 2: col. 1819; Flügel, 6:114; *OALT*, 1:40–41 (no. 1); *Riyazî ilimler*, 1: 385, ç.2.		
15.	Mollā-zāde al-Rūmī (d. ca. 900/1495)	*Sharḥ al-Mulakhkhaṣ*	[43]
	İhsanoğlu, *History*, 2:545–46; *OALT*, 1:58–59.		
16.	ʿAbd al-Salām al-Muhtadī al-Muḥammadī (d. after 918/ 1512-13) = Hoja Īliyā al-Yahūdī	*Ḥāshiya* (to the base text or a commentary?); ʿAbd al-Salām/ Īliyā migrated to the Ottoman Empire from Andalusia and lived during reigns of Sultans Bāyazīd II and Selīm I	[44]
	İhsanoğlu, *History*, 2:546; *OALT*, 1:71.		

	Author	*Title and Description*	
17.	Anonymous (late 8th/14th c.)	*Ḥāshiya ʿalā Sharḥ al-Mulakhkhaṣ,* a gloss on a *Mulakhkhaṣ* commentary (which one?); copied by ʿAlī ibn Fatḥ Allāh al-Maʿdānī al-Isfahānī (=al-Ṣābirī) with the seal of Mehmed II	[45]
	Riyazî ilimler, 1:390, e.6.		
18.	Faṣīḥ al-Dīn ʿAbd al-Karīm al-Niẓāmī al-Nīsābūrī (d. ca. 850/1446)	*Ḥāshiya ʿalā Sharḥ al-Mulakhkhaṣ,* a gloss on a *Mulakhkhaṣ* commentary (which one?); oldest copy dated 9th century H	[46]
	Riyazî ilimler, 1:391, e.10.		
19.	Faṣīḥ al-Dīn Muḥammad ibn ʿAbd al-Karīm Niẓāmī al-Kūhistānī (d. 936-37/1530)	*Sharḥ al-Mulakhkhaṣ;* Faṣīḥ al-Dīn was a student of Qushjī	[47]
	*MAMS*2, 309 (no. 914), A2.		
20.	Anonymous	*Ḥāshiya ʿalā sharḥ al-Mulakhkhaṣ* (which one?)	[48]
	OALT, 1:745.		

B. Translations, Commentaries, Supercommentaries, and Glosses (Persian)

	Author	*Title and Description*	
1.	Muḥammad ibn ʿUmar al-Andiqānī (8th/14th c.)	Persian translation of the *Mulakhkhaṣ;* Ayasofya MS 2592 has a copy date of 796 H	[49]
	Andiqānī; *Riyazî ilimler*, 1:388, e1.1.		
2.	Ḥusayn ibn al-Ḥusayn al-Khwārizmī al-Kubrawī (d. 839/1435-36)	Persian commentary on the *Mulakhkhaṣ* dedicated to Ulugh Beg	[50]
	*MAMS*2, 272 (no. 805), A2; *PL*, 50 (no. 88a), 73 (no. 106[2]).		
3.	Ḥamza ibn Ḥājj ibn Sulaymān (9th/15th c.)	Persian translation by the order of Mehmed II	[51]
	OALT, 1:21, 56–57; *Riyazî ilimler*, 1:388, e1.2.		
4.	Anonymous	*Ḥāshiya ʿalā baḥth al-taḍārīs* (on the Earth's undulations as found in Qāḍīzāde's *Sharḥ*); presented to Sultan Bāyazīd II (r. 886–918/1481–1513)	[52]
	Riyazî ilimler, 1:387–88, d2.4.		

	Author	*Title and Description*	
5.	Maḥmūd ibn Muḥammad ibn Muḥammad al-Qāḍī al-Wālishtānī al-Harawī (15th c.)	A Persian revision of the *Mulakhkhaṣ* compiled for Ghiyāth al-Dīn Aḥmad; Harawī worked at the court of Shāhrukh ibn Tīmūr	[53]
	Andiqānī, 873; *MAMS*2, 282 (no. 832), A1.		
6.	Kamāl al-Dīn Ḥusayn ibn ʿAbd al-Ḥaqq Ardabīlī (fl. 15th/16th centuries)	Persian commentary on *al-Mulakhkhaṣ*	[54]
	Andiqānī, 869 (no. 8).		
7.	Qāḍī Nūr Allāh Shūshtarī (d. 1019/1610)	Persian commentary on *al-Mulakhkhaṣ*	[55]
	Andiqānī, 869 (no. 9).		
8.	Muḥammad Zamān ibn Muḥammad Ṣādiq ibn Abī Yazīd Anbālajī Dihlawī	Persian commentary entitled *Ḥikam al-riyāḍī* and completed 1130/1718-19	[56]
	PL, 50 (no. 88b).		
9.	Mullā Muḥammad Jaʿfar Sharīʿatmadār Astarābādī (1198–1263/1783–1847)	Persian commentary on *al-Mulakhkhaṣ*; Astarābādī was a member of the ʿulamāʾ who traveled (e.g., to Karbalā, Mecca, and Tehran) and had various teaching circles	[57]
	Abada, 92–95 (3.6); Andiqānī, 869 (no. 10).		
10.	Sayyid Muḥammad Taqī ibn Ḥusayn ibn Dildār ʿAlī Naqawī (19th c.)	Persian commentary on *al-Mulakhkhaṣ*	[58]
	Andiqānī, 869 (no. 11).		
11.	Anonymous	Persian commentary on *al-Mulakhkhaṣ*	[59]
	OALT, 2:786.		

C. Other Translations (Turkish and Hebrew)

1.	ʿAbbās Wasīm Efendi (d. 1173/1760)	*Tarjamat kitāb al-Bīrjandī min al-khusūf wa-l-kusūf;* Turkish trans. of ch. 10 of Bīrjandī's *Ḥāshiya* on Qāḍīzāde's *Sharḥ al-Mulakhkhaṣ* dealing with lunar and solar eclipses	[60]
	OALT, 1:446–47 (no. 3); *Riyazî ilimler*, 1:385, c.3.		
2.	Moses Ben Elijah the Greek (late 8th/14th c.)	Hebrew translation of the *Mulakhkhaṣ*	[61]
	Morrison, "The Role of Oral Transmission"; Vajda.		

Glossary

II.3[2] = first appearance and/or definition of term occurs in Book II, Chapter 3, paragraph 2 of edition and translation; ج = جمع (plural); مص = مصدر (verbal noun)

آ

permanent [visibility] (II.2[2])	أَبَدِيّ
aether (Intr.[1])	الأَثِير
end (of Pisces, etc.) (II.2[10])	آخِر (الحُوت)
end of the inhabitable land (II.1[10])	آخِر العِمارَة
beyond (in longitude)	مُتَأَخِّر
the Moderns (I.2[6])	المُتَأَخِّرون
earth (element) (Intr.[1]); land (II.3[3])	أَرْض
the Earth	الأَرْض
Leo (I.5[26])	الأَسَد
astrolabe (II.3[5])	أَسْطُرْلاب
horizon	أُفُق (ج) آفاق
horizon of the right orb (II.2[1])	أُفُق الفَلَك المُسْتَقِيم
horizon of the erect sphere (II.2[1])	أُفُق الكُرَة المُنْتَصِبَة
oblique horizons (II.2[2])	الآفاق المائِلَة
clime (II.1)	إِقْلِيم (ج) أَقالِيم
residents (I.3[9])	أَهْل (ج) أَهالٍ
apogee (I.1[2])	أَوْج
solar apogee	أَوْج الشَّمْس
lunar apogee (I.2[7])	أَوْج القَمَر
first (of Aries)	أَوَّل (الحَمَل)

ب

ocean; sea (II.I[1 and 2])	بَحْر (ج) بِحار
to reappear (after being eclipsed) (I.5[36])	يَبْتَدِئُ
full Moon (Pref.[1])	البَدْر
zodiacal sign (I.3[5])	بُرْج (ج) بُرُوج

S.P. Ragep, *Jaghmīnī's Mulakhkhaṣ*, Sources and Studies in the History of Mathematics and Physical Sciences, DOI 10.1007/978-3-319-31993-3

the zodiacal equator (I.2[3])	مِنطَقة البُرُوج
cold (II.1[10])	بَرْد
simple (bodies) (Intr.[1])	بَسِيط (ج) بَسائِط
sight; vision (I.4[15])	بَصَر (ج) أبْصار
slower motion (I.2[6]); slow (in speed) (I.5[30] and [31])	حركة بطيئة؛ بُطْء السير
to move away	يَبْعُدُ
distance	بُعْد (ج) أبْعاد
farthest distance (I.5[13])	البُعْد الأبْعَد
double elongation (I.5[37])	البُعْد المُضَعَّف
nearest distance (II.5[8])	البُعْد الأقْرَب
mean distance (I.4[9])	البُعْد الأوْسَط
to remain (after subtraction)	يَبْقى
locality	بَلَد (ج) بِلاد
to attain; reach	يَبْلُغ

ت

hill (Intr.[1])	تَلّ (ج) تِلال
complement of the azimuth (I.3[8])	تَمام السَمْت
colatitude (II.2[7]); latitude's completion (II.2[8])	تَمام العَرْض
complement of the arc (I.4[1])	تَمام القَوْس
complement of obliquity (II.2[6])	تَمام المَيْل
complementary body (I.1[3]); complement	المُتَمِّم

ث

the fixed stars (I.1[10])	الثَوابِت
thickness	ثِخَن
triangle (I.4[5])	مُثلَّث
second	ثانِيَة (ج) ثَوانٍ
Taurus	الثَوْر

ج

mountain (II.1[1])	جَبَل (ج) جِبال
Capricornus	الجَدْي
body	جِرْم (ج) أجْرام
aethereal bodies (Intr.[1])	أجْرام أثِيرِيَّة
the simple bodies	الأجْرام البَسيطَة
the celestial bodies (I.2[2])	الأجْرام السَماوِيَّة
part (I.1[1]); unit (II.3[2])	جُزْء (ج) أجْزاء
solid; body	جِسْم (ج) أجْسام
spherical body (I.1[1])	جِسْم كُرِيّ
to reappear (after being eclipsed) (I.5[36])	يَنْجَلِي (مص) اِنْجِلَاء
to be in conjunction	يَجْتَمِعُ
conjunction	اِجْتِماع

حَرَكَة الكُلّ	motion of the Universe (I.2[2])
حَرَكَة الوَسَط/الحَرَكَة الوُسْطَى	mean motion (I.2[9])
الحَرَكَة اليَوْمِيَّة	the daily motion
مَحْسوس	perceptible (I.4[15])
حاسِب (ج) حُسّاب	mathematician; calculator (II.3[7])
مَحْصور	to be bounded (I.4[3])
يَحْصُلُ	to reach; to obtain; to result (from an arithmetical operation)
حَضِيض	perigee (I.1[2]); epicyclic perigee (I.4[9])
يَنْحَطُّ	to be depressed; to descend
إنْحِطاط	depression (II.2[2])
حَقِيقِيّ	true (II.3[7])
حُكْم (ج) أحْكام	rule (II.3[1])
حَكِيم (ج) حُكَماء	sage (Pref.[1])
حُمْرَة	red; redness
الحَمَل	Aries
(فَلَك) حامِل	deferent (orb) (I.1[6])
(فَلَك) حامِل لِمَرْكَز الحامِل	deferent (orb) of the deferent center (I.3[13])
حَمائِلِي	slanted (II.2[2])
حَمّام (ج) حَمّامات	bath house (II.1[10])
الحُوت	Pisces
مِحْوَر	axis (I.2[8])
يُحِيطُ	to bound; enclose
مُحِيط	circumference; enclosing

جانِب (ج) جَوانِب	side
جَنُوب	south
نُقْطَة الجَنُوب	south point (I.3[7])
مَجاز	crossing point (I.3[12])
الجَوْزاء	Gemini
الجَوْزَهْر / الجَوْزَهَر	lunar nodes; *jawzahar* (I.2[7])
جَوْف (ج) أجْواف	cavity (I.1[2])

ح

حَدّ (ج) حُدُود	definition (I.4[12])
مُحَدَّب	convex (I.1[2])
نُقْطَة المُحاذاة	alignment point (I.5[5])
يَنْحَرِفُ عَن	to depart from
الإنْحِراف	slant (I.5[18])
يَحْتَرِقُ (مص) إحْتِراقُ	to combust (I.5[33])
حَرَكَة (ج) حَرَكات	motion
حَرَكَة الأَوْج	motion of the apogee (I.2[3])
الحَرَكَة الأُولَى	the prime motion (I.2[2])
الحَرَكَة الخاصّة	proper motion (I.2[10])
حَرَكَة الإخْتِلاف	motion of anomaly (I.2[10])
حَرَكَة المَرْكَز	motion of the center (I.2[9])
حَرَكَة الطُول	motion of longitude (I.2[9])
حَرَكَة العَرْض	motion of latitude (I.2[9])

enclosing (I.1[3])	حاوِيَة
enclosed (I.1[3])	مَحْوِيّ
animal (Intr.[1])	حَيَوان (ج) حَيَوانات
vacillating planets (I.2[10])	المُتَحَيِّرَة

خ

wasteland (II.1[1])	خَرِبة
to emerge; to be extended; to result	يَخْرُجُ
to produce or extend (a line, etc.)	يُخْرِجُ
to be etched on (I.3[7])	يُسْتَخْرَجُ في
eccentric (orb) (I.1[3])	الخارِج المَرْكَز
cone (II.3[6])	مَخْرُوط
shadow cone (II.3[6])	مَخْرُوط الظِلّ
conic (II.3[3])	مَخْرُوطي
autumn, fall	خَريف
lunar eclipse (I.5[35])	خُسُوف القَمَر
proper movement (II.2[8])	المَسِير الخاصّ
characteristic	خاصّة (ج) خَواصّ
proper (motion) (I.2[10])	(الحَرَكَة) الخاصّة
proper anomaly (I.5[18])	خاصّة إخْتِلاف
mean proper anomaly	الخاصّة الوُسْطى
line	خَطّ (ج) خُطُوط
dirigent line (I.5[9])	خَطّ المُدِير
(Earth's) equator (I.3[2])	خَطّ الإسْتِواء
east-west line (I.3[6]; II.3[3])	خَطّ المَشْرِق والمَغْرِب
equinox line (II.3[3])	خطّ الاعْتِدال
straight line (II.3[4])	خَطّ مُسْتَقِيم
meridian line (I.3[7]; II.3[3])	خَطّ نِصْف النَهار
mid-heaven line (II.3[5])	خطّ وَسَط السَماء
invisibility (II.2[2])	خَفاء / إخْتِفاء
of permanent invisibility (II.2[2])	أبَديّ الخَفاء
Eternal Islands (II.1[2])	جَزائِر الخالِدات
in reverse	بِالْخِلاف
to differ; to vary	يَخْتَلِفُ
to be invariable	لا يَخْتَلِفُ
difference; variation (Intr.[2]); divergence (I.4[15]); anomaly (I.5[17]); change (II.3[10])	اخْتِلاف (ج) إخْتِلافات
first anomaly (I.5[3])	إخْتِلاف أوَّل
second anomaly (I.5[4]); anomaly of the nearest distance	إخْتِلاف ثانٍ؛ إخْتِلاف البُعْد الأقْرَب
third anomaly (I.5[5])	إخْتِلاف ثالِث
variable movement (I.4[10]; Fig. 7)	إخْتِلاف المَسير
longitudinal anomaly (I.5[1])	إخْتِلاف الطُول
latitudinal anomaly (I.5[14])	إخْتِلاف العَرْض
transit difference (II.3[1])	إخْتِلاف المَمَرّ

the equinoctial circle دائِرَة مُعَدِّل النَهار (I.3[2])
latitude circle (I.3[11]) دائِرَة العَرْض
great circle دائِرَة عَظِيمَة (عُظْمَى) (ج) عِظام (I.3[1])
solstitial colure الدائِرَة المارّة بالأقْطاب الأرْبَعَة (I.3[5])
declination دائِرَة المَيْل (ج) دَوائِر المُيُول circle (I.3[10])
meridian circle (I.3[7]) دائِرَة نِصْف النَهار
Indian circle (II.3[3]) الدائِرَة الهِنْدِيَّة
imaginary circle (I.3[12]) الدائِرَة المُتَوَهَّمَة
circuit مَدار (ج) مَدارات
solar circuit (I.4[20]) مَدار الشَمْس
parallel مَدار عَرْضِيّ (ج) مَدارات العَرْض of latitude (I.3[4])
day-circuit مَدار يَوْمِيّ (ج) مَدارات يَوْمِيَّة (I.3[2])
epicycle تَدْوِير (ج) تَداوِير/تَدْوِيرات (I.1[5])
dirigent (I.1[7]) مُدِير
round (Intr.[1]) إسْتِدارَة
wheel-like (II.2[1]) دُولابِيّ

ذ

apex (of epicycle) ذِرْوَة/ذُرْوَة (ج) ذُرىً (I.4[8])
apparent apex (I.5[9]) الذِرْوَة المَرْئِيَّة
mean apex (I.5[9]) الذِرْوَة الوُسْطى

parallax إخْتِلاف المَنْظَر (ج) المَناظِر (I.4[15]); divergence of sight
without variation غَيْر اخْتِلاف
vacuum (Intr.[2]) خَلاء

د

equal degrees (I.4[5]) دَرَج سَوَاء
degree دَرَجَة (ج) دَرَجات
degree دَرَجَة طُلُوع الكَوكَب of rising of the star (II.3[1])
degree دَرَجَة مَمَرّ الكَوكَب of transit of the star (II.3[1])
in one stroke (II.2[6]) دَفْعَةً
minute (II.3[9]) دَقِيقَة (ج) دَقائِق
Aquarius الدَلْو
to revolve (I.3[13]); to rotate (II.3[8]) يَدُورُ
rotation (I.3[2]); turning (II.2[1]) دَوْر
rotation (I.5[8]) دَوران
rotation (I.2[2]) دَوْرَة
circle دائِرَة (ج) دَوائِر
horizon circle (I.3[6]) دائِرَة الأُفُق
circle of the initial دائِرَة أوّل السُمُوت azimuth (prime vertical) (I.3[9])
zodiacal circle (I.3[3]) دائِرَة البُرُوج
altitude circle (I.3[8]) دائِرَة الارْتِفاع
azimuth circle (I.3[8]) دائِرَة السَمْتِيَّة
small circle دائِرَة صَغِيرَة (صُغْرَى) (I.3[1])

tail (a node) (I.2[4])	ذَنَب

ر

apex; head (a node) (I.2[4]); tip (of a line) (I.4[6]); overhead (II.2[1])	رأس (ج) رُؤوس
head (of Gemini, etc.)	رأس (الجَوْزاء)
apparent	مَرْئِيّ
almuri (Capricorn marker on astrolabe rete)	المُرِيّ (مُرْئِ / مُورِى)
quarter mark (I.3[5]); one-fourth (II.3[3])	رُبْع (ج) أرْباع
spring (the season)	رَبِيع
quadrature (I.5[37])	تَرْبِيع
arrangement (Intr.[2])	تَرْتِيب
retrogradation (I.5[29])	رُجُوع
retrograding (I.5[30])	راجِع
spinning (II.2[12])	رَحَوِيّ
sundial (I.3[7])	رُخامَة (ج) رُخامات
to turn back; return (I.5[37])	يَرُدُّ (إلى)
to trace; to be traced	يَرْسُمُ / يُرَسِّمُ / يَرْتَسِمُ
to observe	يَرْصُدُ
to be elevated	يَرْتَفِعُ
altitude (I.4[11])	ارْتِفاع
thinness	رِقّة
composite (Intr.[1])	مُرَكَّب
composite (bodies) (Intr.[1])	مُرَكَّبات
center	مَرْكَز (ج) مَراكِز
eccentricity (lit., what is between the two centers) (I.4[9])	ما بَيْن المَرْكَزَيْن
eccentric center	مَرْكَز الخارِج
solar center	مَرْكَز الشَمْس
equant center (I.5[5])	مَرْكَز مُعَدِّل المَسِير
center of the World	مَرْكَز العالَم
lunar center	مَرْكَز القَمَر
embedded (I.1[4])	مَرْكُوز

ز

Saturn	زُحَل
unit of time (II.3[8])	زَمَن (ج) أزْمان
period (of time)	زَمان (ج) أزْمِنَة
duration of daytime (nighttime) (II.3[8])	زَمان النهار (الليل)
Venus	الزُهَرَة
angle	زاوِيَة (ج) زَوايا
angle of anomaly (I.5[2]; commentary Fig. C1)	زاوِيَة الاخْتِلاف
angle of the equation (I.4[6])	زاوِيَة التَعْدِيل
right angle (I.3[8])	زاوِيَة قائِمَة (ج) زَوايا قائِمَة
astronomical handbook; *zīj* (I.2[10])	زِيْج (ج) زِيجات
to increase; to add	يَزِيدُ (مص) زِيادَة
waxing Moon (I.5[35])	الزِيادَة
exceeds	زائِد

to depart	يزُولُ / يُزايِلُ

س

coast (II.1[2])	ساحِل (ج) سَواحِل
Cancer	السَرَطان
rapid (I.2[2]); swift (I.5[30])	سَريع
surface (Intr.[1])	سَطْح (ج) سُطُوح
lower	أَسْفَل / سُفْلِيّ (ج) أسافِل
the two lower planets (I.5[18])	السُفْلِيّان
residents (II.1[10])	سُكّان
inhabited zone or region (I.5[36])	مَساكِن
azimuth (I.3[8])	سَمْت (ج) سُمُوت
complement of azimuth (I.3[8])	تَمام السَمْت
zenith (I.3[6])	سَمْت الرأس؛ سَمْت رؤوس
qibla azimuth (I.4[19]); *qibla* bearing (II.3[4])	سَمْت القِبْلَة
nadir (I.3[6])	سَمْت القَدَم
arc of azimuth (I.3[8])	قَوْس السَمْت
azimuth point (I.3[8])	نُقْطَة السَمْت
the ascendant azimuth (I.4[18])	سَمْت من الطالِع
circle of the initial azimuth (prime vertical) (I.3[9])	دائِرَة أوّل السُمُوت
alignment (I.2[8])	مُسامَتَة
the celestial (bodies) (I.2[2])	الأجْرام السَماوِيّة
year (II.3[9])	سَنَة (ج) سِنون
solar year (I.2[6])	سَنَة شَمْسِيّة
lunar year (I.2[6])	سَنَة قَمَرِيّة
Virgo	السُنْبُلَة
blackness (I.5[36])	سَواد
hour (II.3[6])	ساعَة (ج) ساعات
seasonal (temporal) hours (II.3[8])	ساعات زَمانِيّة
equal hours (II.3[8])	ساعات مُسْتَوِيَة
regularized hours (II.3[8])	ساعات مُعْتَدِلَة
unequal (distorted) hours (II.3[8])	ساعات مُعَوَّجَة
to equal; be equal to	يُساوِي
to level (land) (II.3[3])	يُسَوِّي
are equal to one another I.5[23]	مُتَساوِيَة
regular (order) (II.2[9])	مُسْتَوٍ
speed (I.5[30])	سَيْر
movement (I.4[10]); course (I.2[10])	مَسِير
proper movement (II.2[8])	المَسير الخاصّ
mean motion (II.3[10])	مَسير وَسَط
(wandering) planets (I.3[12])	السَيّارَة

ش

is similar to	شَبِيه بِـ
uniform (I.1[1])	مُتَشابِهَة
winter	شِتاء
the law (II.3[8])	الشَرْع
east	شَرْق
east-west line (I.3[6])	خَطّ المَشْرِق والمَغْرِب

east point (I.3[6])	نُقْطَة المَشْرِق
common point	نُقْطَة مُشْتَرَكَة
Jupiter	المُشْتَرِي
barleycorn (Intr.[1])	شَعِير
dusk (II.3[6])	شَفَق
plumb-line (II.3[3])	شاقول
figure; form; shape	شَكْل (ج) أشْكال
(geological) formations (Intr.[1])	تَشْكيلات
Sun	الشَمْس
north	شَمال
north point (I.3[7])	نُقْطَة الشَمال
month (II.3[10])	شَهْر (ج) شُهور
proposal (Pref.[1])	إشارَة

ص

digit (II.3[2])	إصْبَع (ج) أصابِع
desert (II.1[1]	صَحْراء (ج) صَحارى
ascending (I.4[10])	صاعِد
the Slavs (II.1[10])	الصَقالِبَة
worshipper (II.3[4])	مُصَلِّي
illustration	صُورَة (ج) صُوَر
to conceive (I.1[9])	يَتَصَوَّرُ
summer	صَيْف

ض

vice versa (I.2[11])	بِالضِدّ
multiplied by (from an arithmetical operation) (II.3[10])	ضَرَبَ في
irregular (Intr.[1])	مُضَرَّس
undulations (Intr.[1])	تَضارِيس
side	ضِلْع (ج) أضْلاع
light (I.5[36])	ضَوء
light (I.5[36])	ضِياء
luminous [part] (I.5[36])	مُضِيء
to become illuminated (I.5[36])	يَسْتَضِيءُ

ط

natural state (Intr.[1])	طَبْع
a nature (Intr.[1])	طَبِيعَة (ج) طَبائِع
to coincide (with)	يَنْطَبِقُ (على)
coincidence (I.5[21])	اِنْطِباق
coincident; coinciding (with)	مُنْطَبِقَة (على)
endpoint; edge	طَرَف (ج) أطْراف
path (of the Sun) (I.5[36])	طَرِيقَة (الشَمْس)
to rise; to ascend	يَطْلُعُ (مص) طُلُوع
ascendant (II.3[1])	الطالِع
rising place (I.4[16])	مَطْلَع
co-ascension (I.4[3])	مَطالِع
co-ascension of the zodiacal arc (I.4[5])	مَطالِع قَوْس البُرُوج
co-ascension of a [discrete] part (I.4[3])	مَطالِع الجُزء
length (II.2[8]); longitude (I.3[3])	طُول
longitude of a locality (I.4[2])	طُول البَلَد

equinoctial (I.2[2]) مُعَدِّل النَهار
the equinoctial circle (I.3[2]) دائِرَة مُعَدِّل النَهار
to balance (I.3[2]); to equalize (I.5[9]) يَعْتَدِلُ
equinox الاِعْتِدال
equinox (line) (I.3[6]; II.3[3]) (خطّ) الاِعْتِدال
equinox (point) (I.3[5]) (نُقْطَة) الاِعْتِدال
autumnal equinox (I.3[5]) الاِعْتِدال الخَرِيفيّ
vernal equinox (I.3[5]) الاِعْتِدال الرَبيعيّ
regularized (II.3[8]) مُعْتَدِل
mineral (Intr.[1]) مَعْدِن (ج) مَعادِن / مَعْدِنِيّات
latitude عَرْض
local latitude (I.4[11]) عَرْض البَلَد
colatitude (II.2[7]) تَمام العَرْض
latitude circle (I.3[11]) دائِرَة العَرْض
parallel of latitude (I.3[4]) مَدار عَرْضِيّ (ج) مَدارات العَرْض
characteristics عَرَض (ج) أعْراض
afternoon [ʿaṣr] prayer (II.3[2]) عَصْر
waterwheel with buckets (II.2[1]) عُصْمور (ج) عصامير
Mercury عُطارِد
Scorpius العَقْرَب
reversed (shadow) (II.3[2]) المَعْكُوس
science; discipline عِلْم (ج) عُلُوم

ظ

shadow (I.5[36]) ظِلّ (ج) أظْلال
الظِلّ الأوّل؛ الظِلّ المَعْكُوس؛ الظِلّ المُنْتَصِب
first shadow; reversed shadow [*umbra versa*]; erect shadow (II.3[2])
second shadow الظِلّ الثاني؛ الظِلّ المُسْتَوِي (II.3[2]); planar shadow [*umbra recta*] (II.2[4])
shadow cone (II.3[6]) مَخْرُوط الظِلّ
dark (I.5[36]) مُظْلِم
to appear; to be visible يَظْهَرُ
noon [*ẓuhr*] prayer (II.3[2]) ظُهْر
visibility ظُهُور
permanent visibility (II.2[2]) أبَدِيّ الظُهُور
visible ظاهِر (ج) ظواهر

ع

number عَدَد (ج) أعداد
equation (I.4[6]) تَعْدِيل
equation of daylight (I.4[5]) تَعْدِيل النَهار
equation of the time [nychthemeron] (II.3[7]) تَعْدِيل الأيّام بِلَيالِيها
angle of the equation (I.4[6]) زاوِيَة التَعْدِيل
arc of the equation (I.4[6]) قَوْس التَعْدِيل
equant (center or orb) (I.5[5]) (مَرْكَز / فَلَك) مُعَدِّل المَسِير

ʿilm al-hay'a (Pref.[1]) عِلْم الهَيْئَة
the World (Intr.[2]) العالَم
to mark (II.3[3]) يُعَلِّمُ
the upper planets (I.1[5]) (الكَواكِب) العَلَوِيَّة
perpendicular (II.3[2]) عَمُود
habitable land; habitation (II.1[2]) عِمارَة (ج) عِمارات
inhabited (quarters) (II.1[title and [1]) المَعْمُور
element (Intr.[1]) عُنْصُر (ج) عَناصِر / عُنْصُرِيّات
the rete [lit., spider] (of astrolabe) (II.3[5]) العَنْكَبوت
return عَوْدَة

غ

to set يَغْرُبُ (مص) غُرُوب
west غَرْب
west point (I.3[6]) نُقْطَة المَغْرِب
embedded (I.1[4]) مُغْرَق
thickness (I.1[1]) غِلَظ
maximum (I.3[5]) غايَة
maximum altitude (I.4[14]) غايَة ارْتِفاع
maximum equation (I.4[10]) غايَة التَعْدِيل
maximum declination (I.4[12]) غايَة المَيْل
does not change (I.5[19]) لا يَتَغَيَّرُ

ف

dawn (II.3[6]) فَجْر
to be assumed يُفْرَضُ
given or designated مَفْرُوض
departure (II.3[7]) مُفارَقَة
to separate (I.3[6]) يَفْصِلُ
chapter; season فَصْل (ج) فُصُول
excess; difference فَضْل
orb فَلَك (ج) أفْلاك / فَلَكِيّات
the celestial orb (I.3[6]) الفَلَك
the zodiacal orb (I.2[6]) فَلَك البُرُوج
orb of the fixed stars (Intr.[2]) فَلَك الثَوابِت
solid orb (I.1[1]) فَلَك مُجَسَّم
atlas orb فَلَك الأَطْلَس
equant orb (I.5[5]) فَلَك مُعَدِّل لِلْمَسير
the greatest orb (Intr.[2]) فَلَك الأَعْظَم
the highest orb (I.3[8]) الفَلَك الأَعْلى
orb of orbs (Intr.[2]) فَلَك الأَفْلاك
the right orb (I.3[2]) الفَلَك المُسْتَقِيم
difference تَفاوُت

ق

cupola of the Earth (II.1[1]) قُبَّة الأَرْض
qibla قِبْلَة
to face; to be in opposition to يُقابِلُ (مص) مُقابَلَة
opposite; facing (one another) متقابِلَة
amount قَدْر (ج) أَقْدار

amount; measure مِقْدار (ج) مَقادِير
to determine; to measure يُقَدِّرُ
foot (II.3[2]) قَدَم (ج) أَقْدام
introduction (Pref.[2]) مُقَدِّمَة
to be in advance of يَتَقَدَّمُ
to approach يَقْرُبُ
closeness; nearness قُرْب
approximately بِالتَقْرِيب
to be in conjunction with يُقارِنُ
conjunction مُقارَنَة
to divide يَقْسِمُ / يُقَسِّمُ
division; part قِسْم (ج) أَقْسام
pole (I.2[2]) قُطْب (ج) أَقْطاب
diameter (I.1[4]) قُطْر (ج) أَقْطار
radius (I.5[3]) نِصْف قُطْر (ج) أَنْصاف أَقْطار
to be in opposition; in alignment يَتَقاطَرُ
to cut off or describe (e.g. an arc); to intersect; to traverse يَقْطَعُ
segment قِطْعَة (ج) قِطَع
to intersect يُقاطِعُ (مص) مُقاطَعَة
to intersect one another يَتَقاطَعُ
intersection point نُقْطَة تَقاطُع
concave (Intr.[1]) مُقَعَّر
solstice (point) (نُقْطَة) الِانْقِلاب
winter solstice (point) (I.3[5]) (نُقْطَة) الِانْقِلاب الشَتْوِيّة
summer solstice (point) (I.3[5]) (نُقْطَة) الِانْقِلاب الصَيْفِيّة

the Moon القَمَر
almucantar (I.3[6]) مُقَنْطَرَة (ج) مُقَنْطَرات
altitude almucantar (II.3[5]) مُقَنْطَرَة الاِرْتِفاع
arc (I.4[1]) قَوْس (ج) قُسِيّ / قِسِيّ
Sagittarius القَوْس
arc of the azimuth (I.3[8]) قَوْس السَمْت
arc of *qibla* bearing (II.3[4]) قَوْس سَمْت القِبْلَة
arc of the equation (I.4[6]) قَوْس التَعْدِيل
arc of night (I.4[20]) قَوْس اللَيْل
arc of daylight (I.4[20]) قَوْس النَهار
part (main division of *Mulakhkhaṣ*) مَقالَة
to be perpendicular; to stand erect (I.4[10]) يَقُومُ على
people (II.1[10]) قَوْم
true position: true position of the Sun (I.4[6]); true position of the planets (I.4[7]) تَقْوِيم: تَقْوِيم الشَمْس؛ تَقْوِيم الكَواكِب
to be stationary; station (I.5[29]) يُقِيمُ (مص) إقامَة
being stationary مُقِيم
first station; second station (I.5[30]) المُقام الأوّل؛ المُقام الثاني
to undergo direct motion (I.5[30]) يَسْتَقِيمُ (مص) اِسْتِقامَة

م

comparable; is equivalent to; is equal to; the same as; similar to	مِثْل (ج) أمْثال
parecliptic (orb) (I.1[3])	(الفَلَك) المُمَثَّل
Almagest (II.1[2])	المَجِسْطِي
new Moon (I.5[35])	المُحاق
to pass through; to transit	يَمُرُّ
transit; traverse	مُرُور
transiting	مَمَرّ
pasture (II.1[1])	مَرْج (ج) مُروج
Mars	المِرّيخ
to be contiguous with; to be tangent to; to touch	يُماسُّ
tangent	مُماسّ
tangency	تَماسّ
Mihrjān (II.2[2])	مِهْرجان
water (Intr.[1])	ماء (ج) مِياه
to incline	يَمِيلُ
inclination; declination	مَيْل (ج) مُيُول
first declination (I.3[10]; I.4[12])	مَيْل أوّل
second declination (I.3[10]; I.4[12])	مَيْل ثانٍ
total obliquity or complete declination; maximum declination (I.4[12])	المَيْل الكُلّيّ / غاية المَيْل
declination circle (I.3[10])	دائِرَة المَيْل (ج) دَوائِر المُيُول

having direct motion; being straight	مُسْتَقِيم
the right (erect) orb (I.3[2])	الفَلَك المُسْتَقِيم
gnomon (II.3[2])	مِقْياس

ك

sphere	كُرَة (ج) كُرات
enclosing sphere (Intr.[2])	الكُرَة المُحِيطَة
erect sphere (II.2[1])	الكُرَة المُنْتَصِبَة
spherical (Intr.[1])	كُرِيّ
solar eclipse (I.5[35])	كُسُوف الشَمْس
Kaʿba (II.3[4])	كَعْبَة
the Universe (I.2[2])	الكُلّ
quantity	كَمِّيَّة (ج) كَمِّيّات
opaque (I.5[36])	كَمِد
full Moon (I.5[35])	الكَمال
star; planet (I.1[6])	كَوْكَب (ج) كَواكِب
vacillating planets (I.2[10])	الكَواكِب المُتَحَيِّرَة
wandering planets	الكَواكِب السَيّارَة
position; place	مَكان
manner (I.1[9])	كَيْفِيَّة

ل

twist (for latitude) (I.5[18])	الالْتِواء
night; nighttime	لَيْل (ج) لَيالٍ
arc of night (I.4[20])	قَوْس اللَيْل

declination/inclined orbs (I.3[12])	الأفْلاك المائِلة

ن

plants (Intr.[1])	نَبات (ج) نَباتات / نَباتِيّات
astrologer (II.3[7])	مُنَجِّم (ج) مُنَجِّمون
ratio	نِسْبَة (ج) نِسَب
radius	نِصْف قُطْر (ج) أنْصاف أقْطار
noon	نِصْف النَهار
meridian (line or circle) (I.3[7])	(خَطّ / دائِرَة) نِصْف النَهار
to bisect	يُنَصِّفُ
midpoint (I.4[9]); midway (I.5[7]); middle (II.1[1])	مُنْتَصَف
planetary sector (I.4[9])	نِطاق (ج) نِطاقات
equator	مِنْطَقَة (ج) مَناطِق
zodiacal equator (I.2[3])	مِنْطَقَة البُرُوج
perspective (I.4[15])	مَنْظَر
parallax (I.4[15])	إِخْتِلاف المَنْظَر
to subtract from	يَنْقُصُ مِن
subtraction	نُقْصان
waning Moon (I.5[35])	النُقْصان
point	نُقْطَة (ج) نُقَط
to shift (I.3[8])	يَنْتَقِلُ
reversed (order) (II.2[9])	مَنْكُوس
daylight; daytime (II.2[1])	نَهار (ج) أنْهُر
duration of daylight (II.3[8])	زَمان النَهار
arc of daylight (I.4[20])	قَوْس النَهار
limit (I.4[12])	نِهايَة
to terminate (I.4[6]); to reach (I.4[10])	يَنْتَهِي / يَنْتَهِي إلى
fire (Intr.[1])	نار
light (II.3[6]	نُور (ج) أنْوار
the two luminaries (II.3[10])	النَيِّران
Nayrūz [II.2[2])	نَيْروز

ه

descending (I.4[10])	هابِط
crescent Moon (I.5[36])	هِلال
air (Intr.[1])	هَواء
configuration; shape	هَيْئَة (ج) هَيْئات

و

to subtend (I.4[6])	يُوتِرُ
face (of a celestial body)	وَجْه
direction	جِهَة (ج) جِهات
to face (I.5[36]); to be facing (II.3[4])	يُواجِهُ (مص) مُواجَهَة
wild animal (II.1[10])	وَحْش (ج) وُحوش
slope (I.5[18]) (of latitude)	الوِراب
beyond	وَراء
Libra	المِيزان
parallel (I.2[8])	مُوازٍ
parallel to (II.1[3])	على مُوازاة
to be parallel to one another (I.1[1])	مُتَوازٍ

وَسَط (ج) أوْساط middle; mean [motion]
midpoint [climes] (II.1[3])
وَسَط السَماء mid-heaven (II.3[5])
وَسَط الشَمْس solar (Sun's) mean (I.4[6])
وَسَط القَمَر Moon's mean motion (I.5[37])
وَسَط الكَوْكَب planet's mean (I.4[7])
سَعَة مَشْرِق ortive amplitude (I.4[5])
سَعَة مَغْرِب occasive amplitude (I.4[16])
يَصِفُ؛ وَصْف to describe; description
يَصِلُ إلى to reach
يَتَّصِلُ بِـ ؛ مُتَّصَل بِـ to be connected with; to be contiguous with
وَضْع (ج) أوْضاع position
مَوْضِع (ج) مَواضِع location; place; position
مُوافِقَة corresponding to (I.5[30])
وَقْت (ج) أوْقات time
توالٍ sequence
على تَوالِي in the sequence of
لا على تَوالِي / إلى خِلاف تَوالِي in the counter-sequence of
وَهْدَة (ج) وِهاد valley (Intr.[1])
مَوْهوم imagined
يَتَوَهَّمُ to imagine
مُتَوهَّم imaginary

ي

يوم day
اليَوْم بِلَيْلَتِه nychthemeron (I.2[3])
يَوْمِيّ daily

Bibliography

Abada, Abdelfettah. "The 'Ulamā' of Iran in the 19th Century Hagiographical Literature." B.A. thesis, University of Durham, UK, 1989.

Abdukhalimov, Bahrom. "Aḥmad al-Farghānī and His Compendium of Astronomy." *Journal of Islamic Studies* 10, no. 2 (1999): 142–58.

Ahlwardt, Wilhelm. *Die Handschriften-Verzeichnisse der Königlichen Bibliothek zu Berlin. Verzeichniss der Arabischen Handschriften.* 10 vols. Berlin, 1887–99.

Aiton, E. J. "Peurbach's *Theoricae novae planetarum*: A Translation with Commentary." *Osiris*, 2nd series, 3 (1987): 4–43.

Akpınar, Cemil. "Harakī." In *İslam Ansiklopedisi*, 16:94–96. Istanbul: Turkiye Diyanet Vakfi (TDV), 1997.

Andiqānī, Muḥammad ibn 'Umar. *Tarjumah-yi al-Mulakhkhaṣ fī al-Hay'a.* Edited by Arash Abutorabi Hamedani. In *Nāmah-yi Ma'ānī, Yādnāmah-yi Ustād Aḥmad Gulčīn Ma'ānī (Memoirs of Master Aḥmad Gulčīn Ma'ānī)*, edited by Behrouz Imani, 866–916. Tehran: Tehran, Library, Museum and Document Center of Iran Parliament, 1383 H. Sh./2004-5.

Asper, Markus and Anna Maria Kanthak, eds. *Writing Science: Medical and Mathematical Authorship in Ancient Greece.* Science, Technology, and Medicine in Ancient Cultures. Berlin/Boston: Walter de Gruyter, 2013.

al-Baghdādī, Ismā'īl Bāshā. *Hadiyyat al-'ārifīn.* 2 vols. Istanbul, 1951 (vol. 1) and 1955 (vol. 2). Reprint, Baghdad and Beirut: Makabat al-muthannā, n.d.

Barker, Peter. "The Reality of Peurbach's Orbs: Cosmological Continuity in Fifteenth and Sixteenth Century Astronomy." In *Change and Continuity in Early Modern Cosmology*, edited by Patrick J. Boner, 7–32. Archimedes 27: New Studies in the History and Philosophy of Science and Technology. Dordrecht; New York: Springer, 2011.

Barthold, Vassiliy Viladimirovic. *Turkestan down to the Mongol Invasion.* 2nd ed. Translated from the original Russian and revised by the author with the assistance of H. A. R. Gibb. London: Luzac and Co., 1958.

al-Battānī, *Zīj*. See Nallino, Carlo A. *Al-Battānī Sive Albatenii Opus Astronomicum.*

BEA: The Biographical Encyclopedia of Astronomers. Edited by Thomas Hockey et al. 2 vols. New York: Springer-Verlag 2007.

S.P. Ragep, *Jaghmīnī's Mulakhkhaṣ*, Sources and Studies in the History of Mathematics and Physical Sciences, DOI 10.1007/978-3-319-31993-3

Berggren, J. Lennart. *Episodes in the Mathematics of Medieval Islam*. New York; Berlin; Heidelberg: Springer-Verlag, 2003.

Berggren, J. Lennart. "Maʾmūn: Abū al-ʿAbbās ʿAbdallāh ibn Hārūn al-Rashīd." In *The Biographical Encyclopedia of Astronomers*, edited by Thomas Hockey et al., 2:733. New York: Springer-Verlag, 2007.

Berggren, J. Lennart and Alexander Jones. *Ptolemy's Geography: An Annotated Translation of the Theoretical Chapters*. Princeton, NJ: Princeton University Press, 2000.

al-Bīrūnī, Abū Rayḥān Muḥammad ibn Aḥmad. *India*. Edition of Arabic text by Edward C. Sachau as *Alberuni's India: An Account of the Religion, Philosophy, Literature, Geography, Chronology, Astronomy, Customs, Laws and Astrology of India About A.D. 1030*. London: Trübner, 1887. Translation with notes by Edward C. Sachau as *Alberuni's India*. 2 vols. London: Trübner & Co., 1888. Reprint, London: Trübner & Co., 1910.

al-Bīrūnī, Abū Rayḥān Muḥammad ibn Aḥmad. *Kitāb al-Tafhīm li-awāʾil ṣināʿat al-tanjīm* (=The Book of Instruction in the Elements of the Art of Astrology). Translated by R. Ramsay Wright, with the facsimile reproduction of an Arabic MS. London: Luzac and Co., 1934. Persian text, edited by Jalāl al-Dīn Humāʾī. Tehran: Intishārāt-i Bābak, 1362 H. Sh./1983-84.

al-Bīrūnī, Abū Rayḥān Muḥammad ibn Aḥmad. *Al-Qānūn al-Masʿūdī*. 3 vols. Hyderabad: Dāʾirat al-maʿārif al-ʿUthmāniyya, 1954–56.

Black, Crofton. *Transformation of Knowledge: Early Manuscripts from the Collection of Lawrence J. Schoenberg*. London: Paul Holberton, 2006.

Boilot, D. J. "L'oeuvre d'al-Beruni, Essai bibliographique." *Mélanges de l'Institut Dominicain d'Études Orientales* 2 (1955): 161–256.

Bolt, Marvin. "ʿAlī ibn ʿĪsā al-Asṭurlābī." In *The Biographical Encyclopedia of Astronomers*, edited by Thomas Hockey et al., 1:34. New York: Springer-Verlag, 2007.

Bolt, Marvin. "Marwarrūdhī: Khālid ibn ʿAbd al-Malik al-Marwarrūdhī." In *The Biographical Encyclopedia of Astronomers*, edited by Thomas Hockey et al., 2:740. New York: Springer-Verlag, 2007.

Bosworth, C. Edmund. *The Islamic Dynasties: A Chronology and Genealogical Handbook*. Edinburgh: Edinburgh University Press, 1967.

Bosworth, C. Edmund. "The Political and Dynastic History of the Iranian World (A.D. 1000–1217)." In *The Cambridge History of Iran*, vol. 5, *The Saljuq and Mongol Periods*, edited by J. A. Boyle, 1–202. Cambridge: Cambridge University Press, 1968.

Bosworth, C. Edmund. "K͟hʷārazm" and "K͟hʷārazm-S͟hāhs." *Encyclopaedia of Islam*, 2nd ed., 4:1060–68. Leiden: E. J. Brill, 1978.

Bosworth, C. Edmund. "The Persian Contribution to Islamic Historiography in the Pre-Mongol Period." In *The Persian Presence in the Islamic World*, edited by Richard G. Hovannisian and Georges Sabagh, 218–36. Cambridge: Cambridge University Press, 1998.

Bosworth, C. Edmund and M. S. Asimov, eds. *History of Civilizations of Central Asia*. Vol. 4, *The Age of Achievement: A.D. 750 to the End of the Fifteenth Century. Part Two: The Achievements*. Paris: UNESCO Publ., 2000.

Bowen, Alan C. "Three Introductions to Celestial Science in the First Century BC." In *Writing Science: Medical and Mathematical Authorship in Ancient Greece*, edited by Markus Asper in collaboration with Anna-Maria Kanthak, 299–329. Science, Technology, and Medicine in Ancient Cultures. Berlin; Boston: Walter de Gruyter, 2013.

Boyle, John A. *The History of the World Conqueror.* See Juwaynī.

Boyle, John A. "Dynastic and Political History of the Īl-Khāns." In *The Cambridge History of Iran*, vol. 5, *The Saljuq and Mongol Periods*, edited by J. A. Boyle, 303–421. Cambridge: Cambridge University Press, 1968.

Brentjes, Sonja. "The Prison of Categories—'Decline' and its Company." In *Islamic Philosophy, Science, Culture, and Religion: Studies in Honor of Dimitri Gutas*, edited by Felicitas Opwis and David Reisman, 131–56. Leiden: E. J. Brill, 2012.

Brockelmann, Carl. *Geschichte der arabischen Litteratur.* 2 vols. plus 3 supplements. Weimar: Verlag von Emil Felber (vol. 1 [1898]); Berlin: Verlag von Emil Felber (vol. 2 [1902]; Leiden: E. J. Brill (suppl. 1 [1937]; suppl. 2 [1938]; suppl. 3 [1942]). Reprinted with a new introduction by Jan Just Witkam. Leiden: E. J. Brill, 1996.

Browne, Edward G. *Arabian Medicine, Being the Fitzpatrick Lectures Delivered at the College of Physicians in November 1919 and November 1920.* Cambridge: Cambridge University Press, 1921.

Browne, Edward G. *A Supplementary Hand-List of the Muhammadan Manuscripts, Including All Those Written in the Arabic Character, Preserved in the Libraries of the University and Colleges of Cambridge.* Cambridge: Cambridge University Press, 1922.

Burnett, Charles. "The Certitude of Astrology: The Scientific Methodology of al-Qabīṣī and Abū Maʿshar." *Early Science and Medicine* 7, no. 3 (2002): 198–213.

Byrne, James Steven. "A Humanist History of Mathematics? Regiomontanus's Padua Oration in Context." *Journal of the History of Ideas* 67, no. 1 (Jan., 2006): 41–61.

Calvo, Emilia. "Khāzin: Abū Jaʿfar Muḥammad ibn al-Ḥusayn al-Khāzin al-Khurāsānī." In *The Biographical Encyclopedia of Astronomers*, edited by Thomas Hockey et al., 1:628–29. New York: Springer-Verlag, 2007.

Carmody, Francis J. *Arabic Astronomical and Astrological Sciences in Latin Translation: A Critical Bibliography.* Berkeley: University of California Press, 1956.

Carmody, Francis J. *The Astronomical Works of Thabit b. Qurra.* Berkeley: University of California Press, 1960.

Chalisova, Natalia. "Waṭwāṭ, Rašid al-Din." *Encylopædia Iranica.* Online edition. NewYork, 1996–. Last updated June 28, 2011. http://www.iranicaonline.org/articles/watwat-rasid-al-din

Chérif, Tarabein. "Contribution à l'histoire de la pharmacie arabe. Étude particulière du manuscrit intitulé: Al-Nadjibiate Al-Samarkandiate." Ph.D. dissertation, Strasbourg University, 1952.

Dallal, Ahmad S. *Islam, Science, and the Challenge of History.* New Haven: Yale University Press, 2010.

Delambre, J. B. *Histoire de l'astronomie du Moyen Age*, Chapter 2, "Albategnius," 10–62. Paris, 1819. Reprinted as "*Albategnius*. Histoire de l'astronomie du Moyen Age in *Islamic Mathematics and Astronomy*, edited by Fuat Sezgin, 68:11–64. Frankfurt am Main, 1998.

de Slane, Baron William MacGuckin. *Catalogue des manuscrits arabes/par le baron de Slane.* Paris: Imprimerie nationale, 1883–95.

DeYoung, Gregg. "Farghānī: Abū al-ʿAbbās Aḥmad ibn Muḥammad ibn Kathīr al-Farghānī." In *The Biographical Encyclopedia of Astronomers*, edited by Thomas Hockey et al., 1:357. New York: Springer-Verlag, 2007.

Dicks, D. R. "Geminus." In *Dictionary of Scientific Biography*, edited by Charles Coulston Gillispie, 5:344–47. New York: Charles Scribner's Sons, 1972.

Dietrich, Albert. "al-Suwaydī." *Encyclopaedia of Islam*, 2nd ed., 9:909–10. Leiden: E. J. Brill, 1997.

Dold-Samplonius, Yvonne. "Al-Khāzin, Abū Jaʿfar Muḥammad Ibn Al-Ḥasan Al-Khurāsānī." In *Dictionary of Scientific Biography*, edited by Charles Coulston Gillispie, 7:334–35. New York: Charles Scribner's Sons, 1973.

Dold-Samplonius, Yvonne. "Al-Sijzī." In *Encyclopaedia of the History of Science, Technology, and Medicine in Non-Western Cultures,* edited by Helaine Selin, 159–60. Dordrecht: Kluwer Academic Publishers, 1997.

DSB: *Dictionary of Scientific Biography.*

Eastwood, Bruce. *Ordering the Heavens: Roman Astronomy and Cosmology in the Carolingian Renaissance*. History of Science and Medicine Library. Leiden: E. J. Brill, 2007.

EI2: *Encyclopaedia of Islam,* 2nd edition.

EI3: *Encyclopaedia of Islam,* 3rd edition.

Elgood, Cyril. "Tibb-ul-Nabbi or Medicine of the Prophet." *Osiris* 14 (1962): 33–192.

Evans, James and J. Lennart Berggren. *Geminos's Introduction to the Phenomena: A Translation and Study of a Hellenistic Survey of Astronomy.* Princeton, NJ: Princeton University Press, 2006.

Faqīh Imāmī, Aḥmad. *Fihrist-i Nuskha-hā-yi Khaṭṭī-i Maktab al-Zahrāʾ*. Qum: Majmaʿ-i Dhakhāʾir-i Islāmī, 1385 H. Sh./2006-7.

Farghānī, Aḥmad ibn Muḥammad ibn Kathīr. *Jawāmiʿ ʿilm al-nujūm wa-uṣūl al-ḥarakāt al-samāwīya*. Edited by Jacob Golius and reprinted by Fuat Sezgin. Frankfurt am Main: Institut für Geschichte der Arabisch-Islamischen Wissenschaften an der Johann Wolfgang Goethe Universität, 1986.

Fazlıoğlu, İhsan. "Kamāl al-Dīn al-Turkmānī: Kamāl al-Dīn Muḥammad ibn Aḥmad ibn ʿUthmān ibn Ibrāhīm ibn Muṣṭafā al-Māridīnī." In *The Biographical Encyclopedia of Astronomers*, edited by Thomas Hockey et al., 1:609. New York: Springer-Verlag, 2007.

Fazlıoğlu, İhsan. "The Samarqand Mathematical-Astronomical School: A Basis for Ottoman Philosophy and Science." *Journal for the History of Arabic Science* 14 (2008): 3–68.

Feldhay, Rivka and F. Jamil Ragep, eds. *Before Copernicus: The Cultures and Contexts of Scientific Learning in the Fifteenth Century*. Montreal: McGill-Queens University Press, in press.

Fellmann, Irene. *Das Aqrābāḏīn al-Qalānisī: Quellenkritische und begriffsanalytische Untersuchungen zur arabisch-pharmazeutischen Literatur*. Beirut: Orient-Institut der Deutschen Morgenländischen Gesellschaft; Wiesbaden: In Komission bei Franz Steiner, 1986.

Fihrist. See Ibn al-Nadīm.

Flügel, Gustavus. *Lexicon Bibliographicum et Encyclopædicum a Mustafa ben Abdallah. Katib Jelebi dicto et nomine Haji Khalfa celebrato compositum.* Ad codicum vindobonensium parisiensium et berolinensis fidem primum edidit latine vertit et commentario indicibusque instruxit Gustavus Fluegel. 7 vols. Leipzig and London, 1835–58.

GAL. See Brockelmann, *Geschichte der arabischen Litteratur*.

GAS, 5. See Sezgin, *Geschichte des arabischen Schrifttums*, vol. 5.

GAS, 6. See Sezgin, *Geschichte des arabischen Schrifttums*, vol. 6.

Gelder, Geert Jan van. "Muḥdat͟hūn." *Encyclopaedia of Islam*, 2nd ed., 12:637–40. Leiden: E. J. Brill, 2004.

Gelder, Geert Jan van. "Ancients and Moderns." *Encyclopaedia of Islam, THREE*. Edited by Gudrun Krämer, Denis Matringe, John Nawas, Everett Rowson. Brill Online, 2015. 19 Jan., 2015 http://referenceworks.brillonline.com/entries/encyclopaedia-of-islam-3/ancients-and-moderns-SIM_0040.

Ghalandari, Hanif. "A Survey of the Works of '*Hay'a*' in the Islamic Period with a Critical Edition, Translation and Commentary of the Treatise *Muntaha al-Idrāk fī Taqāsīm al-Aflāk* written by Bahā' al-Dīn al-K͟haraqī (d. 553 AH/1158 AD)." Ph.D. dissertation, Institute for Humanities and Cultural Studies (IHCS), Tehran, October 2012.

Ghalandari, Hanif. "Chaghmīnī." *The Great Islamic Encyclopedia*, 19:356–57 Tehran, 1390 H. Sh./2012.

Gilbert, Joan Elizabeth. "The Ulama of Medieval Damascus and the International World of Islamic Scholarship." Ph.D. dissertation, University of California, Berkeley, 1977. ProQuest (7812573).

Gilbert, Joan Elizabeth. "Institutionalization of Muslim Scholarship and Professionalization of the 'Ulamā' in Medieval Damascus." *Studia Islamica* 52 (1980): 105–34.

Goldstein, Bernard R. "The Arabic Version of Ptolemy's Planetary Hypotheses." *Transactions of the American Philosophical Society*, n.s., 57, no. 4 (1967): 3–55.

Goldstein, Bernard R. and Alan C. Bowen. "Pliny and Hipparchus's 600-Year Cycle." *Journal for the History of Astronomy* 26, no. 2 (1995): 155–58.

Gottwaldt, Josephus M. E. *Opisanie arabskich rukopisej prinadležavšich bibliotekě Imperatorskago kazangskago universiteta*. Kazan, 1885.

Graßhoff, Gerd. "Natural Law and Celestial Regularities from Copernicus to Kepler." In *Natural Law and Laws of Nature in Early Modern Europe: Jurisprudence, Theology, Moral and Natural Philosophy*, edited by Lorraine Daston and Michael Stolleis, 143–61. Farnham: Ashgate, 2008.

Griffel, Frank. "On Fakhr al-Dīn al-Rāzī's Life and the Patronage He Received." *Journal of Islamic Studies* 13, no. 3 (2007): 313–44.

Griffel, Frank. "The Western Reception of al-Ghazālī's Cosmology from the Middle Ages to the 21st Century." *Dîvân* 16, issue 30 (2011/1): 33–62.
Hadzibegovic, Zalkida. "Compendium of the Science of Astronomy by al-Jaghmīnī Used in Bosnia for Teaching and Learning Planetary Motions." Oral Presentation at the GIREP-EPEC Conference 2007, 1–4. Opatija, Croatia.
Ḥā'irī, 'Abd al-Ḥusayn, ed. *The Majlis Library Catalogue* (*Fihrist-i Kitābkhānah-yi Majlis-i Shūrā-yi Millī, kutub-i khattī*), vol. 10, part 1.Tehran: Majlis-i Shūrā, 1347 H. Sh./1968-69.
Ḥājjī Khalīfa. *Kashf al-ẓunūn 'an asāmī al-kutub wa-'l-funūn.* 2 vols. Edited by Şerefettin Yaltkaya and Kilisli Rifat Bilge. Istanbul, 1941. See also Flügel, Gustavus, *Lexicon.*
HAMA. See Neugebauer, *A History of Ancient Mathematical Astronomy.*
Hartner, Willy. "The Mercury Horoscope of Marcantonio Michiel of Venice: A Study in the History of Renaissance Astrology and Astronomy." In *Vistas in Astronomy*, edited by Arthur Beer, 1:84–138. London: Pergamon Press, 1955. Reprinted in Willy Hartner, *Oriens-Occidens: Ausgewählte Schriften zur Wissenschafts- und Kulturgeschichte. Festschrift zum 60. Geburtstag*, 1:440–95. Hildesheim: George Olms, 1968.
Hartner, Willy. "Asṭurlāb." *Encyclopaedia of Islam*, 2nd ed., 1:722–28. Leiden: E. J. Brill, 1960. Reprinted in Willy Hartner, *Oriens-Occidens: Ausgewählte Schriften zur Wissenschafts- und Kulturgeschichte. Festschrift zum 60. Geburtstag,* 1:312–18. Hildesheim: George Olms, 1968.
Hartner, Willy. "Mediaeval Views on Cosmic Dimensions and Ptolemy's *Kitāb al-Manshūrāt.*" In *Mélanges Alexandre Koyré*, 2 vols., 1:254–822. Paris: Hermann, 1964. Reprinted in Willy Hartner, *Oriens-Occidens: Ausgewählte Schriften zur Wissenschafts- und Kulturgeschichte. Festschrift zum 60. Geburtstag,* 1:319–48. Hildesheim: George Olms, 1968.
Hartner, Willy. "al-Battānī, Abū 'Abd Allāh Muḥammad ibn Jābir ibn Sinān al-Raqqī al-Ḥarrānī al-Ṣābi'." In *Dictionary of Scientific Biography*, edited by Charles Coulston Gillispie, 1:507–16. New York: Charles Scribner's Sons, 1970.
Hashemipour, Behnaz. "Gayhānshenākht: A Cosmological Treatise." In *Sciences, techniques et instruments dans le monde iranien (Xe–XIXe siècle)*, études réunies et présentées par N. Pourjavady et Ž. Vesel, 77–84. Actes du colloque tenu à l'Université de Téhéran (7–9 juin 1998). Tehran, 2004.
Hashemipour, Behnaz. "Buzjānī: Abū al-Wafā' Muḥammad ibn Muḥammad ibn Yaḥyā al-Būzjānī." In *The Biographical Encyclopedia of Astronomers*, edited by Thomas Hockey et al., 1:188–89. New York: Springer-Verlag, 2007.
Hashemipour, Behnaz. "Qaṭṭān al-Marwazī: 'Ayn al-Zamān Abū 'Alī Ḥasan ibn 'Alī Qaṭṭān al-Marwazī." In *The Biographical Encyclopedia of Astronomers*, edited by Thomas Hockey et al., 2:943–44. New York: Springer-Verlag, 2007.
Heiberg, J. L., ed. *Claudii Ptolemaei opera quae exstant omnia.* Vol. 2, *Opera astronomica minora.* Leipzig: Teubner, 1907.
Herrera-Casais, Mónica. "The Nautical Atlases of 'Alī al-Sharafī." *Suhayl* 8 (2008): 223–63.

Hildebrandt, Thomas. "Waren Ǧamāl ad-Dīn al-Afġānī und Muḥammad ʿAbduh Neo-Muʿtaziliten?" *Die Welt des Islams* 42, no. 2 (2002): 207–62.
Hitti, Philip K. et al. *Descriptive Catalogue of the Garrett Collection of Arabic Manuscripts in the Princeton University Library*. Princeton, NJ: Princeton University Press, 1938.
Ibn Abī Uṣaybiʿa. *ʿUyūn al-anbāʾ fī ṭabaqāt al-aṭibbāʾ*. Edited by A. Müller. 2 vols. plus corrections. Cairo: al-Maṭbaʿa al-Wahabiyya, 1299/1882. Königsberg, 1884.
Ibn Abī Uṣaybiʿa. *ʿUyūn al-anbāʾ fī ṭabaqāt al-aṭibbāʾ*. Edited by Nizār Riḍā. Beirut: Dār maktabat al-ḥayāh, 1965.
Ibn al-Akfānī. See Witkam, *De Egyptische Arts.*
Ibn al-Athīr, ʿIzz al-Dīn. *al-Kāmil fī ʾl-Taʾrīkh*. Vol 12. Beirut: Dār ṣādir, 1966.
Ibn al-Athīr, ʿIzz al-Dīn. *The Chronicle of Ibn al-Athīr for the Crusading Period from* al-Kāmil fī'l Taʾrīkh. Part 2, *The Years 541–589/1146–1193: The Age of Nur al-Din and Saladin.* Translated by D. S. Richards. Burlington, VT: Ashgate, 2007.
Ibn al-Athīr, ʿIzz al-Dīn. *The Chronicle of Ibn al-Athīr for the Crusading Period from* al-Kāmil fī'l Taʾrīkh. Part 3, *The Years 589–629/1193–1231: The Ayyūbids after Saladin and the Mongol Menace.* Translated by D. S. Richards. Aldershot, England; Burlington, VT: Ashgate, 2008.
Ibn al-Haytham. *Al-Shukūk ʿalā Baṭlamyūs*. Edition by A. I. Sabra and N. Shehaby. Cairo: Dār al-kutub, 1971.
Ibn Khaldūn. *The Muqaddimah: An Introduction to History*. Translated from the Arabic by Franz Rosenthal. 3 vols. Princeton, NJ: Princeton University Press, 1967.
Ibn al-Nadīm, Muḥammad ibn Isḥāq. *Kitāb al-Fihrist*. 3rd ed. Edited by Riḍā Tajaddud. Beirut: Dār al-Masīra, 1988.Translated as *The Fihrist of al-Nadīm: A Tenth-Century Survey of Muslim Culture* by Bayard Dodge. 2 vols. New York: Columbia University Press, 1970.
Ibn al-Nadīm. *The Fihrist of al-Nadīm. Abul-Farağ Muḥammad ibn Isḥáq, composed at 377 AH.* A Critical Edition by Ayman Fuʾād Sayyid. 2 vols. in 4. London: Al-Furqān Islamic Heritage Foundation, 2009.
İhsanoğlu, Ekmeleddin, ed. *History of the Ottoman State, Society & Civilisation*. 2 vols. Istanbul: IRCICA, 2001 (vol. 1) and 2002 (vol. 2).
Inalcık, Halil. *The Ottoman Empire: The Classical Age 1300–1600*. Translated by Norman Itzkowitz and Colin Imber. London: Weidenfeld and Nicolson, 1973.
Iskandar, A. Z. *A Catalogue of Arabic Manuscripts on Medicine and Science in the Wellcome Historical Medical Library*. London: The Wellcome Historical Medical Library, 1967.
Iskandar, A. Z. "A Study of Al-Samarqandī's Medical Writings." In *Le Muséon Revue d'Études Orientales*, vol. 85, 451–79. Louvain: Orientaliste, 1972.
Islamic Medical Manuscripts at the National Library of Medicine website. Project of the History of Medicine Division of the National Library of Medicine. Text by Emilie Savage-Smith. The Oriental Institute, Oxford University. http://www.nlm.nih.gov/hmd/arabic/bioJ.html
İzgi, Cevat. *Osmanlı Medreselerinde İlim: Riyazî ilimler.* 2 vols. Istanbul: İz, 1997.

al-Jaghmīnī. *Talkhīṣ kitāb Ūqlīdis*. Edited by Ḥusaynī al-Ishkavarī. Qum: Majma'-i Dhakhā'ir-i Islāmī, 1384/2006.

al-Jaghmīnī. *Qānūnča fī al-ṭibb.* Edition and Persian translation by Ismā'īl Nāẓim. Tehran: Tehran University of Medical Sciences, 2012.

Jawāmi'. See Farghānī, *Jawāmi' 'ilm al-nujūm.*

Jones, Alexander, ed. *Ptolemy in Perspective: Use and Criticism of His Work from Antiquity to the Nineteenth Century*. Archimedes 23: New Studies in the History and Philosophy of Science and Technology. Dordrecht; New York: Springer, 2010.

Junayd ibn Maḥmūd Junayd Shīrāzī. *Tazkirah-yi Hazār Mazār: Tarjumah-yi Shadd al-Izār: Mazārāt-i Shīrāz.* Shiraz: Kitābkhānah-yi Aḥmadī, 1364 H. Sh./1985-86.

Juwaynī, 'Alā' al-Dīn 'Aṭā Malik. *Tārīkh-i Jahān-gushā*. 3 vols. Leiden: E. J. Brill, 1912–1937.

Juwaynī, 'Alā' al-Dīn 'Aṭā Malik. *The History of the World Conqueror by 'Ala-ad-Din 'Ata Malik Juvaini. Translated from the text of Mirza Muhammad Qazvini.* Translated by John Andrew Boyle. 2 vols. Manchester: Manchester University Press, 1958.

Jūzjānī, Minhāj Sirāj. *Ṭabaqāt-i Nāṣirī: A General History of the Muhammadan Dynasties of Asia, including Hindūstān, from A.H. 194 (810 A.D.) to A.H. 658 (1260 A.D.) and the Irruption of the Infidel Mughals into Islām.* 2 vols. By Maulānā, Minhāj-ud-Dīn, Abū-'Umar-i-'Usmān. Translated from original Persian manuscripts by Major H. G. Raverty. London: Gilbert and Rivington, 1881.

Kaḥḥāla, 'Umar Riḍā. *Mu'jam al-mu'allifīn: tarājim muṣannifī al-kutub al-'Arabiyya*. 15 vols. Beirut: Dār iḥyā' al-turāth al-'arabī, 1980.

Kātib Çelebī. See Ḥājjī Khalīfa, *Kashf al-ẓunūn.*

Kennedy, Edward S. "A Survey of Islamic Astronomical Tables." *Transactions of the American Philosophical Society*, n.s., 46, pt. 2 (1956): 121–77.

Kennedy, Edward S. "The Sasanian Astronomical Handbook *Zīj-i Shāh* and the Astrological Doctrine of 'Transit' (*Mamarr*)." *Journal of the American Oriental Society* 78, no. 4 (Oct.-Dec., 1958): 246–62. Reprinted in E. S. Kennedy, et al. *Studies in the Islamic Exact Sciences,* edited by David A. King and Mary Helen Kennedy, 164–84. Beirut: American University of Beirut, 1983.

Kennedy, Edward S. *The Planetary Equatorium of Jamshīd Ghiyāth al-Dīn al-Kāshī.* Princeton, NJ: Princeton University Press, 1960.

Kennedy, Edward S. "Al-Bīrūnī (or Bērūnī), Abū Rayḥān (or Abu'l-Rayḥān) Muḥammad Ibn Aḥmad." In *Dictionary of Scientific Biography*, edited by Charles Coulston Gillispie, 2:147–58. New York: Charles Scribner's Sons, 1970.

Kennedy, Edward S. "Al-Bīrūnī on the Muslim Times of Prayer." In *The Scholar and the Saint: Studies in Commemoration of Abū'l-Rayḥān al-Bīrūnī and Jalāl al-Dīn al-Rūmī,* 83–94. New York: New York University Press, 1975. Reprinted in E. S. Kennedy, et al. *Studies in the Islamic Exact Sciences,* edited by David A. King and Mary Helen Kennedy, 299–310. Beirut: American University of Beirut, 1983.

Kennedy, Edward S., trans. and comm. *The Exhaustive Treatise on Shadows by Abu al-Rayḥān Muḥammad b. Aḥmad al-Bīrūnī, Translation & Commentary.* 2 vols. Aleppo: Institute for the History of Science, 1976.

Kheirandish, Elaheh. "Qusṭā ibn Lūqā al-Baʿlabakkī." In *The Biographical Encyclopedia of Astronomers*, edited by Thomas Hockey et al., 2:948–49. New York: Springer-Verlag, 2007.

Kholeif, Fathalla. *A Study on Fakhr al-Dīn al-Rāzī and His Controversies in Transoxiana.* Beirut: Dar El-Machreq Editeurs, 1984.

al-Khwārizmī, Abū ʿAbd Allāh Muḥammad. *Kitāb Mafātīḥ al-ʿulūm*. Edition by G. van Vloten. Leiden: E. J. Brill, 1895.

King, David A. *A Catalogue of the Scientific Manuscripts in the Egyptian National Library*. Vol. 1, *A Critical Hand List of the Scientific Collections—Indexes of Copyists and Owners*. Cairo: General Egyptian Book Organization, 1981.

King, David A. "The Astronomy of the Mamluks." *Isis* 74, no. 4 (Dec., 1983): 531–55. Reprinted in David A. King, *Islamic Mathematical Astronomy*, III.

King, David A. *A Catalogue of the Scientific Manuscripts in the Egyptian National Library*. Vol. 2, *Descriptive Catalogue Arranged Chronologically According to Subjects—Indexes of Authors and Titles*. Cairo: General Egyptian Book Organization, 1986.

King, David A. *Islamic Mathematical Astronomy*. Collected Studies, Volume 231. London: Variorum Reprints, 1986.

King, David A. "Ḳibla." *Encyclopaedia of Islam*, 2nd ed., 5:83–88. Leiden: E. J. Brill, 1986.

King, David A. *A Survey of the Scientific Manuscripts in the Egyptian National Library*. American Research Center in Egypt/ Catalogs, vol. 5. Winona Lake, Indiana: Eisenbrauns, 1986. [Based on King, *Catalogue of the Scientific Manuscripts,* vol. 1 (1981) and vol. 2 (1986).]

King, David A. *Astronomy in the Service of Islam.* Aldershot: Ashgate Variorum Reprints, 1993.

King, David A. "On the Role of the Muezzin and the Muwaqqit in Medieval Islamic Society." In *Tradition, Transmission, Transformation: Proceedings of Two Conferences on Premodern Science Held at the University of Oklahoma*, edited by F. Jamil Ragep and Sally P. Ragep with the assistance of Steven Livesey, 285–346. Leiden: E. J. Brill, 1996.

King, David A. *World-Maps for Finding the Direction and Distance to Mecca: Innovation and Tradition in Islamic Science*. London: Al-Furqān Islamic Heritage Foundation; Leiden: E. J. Brill, 1999. Reprint, Leiden: Koninklijke Brill NV, 2014.

King, David A. and Julio Samsó. "Astronomical Handbooks and Tables from the Islamic World [750–1900]: an Interim Report." *Suhayl* 2 (2001): 9–105.

Klausner, Carla L. *The Seljuk Vezirate: A Study of Civil Administration 1055–1194.* Harvard Middle Eastern Monographs. Cambridge, MA: Harvard University Press, 1973.

Knorr, Wilbur. *Textual Studies in Ancient and Medieval Geometry*. Boston: Birkhäuser, 1989.

Krause, Max. "Stambuler Handschriften islamischer Mathematiker." *Quellen und Studien zur Geschichte der Mathematik, Astronomie und Physik.* Abteilung B, Studien 3 (1936): 437–532. Reprinted in *Islamic Mathematics and Astronomy*, edited by Fuat Sezgin, vol. 83, 755–56. Frankfurt am Main, 1998.

Kunitzsch, Paul. "al-Nudjūm." *Encyclopaedia of Islam*, 2nd ed., 8:97–105. Leiden: E. J. Brill, 1995.

Kunitzsch, Paul. "Ṣūfī: Abū al-Ḥusayn ʿAbd al-Raḥmān ibn ʿUmar al- Ṣūfī." In *The Biographical Encyclopedia of Astronomers*, edited by Thomas Hockey et al., 2:1110. New York: Springer-Verlag, 2007.

Kut, Günay and Nimet Bayraktar. *Yazma Eserlerde Vakıf Mühürleri.* Ankara: Kültür ve Turizm Bakanlığı, 1984.

KZ. See Ḥājjī Khalīfa. *Kashf al-ẓunūn.*

Lane, Edward W. *An Arabic-English Lexicon.* 8 parts. London: Williams and Norgate, 1863–93. Facsimile reprint by Libraire du Liban in Beirut, 1968.

Langermann, Y. Tzvi. "The Book of Bodies and Distances of Ḥabash al-Ḥāsib." *Centaurus* 28, no. 2 (1985): 108–28.

Langermann, Y. Tzvi. *Ibn al-Haytham's "On the Configuration of the World."* New York: Garland (Harvard Dissertations in the History of Science), 1990.

Langermann, Y. Tzvi. "Ibn al-Haytham: Abū ʿAlī al-Ḥasan ibn al-Ḥasan." In *The Biographical Encyclopedia of Astronomers*, edited by Thomas Hockey et al., 1:556–57. New York: Springer-Verlag, 2007.

Langermann, Y. Tzvi. "Kharaqī: Shams al-Dīn Abū Bakr Muḥammad ibn Aḥmad al-Kharaqī [al-Khiraqī]." In *The Biographical Encyclopedia of Astronomers*, edited by Thomas Hockey et al., 1:627. New York: Springer-Verlag, 2007.

Lazard, Gilbert. *La langue des plus anciens monuments de la prose persane.* Études linguistiques. Paris: C. Klincksieck, 1963.

Leclerc, Lucien. *Histoire de la médecine arabe: exposé complet des traductions du grec; Les sciences en Orient, leur transmission à l'Occident par les traductions latines.* 2 vols. Paris: E. Leroux, 1876.

Lloyd, G. E. R. *Methods and Problems in Greek Science: [Selected Papers].* Cambridge: Cambridge University Press, 1991.

MAMS2. See Rosenfeld and İhsanoğlu, *Mathematicians, Astronomers, and Other Scholars of Islamic Civilization and Their Works (7th–19th c.).*

Marquet, Yves. "Ikhwān al-Ṣafāʾ." *Encyclopaedia of Islam*, 2nd ed., 3:1071–76. Leiden: E. J. Brill, 1971.

Matvievskaya, G. P. and B. A. Rosenfeld, *Matematiki i astronomi musulmanskogo srednevekovya i ikh trudi (VIII–XVII vv.)* [Mathematicians and Astronomers of the Muslim Middle Ages and Their Works (VIII–XVII centuries)]. 3 vols. Moscow: Nauka, 1983.

McCluskey, Stephen C. *Astronomies and Cultures in Early Medieval Europe.* Cambridge: Cambridge University Press, 1998.

Mimura, Taro. "The Arabic Original of (ps.) Māshāʾallāh's *Liber de orbe*: Its Date and Authorship." *The British Journal for the History of Science* 48 (2015): 321–52.

Minorsky, V. and rev. by C. E. Bosworth. "Nasā, Nisā." *Encyclopaedia of Islam*, 2nd ed., 7:966–67. Leiden: E. J. Brill, 1993.

Mittenhuber, Florian. "The Tradition of Texts and Maps in Ptolemy's Geography." In *Ptolemy in Perspective: Use and Criticism of His Work from Antiquity to the Nineteenth Century*, edited by Alexander Jones, 95–119. Archimedes 23: New Studies in the History and Philosophy of Science and Technology. Dordrecht; New York: Springer, 2010.

Mogenet, Joseph. *L'introduction à l'Almageste.* Mémoires de la classe des lettres, Collection in-8°, 2e Série, vol. 51, fasc. 2. Bruxelles: Palais des Académies, 1956.

Montgomery, James E. "Ibn Faḍlān and the Rūsiyyah." *Journal of Arabic and Islamic Studies* 3 (2000): 1–25.

Morelon, Régis. *Thābit ibn Qurra: Œuvres d'astronomie.* Paris: Les Belles Lettres, 1987.

Morelon, Régis. "Ṯābit b. Qurra and Arab Astronomy in the 9th Century." *Arabic Sciences and Philosophy* 4, no. 1 (1994): 111–39.

Morelon, Régis. "Eastern Arabic Astronomy between the Eighth and the Eleventh Centuries." In *Encyclopedia of the History of Arabic Science*, vol. 1, *Astronomy—Theoretical and Applied*, edited by Roshdi Rashed, 20–57. London: Routledge, 1996.

Morelon, Régis. "General Survey of Arabic Astronomy." In *Encyclopedia of the History of Arabic Science*, vol. 1, *Astronomy—Theoretical and Applied*, edited by Roshdi Rashed, 1–19. London: Routledge, 1996.

Morrison, Robert. "The Reception of Early-Modern European Astronomy by Ottoman Religious Scholars." *Archivum Ottomanicum* 21 (2003): 187–95.

Morrison, Robert. "The Role of Oral Transmission for Astronomy Among Romaniot Jews." In *Texts in Transit in the Medieval Mediterranean*, edited by Y. Tzvi Langermann and Robert G. Morrison, 10–28. University Park, PA: Penn State University Press, 2016.

Murphy, Jane Holt. "Improving the Mind and Delighting the Spirit: Jabarti and the Sciences in Eighteenth-Century Ottoman Cairo." Ph.D. dissertation, Princeton University, 2006.

Mūsā, ʿAlī. *Majisṭī Abī al-Wafāʾ al-Būzjānī.* Beirut: Markaz Dirāsāt al-Waḥda al-ʿArabiyya, 2010.

Nallino, Carlo A. "Zu Ǵaġmînî's Astronomie." *Zeitschrift der Deutschen Morgenländischen Gesellschaft* 48 (1894): 120–22. Reprinted in *Islamic Mathematics and Astronomy*, edited by Fuat Sezgin, vol. 77. Miscellaneous Texts and Studies on Islamic Mathematics and Astronomy, II, 302–4. Frankfurt am Main, 1998.

Nallino, Carlo A. *Al-Battānī Sive Albatenii Opus Astronomicum.* 3 vols. Milan: Pubblicazioni Del Reale Osservatorio di Brera, 1899–1907 (vol. 1 [1903], vol. 2 [1907], and vol. 3 [1899]).

Nallino, Carlo A. "Sun, Moon, and Stars (Muhammadan)." In *Encyclopædia of Religion and Ethics*, edited by James Hastings, with the assistance of John A. Selbie and Louis H. Gray, 12:88–101. Edinburgh: T. & T. Clark; New York: Charles Scribner's Sons, 1921.

Nallino, Carlo A. "Al-Battānī." *Encyclopaedia of Islam*, 2nd ed., 1:1104–5. Leiden: E. J. Brill, 1960.

al-Nasawī, Muḥammad ibn Aḥmad. *Historie du sultan Djelal ed-Din Mankobirti, prince du Kharezm par Mohammed en-Nesawi*. Translated into French from the Arabic by Octave Houdas. Paris: E. Leroux, 1895.

al-Nasawī, Muḥammad ibn Aḥmad. *Sīrat al-Sulṭān Jalāl al-Dīn Mankubirtī li-Muḥammad ibn Aḥmad al-Nasawī*. Edited by Ḥāfiẓ Aḥmad Ḥamdī. Cairo: Dār al-fikr al-ʿarabī, 1953.

Neugebauer, Otto. *A History of Ancient Mathematical Astronomy*. 3 parts. Berlin; New York: Springer-Verlag, 1975.

Niazi, Kaveh Farzad. "A Comparative Study of Quṭb al-Dīn Shīrāzī's Texts and Models on the Configuration of the Heavens." Ph.D. dissertation, Columbia University, 2011. ProQuest (3479090).

Niazi, Kaveh. *Quṭb al-Dīn Shīrāzī and the Configuration of the Heavens: A Comparison of Texts and Models*. Archimedes 35: New Studies in the History and Philosophy of Science and Technology. Dordrecht: Springer, 2014.

OALT: Osmanlı Astronomi Literatürü Tarihi (History of Astronomy Literature during the Ottoman Period). Edited by Ramazan Şeşen, Cevat İzgi, Cemil Akpinar, İhsan Fazlıoğlu. General Editor Ekmeleddin İhsanoğlu. 2 vols. Istanbul: IRCICA, 1997.

Osmanlı Matematik Literatürü Tarihi. Edited by Ramazan Şeşen and Cevat İzgi. General Editor Ekmeleddin İhsanoğlu. 2 vols. Istanbul: IRCICA, 1999.

Palmer, Edward Henry. *A Descriptive Catalogue of the Arabic, Persian and Turkish Manuscripts in the Library of Trinity College, Cambridge*. Cambridge: Deighton Bell and Co., 1870.

Pedersen, Olaf. *A Survey of the Almagest*. Odense: Odense University Press, 1974.

Pedersen, Olaf. "The *Corpus Astronomicum* and the Traditions of Medieval Latin Astronomy: A Tentative Interpretation." In *Colloquia Copernicana III*, edited by Owen Gingerich and Jerzy Dobrzycki, 57–96. Wrocław: Ossolineum, 1975.

Pellat, Charles. "Anwāʾ." *Encyclopaedia of Islam*, 2nd ed., 1:523–24. Leiden: E. J. Brill, 1960.

Pertsch, Wilhelm. *Die orientalischen Handschriften der Herzoglichen Bibliothek zu Gotha*, vol. 3, part 3. Gotha: Friedrich Andreas Perthes, 1881.

Pingree, David. "The Fragments of the Works of Yaʿqūb Ibn Ṭāriq." *Journal of Near Eastern Studies* 27, no. 2 (Apr., 1968): 97–125.

Pingree, David. "*ʿIlm al-hayʾa*." *Encyclopaedia of Islam*, 2nd ed., 3:1135–38. Leiden: E. J. Brill, 1971.

Pingree, David. "The Greek Influence on Early Islamic Mathematical Astronomy." *Journal of the American Oriental Society* 93, no. 1 (Jan.-Mar., 1973): 32–43.

Pingree, David. Māshāʾ allāh." In *Dictionary of Scientific Biography*, edited by Charles Coulston Gillispie, 9:159–62. New York: Charles Scribner's Sons, 1974.

Pingree, David. "Māshāʾ allāh: Some Sasanian and Syriac Sources." In *Essays on Islamic Philosophy and Science,* edited by George F. Hourani, 5–14. Albany: State University of New York Press, 1975.

Pingree, David. "Hellenophilia versus the History of Science." *Isis* 83, no. 4 (Dec., 1992): 554–63.

PL. See Storey, *Persian Literature.*

Plofker, Kim. "Ya'qūb ibn Ṭāriq." In *The Biographical Encyclopedia of Astronomers*, edited by Thomas Hockey et al., 2:1250–51. New York: Springer-Verlag, 2007.

Proclus. *Procli Diadochi Hypotyposis Astronomicarum Positionum*. Edited by Simon Grynaeus. Baseleae: Apud Ioannem Vualder, 1540.

Proclus. *Procli Diadochi Hypotyposis Astronomicarum Positionum*. Edited by Carolus Manitius. Leipzig: B. G. Teubner, 1909. Reprint, Stuttgart: B. G. Teubner, 1974.

Ptolemy. See Toomer, G. J. *Ptolemy's Almagest*.

Qāsimlū, Farīd. "Chaghmīnī." *Encyclopaedia of the World of Islam*, 12:61–63. Tehran, 1387 H. Sh./2008-9.

Qaṭṭān, al-Ḥasan ibn 'Alī. *Gayhān Shinākht*. Chāp-i 1.Tehran: Kitābkhānah, Mūzah va Markaz-i Asnād-i Majlis-i Shūrā-yī Islāmī, 2012.

al-Qifṭī, Jamāl al-Dīn Abū al-Ḥasan 'Alī ibn Yūsuf. *Ta'rīkh al-ḥukamā'*. Edited by J. Lippert. Leipzig: Dieterich'sche Verlagsbuchhandlung, 1903.

Qurbānī, Abū al-Qāsim. *Zindagī-nāmah-yi rīyāḍī-dānān dawrah-yi Islāmī*. Tehran: Markaz-i Nashr-i Dānishgāhī, 1365/1986.

Ragep, F. Jamil. "Duhem, the Arabs, and the History of Cosmology." *Synthese* 83 (1990): 201–14.

Ragep, F. Jamil. *Naṣīr al-Dīn al-Ṭūsī's* Memoir on Astronomy (*al-Tadhkira fī 'ilm al-hay'a*). Edition, Translation, Commentary and Introduction. 2 vols. Sources in the History of Mathematics and Physical Sciences 12. New York: Springer-Verlag, 1993.

Ragep, F. Jamil. "Al-Battānī, Cosmology, and the Early History of Trepidation in Islam." In *From Baghdad to Barcelona. Essays on the History of the Exact Sciences in Honour of Prof. Juan Vernet*, edited by Josep Casulleras and Julio Samsó, 267–98. Barcelona: Universidad de Barcelona. Facultad de Filología, 1996.

Ragep, F. Jamil. "*Hay'a*." In *Encyclopaedia of the History of Science, Technology, and Medicine in Non-Western Cultures,* edited by Helaine Selin, 395–97. Dordrecht: Kluwer Academic Publishers, 1997. 2nd ed., Berlin: Springer, 2008, 1061.

Ragep, F. Jamil. "Freeing Astronomy from Philosophy: An Aspect of Islamic Influence on Science." *Osiris,* 2nd series, 16 (2001): 49–71.

Ragep, F. Jamil. "Ibn al-Haytham and Eudoxus: The Revival of Homocentric Modeling in Islam." In *Studies in the History of the Exact Sciences in Honour of David Pingree*, edited by Charles Burnett, Jan P. Hogendijk, Kim Plofker and Michio Yano, 786–809. Leiden: E. J. Brill, 2004.

Ragep, F. Jamil. "Ḳāḍīzāde Rūmī." *Encyclopaedia of Islam*. 2nd ed., 12:502. Leiden: E. J. Brill, 2004.

Ragep, F. Jamil. "'Alī Qushjī and Regiomontanus: Eccentric Transformations and Copernican Revolutions." *Journal for the History of Astronomy* 36, no. 4 (2005): 359–71.

Ragep, F. Jamil. "On Dating Jaghmīnī and His *Mulakhkhaṣ*." In *Essays in Honour of Ekmeleddin İhsanoğlu*, edited by Mustafa Kaçar and Zeynep Durukal, 461–66. Istanbul: IRCICA, 2006.

Ragep, F. Jamil. "Qāḍīzāde al-Rūmī: Ṣalāḥ al-Dīn Mūsā ibn Muḥammad ibn Maḥmūd al-Rūmī." In *The Biographical Encyclopedia of Astronomers*, edited by Thomas Hockey et al., 2:942. New York: Springer-Verlag, 2007.

Ragep, F. Jamil. "When Did Islamic Science Die (and Who Cares)?" *Newsletter of the British Society for the History of Science* 85 (Feb., 2008): 1–3.

Ragep, F. Jamil. "Astronomy in the Fanārī-Circle: The Critical Background for Qāḍīzāde al-Rūmī and the Samarqand School." In *Uluslararası Molla Fenârî Sempozyumu (4–6 Aralık 2009 Bursa)* (International symposium on Molla Fanārī, 4–6 Dec. 2009, Bursa), edited by Tevfik Yücedoğru, Orhan Koloğlu, U. Murat Kılavuz, and Kadir Gömbeyaz, 165–76. Bursa: Bursa Büyükşehir Belediyesi, 2010.

Ragep, F. Jamil. "Islamic Reactions to Ptolemy's Imprecisions." In *Ptolemy in Perspective: Use and Criticism of His Work from Antiquity to the Nineteenth Century*, edited by Alexander Jones, 121–34. Archimedes 23: New Studies in the History and Philosophy of Science and Technology. Dordrecht; New York: Springer, 2010.

Ragep, F. Jamil. "The *Khilāṣ kayfiyyat tarkīb al-aflāk* of al-Jūzjānī: A Preliminary Description of Its Avicennian Themes." In *Avicenna and his Legacy: A Golden Age of Science and Philosophy,* edited by Y. Tzvi Langermann, 303–8. Turnhout: Brepols, 2010.

Ragep, F. Jamil. "Shīrāzī's *Nihāyat al-Idrāk*: Introduction and Conclusion." *Tarikh-e Elm* (Tehran, Iran) 11 (2013): 41–57.

Ragep, F. Jamil. "Astronomy." *Encyclopaedia of Islam, THREE.* Edited by Gudrun Krämer, Denis Matringe, John Nawas, Everett Rowson. Brill Online, 2014. Reference. McGill University. 03 Mar., 2014 http://referenceworks.brillonline.com/entries/encyclopaedia-of-islam-3/astronomy-COM_22652.

Ragep, *Tadhkira.* See Ragep, F. Jamil. *Naṣīr al-Dīn al-Ṭūsī's* Memoir on Astronomy.

Ragep, F. Jamil and Sally P. Ragep with the assistance of Steven Livesey. *Tradition, Transmission, Transformation: Proceedings of Two Conferences on Premodern Science Held at the University of Oklahoma*. Leiden: E. J. Brill, 1996.

Ragep F. Jamil and Sally P. Ragep. "The Astronomical and Cosmological Works of Ibn Sīnā: Some Preliminary Remarks." In *Sciences, techniques et instruments dans le monde iranien (Xe–XIXe siècle),* études réunies et présentées par N. Pourjavady et Ž. Vesel, 3–15. Actes du colloque tenu à l'Université de Téhéran (7–9 juin 1998). Tehran, 2004.

Ragep, F. Jamil and Taro Mimura, eds. and trans. *Epistles of the Brethren of Purity. On* Astronomia*: An Arabic Critical Edition and English Translation of Epistle 3*. Oxford: Oxford University Press in association with The Institute of Ismaili Studies, 2015.

Ragep, Sally P. "Jaghmīnī: Sharaf al-Dīn Maḥmūd ibn Muḥammad ibn ʿUmar al-Jaghmīnī al-Khwārizmī." In *The Biographical Encyclopedia of Astronomers*, edited by Thomas Hockey et al., 1:584–85. New York: Springer-Verlag, 2007.

Ragep, Sally P. "al-Jaghmīnī." In *New Dictionary of Scientific Biography*, edited by Noretta Koertge, 4:20–22. Detroit: Charles Scribner's Sons/Thomson Gale, 2008.

Ragep, Sally P. "Maḥmūd ibn Muḥammad ibn ʿUmar al-Jaghmīnī's *al-Mulakhkhaṣ fī al-hayʾa al-basīṭa*: An Edition, Translation, and Study." Ph.D. dissertation, McGill University, 2015.

Ragep, Sally P. "The Teaching of Theoretical Astronomy in Pre-modern Islam: Looking Beyond Individual Initiatives." In *Schüler und Meister*, edited by Andreas Speer and Thomas Jeschke, 557–68. *Miscellanea Mediaevalia* 39. Berlin: De Gruyter, 2016.

Rashed, Roshdi. *Entre arithmétique et algèbre: recherches sur l'histoire des mathématiques arabes*. Paris: Les Belles Lettres, 1984.

Rashed, Roshdi, ed. *Encyclopedia of the History of Arabic Science*. 3 vols. London: Routledge, 1996.

Rashed, Roshdi. "The Configuration of the Universe: A Book by al-Ḥasan ibn al-Haytham?" *Revue d'histoire des sciences* 60 (2007/1): 47–63.

Rashed, Roshdi. *Founding Figures and Commentators in Arabic Mathematics: A History of Arabic Science and Mathematics Volume 1*. Edited by Nader El-Bizri. London: Routledge; Beirut: Center for Arab Unity Studies, 2012.

Renaud, H. P. J. "Additions et corrections à Suter "Die Mathematiker und Astronomen der Araber." *Isis* 18, no. 1 (Jul., 1932): 166–83.

Richter-Bernburg, Lutz. "Medical and Veterinary Sciences. Part One: Medicine, Pharmacology and Veterinary Science in Islamic Eastern Iran and Central Asia." In *History of Civilizations of Central Asia,* vol. 4, *The Age of Achievement: A.D. 750 to the End of the Fifteenth Century. Part Two: The Achievements*, edited by C. Edmund Bosworth and M. S. Asimov, 299–317. Paris: UNESCO Publ., 2000.

Richter-Bernburg, Lutz. "Jağmini, Maḥmud." *Encylopædia Iranica*, vol. 14, fasc. 4, 373; originally published Dec. 15, 2008; online version last updated Apr. 10, 2012. http://www.iranicaonline.org/articles/jagmini-mahmud

Riyazî ilimler. See İzgi, Cevat. *Osmanlı Medreselerinde İlim: Riyazî ilimler.*

Roper, Geoffrey, gen. ed. *The World Survey of Islamic Manuscripts*, 4 vols. London: Al-Furqān Islamic Heritage Foundation, 1993.

Rosenfeld, Boris A. and Ekmeleddin İhsanoğlu. *Mathematicians, Astronomers, and Other Scholars of Islamic Civilization and Their Works (7th–19th c.).* Istanbul: IRCICA, 2003.

Rosenthal, Franz. "The Technique and Approach of Muslim Scholarship." *Analecta Orientalia* 24 (1947): 1–74.

Rosenthal, Franz. "Al-Kindī and Ptolemy." *Studi Orientalistici in Onore di Giorgio Levi Della Vida*. Pubblicazioni dell'Istituto per l'Oriente, 52, 2:436–56. Rome, 1956.

Rosenthal, Franz. *The Classical Heritage in Islam*. Berkeley: University of California Press, 1975.

Rudloff, G. and Prof. Dr. Ad. Hochheim. "Die Astronomie des Maḥmûd ibn Muḥammed ibn 'Omar al-Ǵagmînî." *Zeitschrift der Deutschen Morgenländischen Gesellschaft* 47 (1893): 213–75. Reprinted in *Islamic Mathematics and Astronomy*, edited by Fuat Sezgin, vol. 77. Miscellaneous Texts and Studies on Islamic Mathematics and Astronomy, II, 237–301. Frankfurt am Main, 1998.

Rūḥullāhī, Ḥusayn. "Dashtakī, Ghiyāth al-Dīn Manṣūr, ārā va āthār-i 'ilmī." *The Encyclopaedia of the World of Islam,* 17: title 7 (*Ḥāshiya Sharḥ al-Mulakhkhaṣ*). Tehran: Encyclopaedia Islamica Foundation, 1391 H. Sh./2013.
Sabra, A. I. "al-Farghānī, Abu'l-'Abbās Aḥmad ibn Muḥammad ibn Kathīr." In *Dictionary of Scientific Biography*, edited by Charles Coulston Gillispie, 4:541–45. New York: Charles Scribner's Sons, 1972.
Sabra, A. I. "Ibn al-Haytham, Abū 'Alī al-Ḥasan ibn al-Ḥasan." In *Dictionary of Scientific Biography*, edited by Charles Coulston Gillispie, 6:189–210. New York: Charles Scribner's Sons, 1972.
Sabra, A. I. "The Andalusian Revolt Against Ptolemaic Astronomy: Averroes and al-Biṭrūjī." In *Transformation and Tradition in the Sciences: Essays in Honor of I. Bernard Cohen*, edited by Everett Mendelsohn, 133–53. Cambridge: Cambridge University Press, 1984. Reprinted in A. I. Sabra, *Optics, Astronomy and Logic: Studies in Arabic Science and Philosophy*, XV. Aldershot: Ashgate Variorum Reprints, 1994.
Sabra, A. I. "Some Remarks on Al-kindi as a Founder of Arabic Science and Philosophy." In *Dr. Mohammad Abdulhadi Abu Ridah Festschrift*, edited by Abdullah O. Al-Omar, 601–7. Kuwait: Kuwait University Press, 1993.
Sabra, A. I. "Configuring the Universe. Aporetic, Problem Solving, and Kinematic Modeling as Themes of Arabic Astronomy." *Perspectives on Science* 6, no. 3 (1998): 288–330.
Sabra, A. I. "One Ibn al-Haytham or Two? An Exercise in Reading the Bio-bibliographical Sources." *Zeitschrift für Geschichte der Arabisch-Islamischen Wissenschaften* 12 (1998): 1–50.
Sabra, A. I. "Reply to Saliba." *Perspectives on Science* 8, no. 4 (2000): 342–45.
Sabra, A. I. "One Ibn al-Haytham or Two? Conclusion." *Zeitschrift für Geschichte der Arabisch-Islamischen Wissenschaften* 15 (2002/2003): 95–108.
Sabra, A. I. "Ibn al-Haytham: Brief Life of an Arab Mathematician: Died circa 1040." *Harvard Magazine*, 54. Sept.-Oct., 2003.
Sagan, Carl. *Cosmos*. New York: Random House, 1980.
Saliba, George. "The First Non-Ptolemaic Astronomy at the Maraghah School." *Isis* 70, no. 4 (Dec., 1979): 571–76. Reprinted in George Saliba, *A History of Arabic Astronomy: Planetary Theories during the Golden Age of Islam*, 113–18. New York: New York University Press, 1994.
Saliba, George. "Astrology/Astronomy, Islamic." In *Dictionary of the Middle Ages*, edited by Joseph R. Strayer, 1:616–24. New York: Scribner, 1982. Reprinted in George Saliba, *A History of Arabic Astronomy: Planetary Theories during the Golden Age of Islam*, 66–81. New York: New York University Press, 1994.
Saliba, George. "The Development of Astronomy in Medieval Islamic Society." *Arab Studies Quarterly*, 4, no. 3 (Summer 1982): 211–25. Reprinted in George Saliba, *A History of Arabic Astronomy: Planetary Theories during the Golden Age of Islam*, 51–65. New York: New York University Press, 1994.
Saliba, George. "The Role of the Astrologer in Medieval Islamic Society." *Bulletin d'Études Orientales* 44 (1992): 45–67.

Saliba, George. "Early Arabic Critique of Ptolemaic Cosmology: A Ninth-Century Text on the Motion of the Celestial Spheres." *Journal for the History of Astronomy* 25, no. 2 (1994): 115–41.
Saliba, George. *A History of Arabic Astronomy: Planetary Theories during the Golden Age of Islam*. New York: New York University Press, 1994.
Saliba, George. "Persian Scientists in the Islamic World: Astronomy from Maragha to Samarqand." In *The Persian Presence in the Islamic World*, edited by Richard G. Hovannisian and Georges Sabagh, 126–46. Cambridge: Cambridge University Press, 1998.
Saliba, George. "Arabic versus Greek Astronomy: A Debate over the Foundations of Science." *Perspectives on Science* 8, no. 4 (2000): 328–41.
Saliba, George. "Islamic Astronomy in Context: Attacks on Astrology and the Rise of the *Hay'a* Tradition." *Bulletin of the Royal Institute for Inter-Faith Studies* 2, no. 1 (Spring/Summer 2002): 25–46.
Saliba, George. *Islamic Science and the Making of the European Renaissance*. Cambridge, MA: MIT Press, 2007.
Samsó, Julio. "Māshā' Allāh." *Encyclopaedia of Islam*. 2nd ed., 6:710–12. Leiden: E. J. Brill, 1991.
Samsó, Julio. "Al-Battānī." In *Encyclopaedia of the History of Science, Technology, and Medicine in Non-Western Cultures*, edited by Helaine Selin, 91. Dordrecht: Kluwer Academic Publishers, 1997.
Samsó, Julio. ""Battani, Al-." In *Medieval Science, Technology, and Medicine: An Encyclopedia*, edited by Thomas F. Glick, Steven J. Livesey, Faith Wallis, 79–80. New York: Routledge, 2005.
Savadi, Fateme and Sajjad Nikfahm-Khubravan. "The Mean Motion of the Planets in *Zīj-i Īlkhānī* and the Criticisms Regarding It." In *Ustād-i Bashar: Essays on the Life, Times, Philosophy and Scientific Achievements of Khwājah Naṣīr al-Dīn Ṭūsī*, edited by H. Masoumi Hamedani and M. Javad Anvari, 363–470. Tehran: Miras-e Maktoob, 2012.
Sayılı, Aydın. "The Institutions of Science and Learning in the Moslem World." Ph.D. dissertation, Harvard University, 1941.
Sayılı, Aydın. *The Observatory in Islam and Its Place in the General History of the Observatory*. Ankara: Türk Tarih Kurumu Basımevi, 1960.
Schmidl, Petra G. "'Urḍī: Mu'ayyad (al-Milla wa-) al-Dīn (Mu'ayyad ibn Barīk [Burayk] al-'Urḍī al-'Āmirī al-Dimashqī)." In *The Biographical Encyclopedia of Astronomers*, edited by Thomas Hockey et al., 2:1161–62. New York: Springer-Verlag, 2007.
Sezgin, Fuat. *Geschichte des arabischen Schrifttums*, vol. 5, *Mathematik*. Leiden: E. J. Brill, 1974.
Sezgin, Fuat. *Geschichte des arabischen Schrifttums*, vol. 6, *Astronomie bis ca. 430 H*. Leiden: E. J. Brill, 1978.
Sezgin, Fuat. *Geschichte des arabischen Schrifttums*, vol. 7, *Astrologie*. Leiden: E. J. Brill, 1979.
Shabbūḥ, Ibrāhīm, ed. *Fihris al-makhṭūṭāt al-muṣawwara*. Vol. 3, pt. 2. Cairo, 1959.
Shank, Michael H. "The Classical Scientific Tradition in Fifteenth-Century Vienna." In *Tradition, Transmission, Transformation: Proceedings of Two*

Conferences on Premodern Science Held at the University of Oklahoma, edited by F. Jamil Ragep and Sally P. Ragep with the assistance of Steven J. Livesey, 115–36. Leiden: E. J. Brill, 1996.

Shank, Michael H. "Regiomontanus and Astronomical Controversy in the Background of Copernicus." In *Before Copernicus: The Cultures and Contexts of Scientific Learning in the Fifteenth Century*, edited by Rivka Feldhay and F. Jamil Ragep. Montreal: McGill-Queens University Press, in press.

Soucek, Svat. *A History of Inner Asia*. Cambridge: Cambridge University Press, 2000.

Steingass, Francis. *The Student's Arabic-English Dictionary*. London: W. H. Allen, 1884.

Stern, S. M. "Petitions from the Ayyūbid Period." *Bulletin of the School of Oriental and African Studies, University of London* 27, no. 1 (1964): 1–32.

Stoetzer, W. "Prosody (*ʿarūḍ*)." In *Encyclopedia of Arabic Literature*, edited by Julie Scott Meisami and Paul Starkey, 2:619–22. London: Routledge, 1998.

Storey, Charles A. *Persian Literature: A Bio-Bibliographical Survey.* 2 vols. Vol. 2, pt. 1, A. Mathematics. B. Weights and Measures. C. Astronomy and Astrology. D. Geography. London: Luzac and Co., 1972. Vol. 2, pt. 2: E. Medicine. London: Luzac and Co., 1971.

Suter, Heinrich. "Der V. Band des Katalogs der arabischen Bücher der viceköniglichen Bibliothek in Kairo." In *Historisch-literarische Abtheilung der Zeitschrift fur Mathematik und Physik*, edited by Dr. O. Schlömilch and Dr. M. Cantor, vol. 38, no. 1, 1–24; vol. 38, no. 2, 41–57; vol. 38, no. 5, 161–84. Leipzig: Verlag von B. G. Teubner, 1893.

Suter, Prof. Dr. Heinrich. "Zu Rudloff und Hochheim, Die Astronomie des Ǵagmînî." *Zeitschrift der Deutschen Morgenländischen Gesellschaft* 47 (1893): 718–19. Reprinted in *Islamic Mathematics and Astronomy*, edited by Fuat Sezgin, vol. 77. Miscellaneous Texts and Studies on Islamic Mathematics and Astronomy, II, 300–1. Frankfurt am Main, 1998.

Suter, Heinrich. "Zur Frage über die Lebenszeit des Verfassers des Mulaḫḫaṣ fî'l-hei'a, Maḥmûd b. Muḥ. b. 'Omar al-Ǵaġmînî." *Zeitschrift der Deutschen Morgenländischen Gesellschaft* 53 (1899): 539–40. Reprinted in *Islamic Mathematics and Astronomy*, edited by Fuat Sezgin, vol. 77. Miscellaneous Texts and Studies on Islamic Mathematics and Astronomy, II, 305–6. Frankfurt am Main, 1998.

Suter, Heinrich. "Die Mathematiker und Astronomen der Araber und ihre Werke." *Abhandlungen zur Geschichte der mathematischen Wissenschaften mit Einschluss ihrer Anwendungen* 10 (1900). Leipzig: Teubner. Reprint, New York, 1972; and reprint, Frankfurt am Main: Institut für Geschichte der Arabisch-Islamischen Wissenschaften, 1986.

Suter, Heinrich. "Nachträge und Berichtigungen." *Abhandlungen zur Geschichte der mathematischen Wissenschaften mit Einschluss ihrer Anwendungen* 14 (1902): 157–85. Reprint, New York, 1972; and reprint, Frankfurt am Main: Institut für Geschichte der Arabisch-Islamischen Wissenschaften, 1986.

Suter, Heinrich. "al-Djaghmīnī." *Encyclopaedia of Islam, First Edition,* 1:1038. Leiden: E. J. Brill, 1913.

Suter, Heinrich and rev. by J. Vernet. "al-Djaghmīnī." *Encyclopaedia of Islam*. 2nd ed., 2:378. Leiden: E. J. Brill, 1965.

Swerdlow, Noel M. "Science and Humanism in the Renaissance: Regiomontanus's Oration on the Dignity and Utility of the Mathematical Sciences." In *World Changes: Thomas Kuhn and the Nature of Science,* edited by Paul Horwich, 131–68. Cambridge, MA: MIT Press, 1993.

Swerdlow, Noel M. "Ptolemy's Theories of the Latitude of the Planets in the *Almagest*, *Handy Tables*, and *Planetary Hypotheses*." In *Wrong for the Right Reasons*, edited by Jed Z. Buchwald and Allan Franklin, 41–71. Archimedes 11: New Studies in the History and Philosophy of Science and Technology. Dordrecht; New York: Springer, 2005.

Ṭāshkubrīzāde, Aḥmad ibn Muṣṭafā. *Miftāḥ al-saʿāda wa-miṣbāḥ al-siyāda*. 3 vols. Beirut: Dār al-kutub al-ʿilmiyya, 1985.

Taub, Liba. *Ptolemy's Universe: The Natural Philosophical and Ethical Foundations of Ptolemy's Astronomy*. Chicago: Open Court Publishing Company, 1993.

Taub, Liba. "On the Variety of 'Genres' of Greek Mathematical Writing: Thinking about Mathematical Texts and Modes of Mathematical Discourse." In *Writing Science: Medical and Mathematical Authorship in Ancient Greece*, edited by Markus Asper in collaboration with Anna-Maria Kanthak, 333–65. Science, Technology, and Medicine in Ancient Cultures. Berlin/Boston: Walter de Gruyter, 2013.

Tihon, Anne and Raymond Mercier. *Ptolemaiou Procheiroi Kanones: Les Tables Faciles de Ptolémée. Volume 1a: Tables A1-A2* Louvain-la-Neuve: Université catholique de Louvain, Institut Orientaliste, 2011–. [The Tables are given in the original Greek and in English translation. Introductions in French and English. Volume 1b contains an Appendix F, entitled: The Handy tables in Arabic; this appendix contains Arabic text with English translations.]

Toomer, G. J. "The Mathematician Zenodorus." *Greek Roman and Byzantine Studies* 13 (1972): 177–92.

Toomer, G. J. "Ptolemy (or Claudius Ptolemaeus)." In *Dictionary of Scientific Biography*, edited by Charles Coulston Gillispie, 11:186–206. New York: Charles Scribner's Sons, 1975.

Toomer, G. J. "Theon of Alexandria." In *Dictionary of Scientific Biography*, edited by Charles Coulston Gillispie, 13:321–25. New York: Charles Scribner's Sons, 1976.

Toomer, G. J. "Hipparchus." In *Dictionary of Scientific Biography*, edited by Charles Coulston Gillispie, 15:207–24. New York: Charles Scribner's Sons, 1978.

Toomer, G. J. *Ptolemy's Almagest.* Translated and annotated by G. J. Toomer with a Foreword by Owen Gingerich. Princeton, NJ: Princeton University Press, 1998.

Topdemir, Hüseyin. "ʿAbd al-Wājid: Badr al-Dīn ʿAbd al-Wājid [Wāḥid] ibn Muḥammad ibn Muḥammad al-Ḥanafī." In *The Biographical Encyclopedia of Astronomers*, edited by Thomas Hockey et al., 1:5–6. New York: Springer-Verlag, 2007.

Ullmann, Manfred. *Die Medizin im Islam*. Handbuch der Orientalistik. Leiden: E. J. Brill, 1970.

Unat, Yavuz and Ömer Türker, trans. *Cevami ilm en-nucûm ve usûl el-harekât es-samâviyye*. Bilimin ve felsefenin doğulu öncüleri dizisi, 14. Ankara: Kültür ve Turizm Bakanlığı, 2012. [Turkish translation of Qāḍīzāde's *Sharḥ al-Mulakhkhaṣ*. Includes facsimiles of Arabic manuscripts Süleymaniye Library, Carullah 1279, ff. 383b–392b and Ayasofya 2662, ff. 1–69.]

Vajda, Georges. "Une Version Hébräique du *Mulaḫḫaṣ fī l-hay'a* de Maḥmūd b. Muḥammad b. 'Umar al-Čaġmīnī." *Journal Asiatique* 247 (1959): 471–78.

van Brummelen, Glen. "Sijzī: Abū Sa'īd Aḥmad ibn Muḥammad ibn 'Abd al-Jalīl al-Sijzī." In *The Biographical Encyclopedia of Astronomers*, edited by Thomas Hockey et al., 2:1059. New York: Springer-Verlag, 2007.

van Dalen, Benno. "Ta'rīkh: 2. Era Chronology in Astronomical Handbooks." *Encyclopaedia of Islam*, 2nd. ed., 10:264–71. Leiden: E. J. Brill, 2000.

van Dalen, Benno. "Battānī: Abū 'Abd Allāh Muḥammad ibn Jābir ibn Sinān al-Battānī al-Ḥarrānī al-Ṣābi'." In *The Biographical Encyclopedia of Astronomers*, edited by Thomas Hockey et al., 1:101–3. New York: Springer-Verlag, 2007.

van Dalen, Benno. "Yaḥyā ibn Abī Manṣūr: Abū 'Alī Yaḥyā ibn Abi Manṣūr al-Munajjim." In *The Biographical Encyclopedia of Astronomers*, edited by Thomas Hockey et al., 2:1249–50. New York: Springer-Verlag, 2007.

Vesel, Živa. "Ikhwān al-Ṣafā'." Commissioned for *The Biographical Encyclopedia of Astronomers*. Edited by Thomas Hockey et al., 2 vols. New York: Springer-Verlag, 2007.
http://islamsci.mcgill.ca/RASI/BEA/Ikhwan_al-Safa%27_BEA.htm.

Watt, W. Montgomery. "al-Iskandar." *Encyclopaedia of Islam*, 2nd. ed., 4:127. Leiden: Leiden: E. J. Brill, 1978.

Wiedemann, Eilhard and Karl Kohl. "Einleitung zu Werken von al-Charaqi." *Sitzungsberichte der Physikalisch-Medizinischen Sozietät zu Erlangen* 58 and 59 (1926-27): 203–18. Reprinted in Eilhard Wiedemann, *Aufsätze zur arabischen Wissenschafts-gechichte,* 2:628–43. Hildesheim: Georg Olms, 1970.

Williams, Henry Smith. *The Great Astronomers*. New York: Newton Publishing Co., 1932.

Witkam, Jan Just. *De Egyptische Arts Ibn al-Akfānī.* Leiden: Ter Lugt Pers, 1989.

Witkam, Jan Just. "Poverty or richness? Some ideas about the generation of Islamic texts revisited." Paper presented at the Commentary Manuscripts (*al-Makhṭūṭāt al-Shāriḥa*) Conference, Bibliotheca Alexandrina, Alexandria, 7–9 Mar., 2006. Pre-print [15 pp.]: http://www.islamicmanuscripts.info/preprints/.

Witkam, Jan Just. *Inventories of Collections of Oriental Manuscripts of the Library of the University of Leiden*. Vol. 1: Manuscripts Or. 1–Or. 1000. Leiden: Ter Lugt Press, 2007.

Woköck, Ursula. *German Orientalism. The Study of the Middle East and Islam from 1800 to 1945.* Series: Culture and Civilization in the Middle East. General Editor Ian Richard Netton. London/New York: Routledge, 2009.

Yāqūt ibn 'Abd Allāh al-Ḥamawī. *Mu'jam al-buldān li-l-Shaykh al-imām Shihāb al-Dīn Abī 'Abd Allāh Yāqūt ibn 'Abd Allāh al-Ḥamawī al-Rūmī al-Baghdādī.* 5 vols. Beirut: Dār ṣādir, 1955–57.

al-Ziriklī, Khayr al-Dīn. *Kitāb al-A'lām*. 8 vols. Beirut: Dār al-'ilm, 1980.

Manuscripts Cited

EGYPT

Cairo, Dār al-kutub, Taymūr Majāmī' MS 429
Cairo, Dār al-kutub, Taymūr Riyāḍa MS 338

FRANCE

Paris, Bibliothèque nationale de France MS arabe 2330
Paris, Bibliothèque nationale de France MS arabe 2504
Paris, Bibliothèque nationale de France MS arabe 2589
Paris, Bibliothèque nationale de France MS arabe 2865
Paris, Bibliothèque nationale de France MS arabe 4821

GERMANY

Berlin, Staatsbibliothek, Landberg MS 33
Berlin, Staatsbibliothek, MS or. oct. 1511
Gotha, Forschungsbibliothek, MS orient. A 1930

IRAN

Isfahan, Abū al-Barakāt MS 50
Isfahan, Maktabat al-Zahrā MS 144
Qum, Faydiyya MS 01832
Qum, Kitābkhānah-yi Buzurg-i Ḥaḍarat-i Āyat Allāh al-ʿUẓmā Marʿashī Najafī MS 8494
Tehran, Majlis-i Shūrā MS 174
Tehran, Majlis-i Shūrā MS 202
Tehran, Majlis-i Shūrā MS 18045
Tehran, University of Tehran, Central Library and Documentation Center MS 6911

THE NETHERLANDS

Leiden, Leiden University Library MS Or. 204

SLOVAKIA

Bratislava, Bratislava Library, Bašagić's Collection of Islamic Manuscripts, Bratislava, TG 15, Ordinal Number 291

SYRIA

Damascus, Ẓāhiriyya MS 4489

TURKEY

Istanbul, Süleymaniye Library, Ayasofya MS 2579
Istanbul, Süleymaniye Library, Ayasofya MS 2581
Istanbul, Süleymaniye Library, Ayasofya MS 2592
Istanbul, Süleymaniye Library, Ayasofya MS 2608
Istanbul, Süleymaniye Library, Ayasofya MS 2662

Istanbul, Süleymaniye Library, Ayasofya MS 3735
Istanbul, Süleymaniye Library, Ayasofya MS 4832
Istanbul, Süleymaniye Library, Fatih MS 3418
Istanbul, Süleymaniye Library, Laleli MS 2127
Istanbul, Süleymaniye Library, Laleli MS 2128
Istanbul, Süleymaniye Library, Laleli MS 2141
Istanbul, Topkapı Sarayı Müzesi, Ahmet III MS 3308

UK

Cambridge, Cambridge University Library MS Or. 593
London, British Museum MS arab. 426 [Add. 7473]
Oxford, Oxford University, Bodleian Library, Seld. MS 11

USA

Los Angeles, CA, UCLA Library, Caro Minasian Collection MS 33
Los Angeles, CA, UCLA Library, Near Eastern Manuscripts Collection MS Ar. 73
Philadelphia, PA, University of Pennsylvania, Rare Book & Manuscript Library, LJS MS 388
Princeton, NJ, Princeton University, Islamic Manuscripts, Garrett MS 373
Princeton, NJ, Princeton University, Islamic Manuscripts, Garrett MS 502H
Princeton, NJ, Princeton University, Islamic Manuscripts, Garrett MS 3559Y

Parameter Index

Ordering is by increasing numerical value. Sexagesimal notation has been adopted whereby 1;02° is 1 degree, 2 minutes. Comm.=Commentary; ^{p}=parts.

S.P. Ragep, *Jaghmīnī's Mulakhkhaṣ*, Sources and Studies in the History of Mathematics and Physical Sciences, DOI 10.1007/978-3-319-31993-3

Subject Index

The Arabic particle *al-* has been ignored for alphabetizing entries. References of the form I.3[1] (in this case, Book I, Chapter 3, paragraph 1) refer to the text, translation and, if there is any, commentary where a term is defined or explained.

S.P. Ragep, *Jaghmīnī's Mulakhkhaṣ*, Sources and Studies in the History of Mathematics and Physical Sciences, DOI 10.1007/978-3-319-31993-3

O

P

Q

T

Plates

Istanbul, Süleymaniye Manuscript Library, Laleli MS 2141

[reproduced with permission of the Süleymaniye Yazma Eser Kütüphanesi (Süleymaniye Manuscript Library) under the authority of the Türkiye Yazma Eserler Kurumu Başkanlığı (Directorate of the Turkish Institution for Manuscripts)]

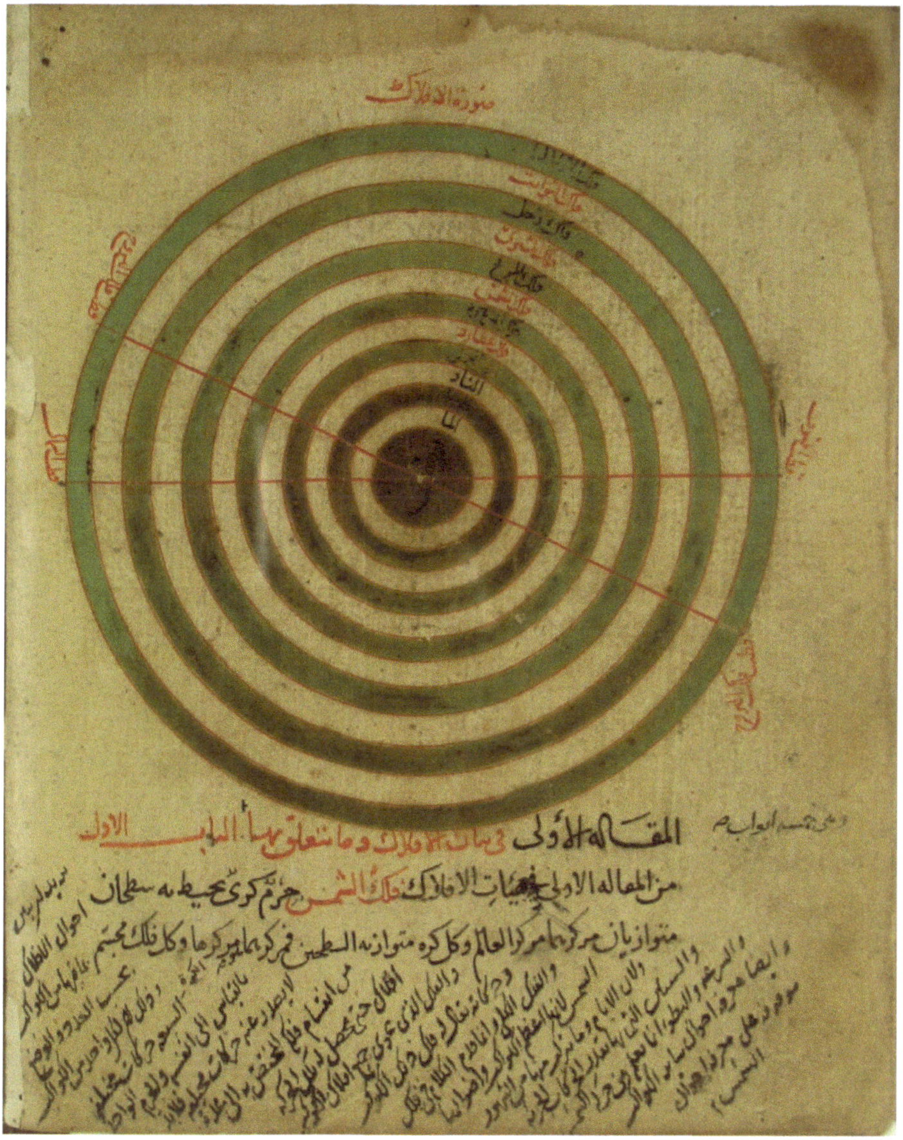

Plate 1. f. 62b: Illustration of the World (Introduction, Figure 1)

Plate 2. f. 64a: Illustration of the Sun's Orbs (Part I, Ch. 1, Figure 2)

Plate 3. f. 64b: Illustration of the Orbs of the Upper Planets and Venus (Part I, Ch. 1, Figure 3)

Plate 4. f. 65a: Illustration of Mercury's Orbs (Part I, Ch. 1, Figure 4)

Plate 5. f. 65b: Illustration of the Moon's Orbs (Part I, Ch. 1, Figure 5)

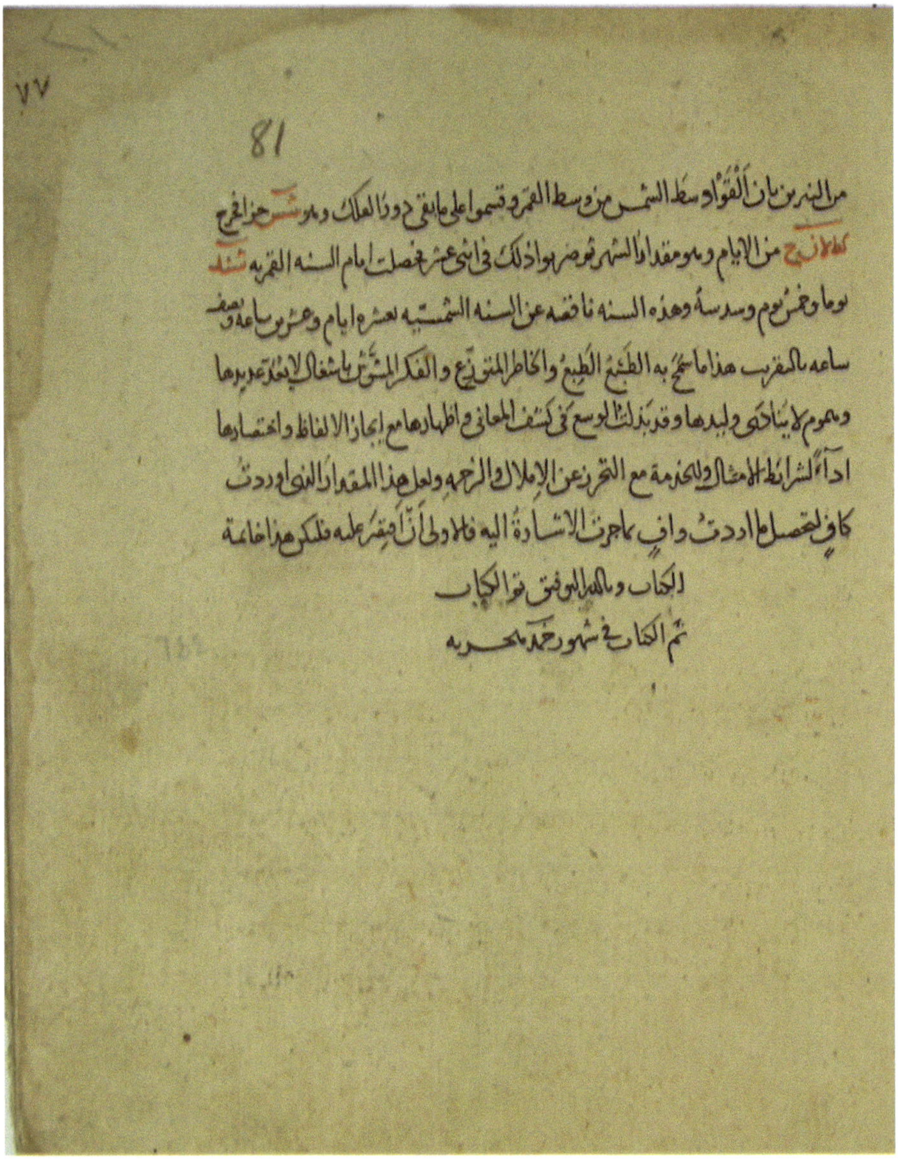

Plate 6. f. 81a: Colophon stating that the book was completed in the months of 644 hijra [1246-47 CE]

MIX
Papier aus verantwortungsvollen Quellen
Paper from responsible sources
FSC® C105338

If you have any concerns about our products,
you can contact us on
ProductSafety@springernature.com

In case Publisher is established outside the EU,
the EU authorized representative is:
Springer Nature Customer Service Center GmbH
Europaplatz 3, 69115 Heidelberg, Germany

Printed by Libri Plureos GmbH
in Hamburg, Germany